Genome Editing in Bacteria

(Part 2)

Edited By

Prakash M. Halami
Department of Microbiology and Fermentation Technology
CSIR- Central Food Technological Research Institute
Mysuru-570020
India

Academy of Scientific and Innovative
Research (AcSIR), Ghaziabad
Uttar Pradesh
India

&

Aravind Sundararaman
Department of Microbiology and Fermentation Technology
CSIR- Central Food Technological Research Institute
Mysuru-570020
India

Genome Editing in Bacteria (Part 2)

Editors: Prakash M. Halami and Aravind Sundararaman

ISBN (Online): 978-981-5223-79-8

ISBN (Print): 978-981-5223-80-4

ISBN (Paperback): 978-981-5223-81-1

© 2024, Bentham Books imprint.

Published by Bentham Science Publishers Pte. Ltd. Singapore. All Rights Reserved.

First published in 2024.

BENTHAM SCIENCE PUBLISHERS LTD.
End User License Agreement (for non-institutional, personal use)

This is an agreement between you and Bentham Science Publishers Ltd. Please read this License Agreement carefully before using the book/echapter/ejournal (**"Work"**). Your use of the Work constitutes your agreement to the terms and conditions set forth in this License Agreement. If you do not agree to these terms and conditions then you should not use the Work.

Bentham Science Publishers agrees to grant you a non-exclusive, non-transferable limited license to use the Work subject to and in accordance with the following terms and conditions. This License Agreement is for non-library, personal use only. For a library / institutional / multi user license in respect of the Work, please contact: permission@benthamscience.net.

Usage Rules:

1. All rights reserved: The Work is the subject of copyright and Bentham Science Publishers either owns the Work (and the copyright in it) or is licensed to distribute the Work. You shall not copy, reproduce, modify, remove, delete, augment, add to, publish, transmit, sell, resell, create derivative works from, or in any way exploit the Work or make the Work available for others to do any of the same, in any form or by any means, in whole or in part, in each case without the prior written permission of Bentham Science Publishers, unless stated otherwise in this License Agreement.
2. You may download a copy of the Work on one occasion to one personal computer (including tablet, laptop, desktop, or other such devices). You may make one back-up copy of the Work to avoid losing it.
3. The unauthorised use or distribution of copyrighted or other proprietary content is illegal and could subject you to liability for substantial money damages. You will be liable for any damage resulting from your misuse of the Work or any violation of this License Agreement, including any infringement by you of copyrights or proprietary rights.

Disclaimer:

Bentham Science Publishers does not guarantee that the information in the Work is error-free, or warrant that it will meet your requirements or that access to the Work will be uninterrupted or error-free. The Work is provided "as is" without warranty of any kind, either express or implied or statutory, including, without limitation, implied warranties of merchantability and fitness for a particular purpose. The entire risk as to the results and performance of the Work is assumed by you. No responsibility is assumed by Bentham Science Publishers, its staff, editors and/or authors for any injury and/or damage to persons or property as a matter of products liability, negligence or otherwise, or from any use or operation of any methods, products instruction, advertisements or ideas contained in the Work.

Limitation of Liability:

In no event will Bentham Science Publishers, its staff, editors and/or authors, be liable for any damages, including, without limitation, special, incidental and/or consequential damages and/or damages for lost data and/or profits arising out of (whether directly or indirectly) the use or inability to use the Work. The entire liability of Bentham Science Publishers shall be limited to the amount actually paid by you for the Work.

General:

1. Any dispute or claim arising out of or in connection with this License Agreement or the Work (including non-contractual disputes or claims) will be governed by and construed in accordance with the laws of Singapore. Each party agrees that the courts of the state of Singapore shall have exclusive jurisdiction to settle any dispute or claim arising out of or in connection with this License Agreement or the Work (including non-contractual disputes or claims).
2. Your rights under this License Agreement will automatically terminate without notice and without the

need for a court order if at any point you breach any terms of this License Agreement. In no event will any delay or failure by Bentham Science Publishers in enforcing your compliance with this License Agreement constitute a waiver of any of its rights.

3. You acknowledge that you have read this License Agreement, and agree to be bound by its terms and conditions. To the extent that any other terms and conditions presented on any website of Bentham Science Publishers conflict with, or are inconsistent with, the terms and conditions set out in this License Agreement, you acknowledge that the terms and conditions set out in this License Agreement shall prevail.

Bentham Science Publishers Pte. Ltd.
80 Robinson Road #02-00
Singapore 068898
Singapore
Email: subscriptions@benthamscience.net

CONTENTS

PREFACE	i
LIST OF CONTRIBUTORS	iii
CHAPTER 1 ADVANCES IN MICROBIAL STUDY FOR CROP IMPROVEMENT	1

Vinay Sharma, Neelam Mishra, Sherin Thomas, Rahul Narasanna, Kalant Jambaladinni, Priscilla Kagolla, Ashish Gautam, Anamika Thakur, Abhaypratap Vishwakarma, Dayanand Agsar, Manish K. Pandey and *Rakesh Kumar*

INTRODUCTION	2
MICROBIAL CONSORTIA	3
Bacteria	3
Fungi	4
Algae	7
BIO-STIMULANTS: INTERACTION OF ROOT EXTRACT WITH SOIL MICROBES	8
AGRONOMICALLY IMPORTANT SOIL MICROBES	10
Nitrogen-Fixing Bacteria	10
Azotobacter chroocochum	10
Azotobacter vinelandii	10
Glucanobacter diazotrophicus	10
Acetobacter xylinum	10
Azospirillum lipoferum	11
Rhizobium sp.	11
Phosphate Solubilizing Microbes	11
Bacillus megaterium	13
Pseudomonas putida	13
Potash Mobilizing Bacteria	13
Frateuria aurentia	13
Plant Growth-Promoting *Rhizobacteria* (PGPR)	14
Bacillus sp.	14
Pseudomonas sp.	14
Biological Control Organisms	14
Metarhizium anisopilae	14
Beauveria bassiana	15
Verticillium lecanii	15
Paecilomyces lilacinus	15
Arthrobotrys spp.	15
Trichoderma viride	15
Microbes for Stress Tolerance	15
Pseudomonas putida	16
Trichoderma harzianum	16
Mycorrihizal Fungi	16
GENOME EDITING OF MICROBES TO BENEFIT CROP PLANTS	16
TRANSFER OF MICROBIAL GENE INTO PLANT SPECIES	17
USE OF MICROBES FOR THE PRODUCTION OF BIOENERGY FROM AGRICULTURE WASTE	22
CONCLUSION	23
ACKNOWLEDGEMENTS	23
REFERENCES	23
CHAPTER 2 GENOME EDITING AGAINST BACTERIAL PLANT PATHOGENS	43

Ashish Warghane, Neha G. Paserkar and *Sumit Bhose*

INTRODUCTION	43
EXISTING GENOME EDITING TECHNIQUES AND ADVANCEMENT UNTIL NOW	44
Meganucleases	45
Zinc-Finger Nucleases (ZFNs)	46
Transcription Activator-Like Effector Nuclease Technique (TALEN)	48
CRISPR/Cas9	51
BACTERIAL GENOME EDITING, APPLICATION, AND ITS SIGNIFICANCE	54
Genome Editing For Plant Disease Resistance Against Bacterial Pathogens	57
CRISPR-Cas9 Mediated Resistance Against Xanthomonas oryzae	57
CRISPR Against "Candidatus liberibacter spp"/Citrus Greening Bacterium	58
CRISPR-Cas9 Mediated Resistance Against Citrus Bacterial Canker	59
CRISPR-Cas9 Mediated Resistance Against Erwinia amylovora	59
Significance of Studying Plant-Pathogen Interaction and Application of Crispr-Cas9 for Insight into the Plant-Pathogen Interaction	60
CONCLUDING REMARKS	60
LIST OF ABBREVIATIONS	61
ACKNOWLEDGEMENTS	61
REFERENCES	62
CHAPTER 3 CRISPR-Cas FOR GENOME EDITING - MOLECULAR SCISSORS FOR COMBATING PATHOGENS	**68**
Poornima Devi C. Ramdev, Divya K. Shankar and B. Renuka	
INTRODUCTION	69
CRISPR-Cas System - Discovery and Function	71
CRISPR/Cas Systems for Gene Editing, Specificity and Molecular Mechanism	74
BACTERIAL VIRULENCE AND BIOMOLECULAR TARGETS	77
APPLICATION OF SYNTHETIC BIOLOGY	87
Controlling Gene Expression with CRISPR	91
CRISPR in the Treatment of Infection	94
CRISPR in Health and Industry	95
FUTURE PROSPECTS	97
CONCLUSION	98
ACKNOWLEDGEMENTS	98
REFERENCES	98
CHAPTER 4 GENOME EDITING OF PLANT GROWTH-PROMOTING MICROBES (PGPM) TOWARDS DEVELOPING SMART BIO-FORMULATIONS FOR SUSTAINABLE AGRICULTURE: CURRENT TRENDS AND PERSPECTIVES	**106**
Sugitha Thankappan, Asish K. Binodh, P. Ramesh Kumar, Sajan Kurien, Shobana Narayanasamy, Jeberlin. B. Prabina and Sivakumar Uthandi	
INTRODUCTION	107
FACTORS INFLUENCING PLANT-MICROBE INTERACTIONS AND THEIR COMPOSITION	108
Biotic Factors	108
Abiotic Factors	109
SIGNIFICANCE OF PLANT-MICROBE INTERACTION IN SUSTAINABLE AGRICULTURE	110
Healthy Plant-microbe Interactions	110
Harmful Plant-microbe Interactions	110
Plant-Pathogen Interaction	111
Microbe Induced Systemic Tolerance [MIST] for Enhanced Crop Resilience	112
Osmolytes Tussles for Stress Resilience	112

Antioxidant Gadgets	114
Root System Architecture	114
Phytohormone Modulation and Cross-Talk: Cues in the Battle	115
Microbial Volatiles [Mvocs] in Plant-microbe Interaction	118
Exopolysaccharide [EPS] Production	119
TOOLS TO EXPLORE PLANT-MICROBE INTERACTIONS	119
CLUSTERED REGULARLY INTERSPACED SHORT PALINDROMIC REPEATS	120
CRISPR/Cas System and Orthologs	120
CRISPR/Cas12 and Cas14	122
RNA-Targeting Endonucleases	122
APPLICATIONS OF MODERN CRISPR-BASED TOOLS	123
CRISPR for Editing Multi-Targets	123
CRISPR-MEDIATED PLANT –MICROBE INTERACTION AND ITS APPLICATION IN AGRICULTURE	123
CRISPR in Understanding Plant-microbe Interactions	123
CRISPR in Understanding Plant Growth Promotion [PGP] and Nutrient Uptake	124
CRISPR in Priming Plant Disease Resistance	125
GE in Bacterial Pathogens	126
GE in Fungal Pathogens	126
Genome Editing for Plant Disease Resistance against Bacterial and Fungal Pathogens	126
GE in Plant-Virus Interactions	128
GE in Unraveling Novel Metabolic Pathways and Metabolome	130
Soil Health	132
Legume- Rhizobium Symbiosis	132
CONCLUSION AND FUTURE PERSPECTIVES	133
ACKNOWLEDGEMENTS	134
REFERENCES	134
CHAPTER 5 APPLICATIONS OF GENOME EDITING IN BIOREMEDIATION	150
Vibhuti Sharma, Rutika Sehgal, Vani Angra and *Reena Gupta*	
INTRODUCTION	150
BIOREMEDIATION	152
General Principle of Bioremediation	153
Types of Bioremediation	154
In-situ Bioremediation	155
Ex-situ Bioremediation	156
Landfarming	156
Biopile	157
Windrows	157
Bioreactor	157
Why Bioremediation is Important?	158
Bioremediation Process: Mechanism	159
GENOME EDITING IN BIOREMEDIATION	159
ROLE OF GENOME EDITING IN BIOREMEDIATION	160
GENOMIC TOOLS USED FOR BIOREMEDIATION OF CONTAMINANTS	160
CRISPR-Cas9	161
Type 1 CRISPR-Cas System	162
Type 2 CRISPR-Cas System	162
Type 3 CRISPR-Cas System	162
TALENs	163
ZFNs	163

GENETIC ENGINEERING OF MICROORGANISMS	164
Advantages of GEMs in Bioremediation	165
Genetically Engineered Bacteria for Bioremediation	165
Production of Biosurfactants	165
Optimizing Biocatalysts	166
Genetically Engineered Fungi for Bioremediation	166
Genetically Engineered Plants for Bioremediation	166
GENOME EDITING TECHNOLOGIES USED FOR THE MODIFICATION OF MICROORGANISMS	167
Rational Designing	167
Genome Shuffling	167
Family Shuffling	167
GENOMIC STRATEGIES AND OMICS APPROACHES USED IN THE PROCESS OF BIOREMEDIATION	168
Metagenomics in Bioremediation	170
Metatranscriptomics and Proteomics in Bioremediation	171
DISADVANTAGES OF GENOMIC APPROACHES	171
FACTORS AFFECTING BIOREMEDIATION	172
Environmental Factors	172
Temperature	172
Oxygen Concentration	172
pH	173
Nutrients Availability	173
Toxic Compounds	173
Biological Factors	173
FUTURE PERSPECTIVE	173
CONCLUSION	174
ACKNOWLEDGEMENTS	174
REFERENCES	174
CHAPTER 6 GENOME EDITING AND GENETICALLY ENGINEERED BACTERIA FOR BIOREMEDIATION OF HEAVY METALS	184
Nirmala Akoijam and S.R. Joshi	
INTRODUCTION	184
ENVIRONMENTAL POLLUTION: CAUSES AND IMPACTS	186
Heavy Metal Tolerance in Plants and Microbes	188
Existing Tools to Combat Heavy Metal Pollution	189
In situ Remediation	190
Ex situ Remediation	191
BIOREMEDIATION AS AN ALTERNATIVE AND ENVIRONMENT-FRIENDLY TECHNIQUE	193
Biopile	194
Windrows	194
Bioreactors	194
Bioventing	195
Biosparging	195
Bioslurping	195
Phytoremediation	195
Phytostabilization	196
Phytovolatilization	196
Phytoextraction	197

Phytofiltration	197
MICROBES-ASSISTED BIOREMEDIATION	198
Microbial Biosorption of Metals	199
Intracellular Sequestration by Protein Binding	200
Extracellular Sequestration	200
Permeability Barrier	201
Microbial Methylation of Metals	201
Microbial Reduction of Metals	202
GENETICALLY ENGINEERED BACTERIA FOR BIOREMEDIATION OF HEAVY METALS	202
Overexpression of Gene or Operon Involved in Metal Detoxification Pathways	203
Expression of Transport Proteins and Efflux Pumps	204
Genome Editing by CRISPR-Cas Technology	204
CONCLUSION AND FUTURE PROSPECTS	206
ACKNOWLEDGMENTS	207
REFERENCES	207

CHAPTER 7 DESIGNING THE METABOLIC CAPACITIES OF ENVIRONMENTAL BIOPROCESSES THROUGH GENOME EDITING ... 222

Ashish Kumar Singh, Bhagyashri Poddar, Rakesh Kumar Gupta, Suraj Prabhakarrao Nakhate, Vijay Varghese, Anshuman A. Khardenavis and *Hemant J. Purohit*

INTRODUCTION	223
CRISPR AND BIOREMEDIATION	224
Conventional Practices for Improvement of Bioremediation Efficiencies	225
Recent Advances in Gene Editing for Enhanced Bioremediation	226
CRISPR Based Genetic Manipulation in Nitrogen Metabolizing Bacteria	229
CRISPR AND METHANOTROPHS	231
Transformation Efficiencies in Methanotrophs	231
CRISPR Based Genetic Manipulation in Methanotrophs	231
CRISPR AND ANAEROBIC DIGESTION	232
CRISPR Based Genetic Manipulation of Hydrolytic Bacteria	232
CRISPR Based Genetic Manipulation of Acidogenic Bacteria	233
CRISPR Based Genetic Manipulation of Methanogenic Bacteria	233
CRISPR/Cas9 Assisted Strain Built-Up and Introduction of Novel Capabilities By Parallel Metabolic Pathway Engineering (PMPE)	234
Strategies For Improving The Efficiency of CRISPR/Cas9	235
CRISPR and Volatile Fatty Acid (VFA) Production	236
CRISPR Based Genetic Manipulation of VFA Producing Bacteria	236
LIMITATIONS OF CRISPR AND STRATEGIES TO OVERCOME THE DRAWBACKS	238
FUTURE PERSPECTIVES	239
CONCLUSION	240
ACKNOWLEDGEMENTS	240
REFERENCES	240

CHAPTER 8 GENETIC ENGINEERING OF METHANOTROPHS: METHODS AND RECENT ADVANCEMENTS ... 247

Eleni N. Moutsoglou and *Rajesh K. Sani*

INTRODUCTION	247
Methanotrophs	247
Industrial Use of Methanotrophs	248
Why Genetically Engineered Methanotrophs?	248

METHODS OF GENETIC ENGINEERING	249
In General	249
In Methanotrophs	249
Conjugation	250
Electroporation	250
CRISPR	250
Vectors Used in Methanotrophs	251
Genome-Scale Metabolic Models	252
SPECIFIC EXAMPLES OF GENETIC ENGINEERING IN METHANOTROPHS	252
EXISTING KNOWLEDGE GAPS	257
PROSPECTS	258
CONCLUSION	259
ACKNOWLEDGMENTS	259
REFERENCES	259
CHAPTER 9 GENOME EDITING IN *CYANOBACTERIA*	262
Bathula Srinivas and Prakash M. Halami	
INTRODUCTION	262
Cyanobacteria as a Host for the Heterologous Expression	264
Shuttle Vectors Used in *Cyanobacteria*	265
Markerless Selection for the Analysis of Transformants in *Cyanobacteria*	265
The Role of CRISPR/Cas in Editing the Genome of the *Cyanobacteria*	266
Applications of CRISPR-Cas9/Cas12a Engineering Tools in *Cyanobacteria*	269
CONCLUDING REMARKS	270
REFERENCES	270
CHAPTER 10 GENOME EDITING IN STREPTOMYCES	278
Johns Saji, Jibin James, Ramesh Kumar Saini and Shibin Mohanan	
INTRODUCTION	278
DIFFERENT GENOME EDITING TECHNIQUES	280
PCR-Targeting System	280
Cre-loxP Recombination System	283
I-SceI Meganuclease-Promoted Recombination System	284
Genome Editing with the CRISPR/Cas Systems	285
CRISPR/Cas9 and HRD-mediated Genome Editing	288
CRISPR/Cas9 and NHEJ Mediated Genome Editing	291
Heterologous Expression and BGCS Cloning using CRISPR/Cas9	293
Genome Editing with the Assistance of Cpf1	294
Transcriptional Repression using dCas (CRISPRi)	294
Editors Based on Cas9 Variants (dCas9 or Cas9n)	295
Genome Editing using Multiplex Automated Genome Editing (MAGE) Tool	297
Multiplex Genome Editing of Streptomyces Species using Engineered CRISPR/Cas System	297
Multiplex Genome Editing using Engineered CRISPR/Cas9 System.	297
Multiplex Genome Editing using Engineered CRISPR/Cas12a System	298
CRISPR-Cpf1 Assisted Multiplex Genome Editing and Transcriptional Repression in Streptomyces	299
CONCLUSION	299
ACKNOWLEDGEMENTS	300
REFERENCES	300
SUBJECT INDEX	307

PREFACE

In the vast landscape of scientific exploration, genetic engineering stands as a beacon illuminating pathways in both basic research and industrial biotechnology. At its heart lie metabolic and genomic manipulations that coax microorganisms to yield invaluable products, sparking innovations that redefine possibilities.

The saga of genetic inquiry into microorganisms hinges upon accessibility to their genomes and the arsenal of molecular tools at our disposal. Early genetic methods for genome editing in bacterial species, rooted in culture and transformation, were painstakingly laborious, often reliant on introducing resistance markers that hindered the pursuit of precise edits such as single amino acid mutations.

Yet, the tide turned with the groundbreaking discovery of CRISPR-Cas technology, unraveling the adaptive immune system of prokaryotes and unfurling vistas of targeted genetic engineering in these organisms. In this tome, we delve into the cutting edge gene editing, exploring diverse strategies employed in prokaryotic genetic manipulation.

This book embarks on a journey that traverses historical perspectives of genome editing, its application in probiotics, and its relevance in agricultural and environmental microbiology. It endeavors to consolidate and update the compendium of knowledge and research in bacterial applications across industries like food and pharmaceuticals, illuminating gene regulation for metabolic engineering through genome editing tools.

Our heartfelt gratitude extends to the esteemed contributing authors who embraced our call to enrich this compendium. Each chapter bears the mark of dedication and expertise, a testament to their profound contributions to bacteriology and molecular biology.

The Bentham Science Group's commitment to publication has facilitated the realization of this comprehensive endeavor, offering a resource intended for researchers, students, teachers, scientists, and enterprising minds intrigued by bacterial metabolic engineering.

In the vast ocean of scientific literature, this book, "Gene Editing in Bacteria," stands as a pioneering compilation, weaving together diverse applications of bacteria across the tapestry of biotechnology.

I dedicate this book to the pioneers of indigenous knowledge in molecular biology and genetic engineering. They not only laid the foundation for an ocean of knowledge but also kindled the flame that propels our relentless pursuit of understanding genome editing techniques in bacteria.

Prakash M. Halami
Department of Microbiology and Fermentation Technology
CSIR- Central Food Technological Research Institute
Mysuru-570020
India

Academy of Scientific and Innovative
Research (AcSIR), Ghaziabad
Uttar Pradesh
India

&

Aravind Sundararaman
Department of Microbiology and Fermentation Technology
CSIR- Central Food Technological Research Institute
Mysuru-570020
India

List of Contributors

Anshuman A. Khardenavis	Academy of Scientific and Innovative Research (AcSIR), Ghaziabad-201002, India Environmental Biotechnology and Genomics Division, CSIR-NEERI, Nehru Marg, Nagpur-440020, Maharashtra, India
Aravind Sundararaman	Department of Microbiology and Fermentation Technology, CSIR- Central Food Technological Research Institute, Mysuru-570020, India
Ashish Gautam	Department of Life Science, Central University of Karnataka (CUK), Kalaburagi, India
Anamika Thakur	Department of Biotechnology, Dr. Y.S. Parmar University of Horticulture and Forestry, Nauni, Solan, Himachal Pradesh, India
Abhaypratap Vishwakarma	Department of Botany, Deshbandhu College, University of Delhi, New Delhi, India
Ashish Warghane	School of Applied Sciences and Technology, Gujarat Technological University, Chandkheda, Ahmedabad, Gujarat, India
Asish K. Binodh	Centre for Plant Breeding and Genetics, Tamil Nadu Agricultural University, Coimbatore-641003, Tamil Nadu, India
Ashish Kumar Singh	Academy of Scientific and Innovative Research (AcSIR), Ghaziabad-201002, India Environmental Biotechnology and Genomics Division, CSIR-NEERI, Nehru Marg, Nagpur- 440020, Maharashtra, India
Bhagyashri Poddar	Academy of Scientific and Innovative Research (AcSIR), Ghaziabad-201002, India Environmental Biotechnology and Genomics Division, CSIR-NEERI, Nehru Marg, Nagpur- 440020, Maharashtra, India
B. Renuka	Promic Svasthya Private Limited, Mysore-570028, Karnataka, India
Bathula Srinivas	Department of Biotechnology, School of Herbal Studies and Naturo Sciences, Dravidian University, Kuppam-517426, India
Dayanand Agsar	Department of Microbiology, Gulbarga University (GU), Kalaburagi, India
Divya K. Shankar	Department of Studies in Microbiology, Pooja Bhagavat Memorial Mahajana PG Centre, Myosre-570016, Karnataka, India
Eleni N. Moutsoglou	Department of Chemical and Biological Engineering, South Dakota School of Mines and Technology, Rapid City, SD 57701, USA BuG ReMeDEE Consortium, Rapid City, SD 57701, USA
Hemant J. Purohit	Environmental Biotechnology and Genomics Division, CSIR-NEERI, Nehru Marg, Nagpur-440020, Maharashtra, India
Jibin James	Department of Botany, Nirmala College, Muvattupuzha, Ernakulam, Kerala, India
Johns Saji	Department of Botany, Nirmala College, Muvattupuzha, Ernakulam, Kerala, India

Jeberlin. B. Prabina	Department of Soil Science, Agricultural College and Research Institute, Killikulam, Vallanad Post, Tuticorin Dt -628252, Tamil Nadu, India
Kalant Jambaladinni	Department of Life Science, Central University of Karnataka (CUK), Kalaburagi, India
Manish K. Pandey	International Crops Research Institute for the Semi-Arid Tropics (ICRISAT), Hyderabad, India
Neelam Mishra	Department of Microbiology, Gulbarga University (GU), Kalaburagi, India
Neha G. Paserkar	Department of Plant Science, McGill University, Quebec H9X 3V9, Canada
Nirmala Akoijam	Department of Biotechnology & Bioinformatics, North-Eastern Hill University, Shillong, India
Prakash M. Halami	Department of Microbiology & Fermentation Technology, CSIR- Central Food Technological Research Institute, Mysore – 570020, India Academy of Scientific and Innovative Research (AcSIR), Ghaziabad, Uttar Pradesh, India
Priscilla Kagolla	Department of Life Science, Central University of Karnataka (CUK), Kalaburagi, India
P. Ramesh Kumar	School of Agricultural Sciences, Karunya Institute of Technology and Sciences (Deemed to be University), Coimbatore-641114, Tamil Nadu, India
Poornima Devi C. Ramdev	Department of Microbiology, Yuvaraja's College (Autonomous), Myosre-570005, India
Rutika Sehgal	Department of Biotechnology, Himachal Pradesh University, Summerhill, Shimla-171005, India
Reena Gupta	Department of Biotechnology, Himachal Pradesh University, Summerhill, Shimla-171005, India
Rakesh Kumar Gupta	Academy of Scientific and Innovative Research (AcSIR), Ghaziabad-201002, India Environmental Biotechnology and Genomics Division, CSIR-NEERI, Nehru Marg, Nagpur- 440020, Maharashtra, India
Rajesh K. Sani	Department of Chemical and Biological Engineering, South Dakota School of Mines and Technology, Rapid City, SD 57701, USA BuG ReMeDEE Consortium, Rapid City, SD 57701, USA
Ramesh Kumar Saini	Department of Crop Science, Konkuk University, Seoul, Korea
Rakesh Kumar	Department of Life Science, Central University of Karnataka (CUK), Kalaburagi, India
Rahul Narasanna	Department of Life Science, Central University of Karnataka (CUK), Kalaburagi, India
Sherin Thomas	Department of Biosciences & Bioengineering, Indian Institute of Technology Bombay, Mumbai, Maharashtra, India
Sumit Bhose	Sea6Energy Pvt. Ltd., Bangalore, Karnataka, India
Sugitha Thankappan	School of Agricultural Sciences, Karunya Institute of Technology and Sciences (Deemed to be University), Coimbatore-641114, Tamil Nadu, India

Sivakumar Uthandi	Department of Microbiology, Tamil Nadu Agricultural University, Coimbatore-641003, Tamil Nadu, India
Sajan Kurien	School of Agricultural Sciences, Karunya Institute of Technology and Sciences (Deemed to be University), Coimbatore-641114, Tamil Nadu, India
Shobana Narayanasamy	Department of Microbiology, Tamil Nadu Agricultural University, Coimbatore-641003, Tamil Nadu, India
S.R. Joshi	Department of Biotechnology & Bioinformatics, North-Eastern Hill University, Shillong, India
Shibin Mohanan	Department of Botany, Nirmala College, Muvattupuzha, Ernakulam, Kerala, India
Suraj Prabhakarrao Nakhate	Academy of Scientific and Innovative Research (AcSIR), Ghaziabad-201002, India Environmental Biotechnology and Genomics Division, CSIR-NEERI, Nehru Marg, Nagpur- 440020, Maharashtra, India
Vinay Sharma	International Crops Research Institute for the Semi-Arid Tropics (ICRISAT), Hyderabad, India Department of Genetics and Plant Breeding, Ch. Charan Singh University, Meerut, India
Vibhuti Sharma	Department of Biotechnology, Himachal Pradesh University, Summerhill, Shimla-171005, India
Vani Angra	Department of Biotechnology, Himachal Pradesh University, Summerhill, Shimla-171005, India
Vijay Varghese	Environmental Biotechnology and Genomics Division, CSIR-NEERI, Nehru Marg, Nagpur- 440020, Maharashtra, India

CHAPTER 1

Advances in Microbial Study for Crop Improvement

Vinay Sharma[1,2], Neelam Mishra[3,‡], Sherin Thomas[4,‡], Rahul Narasanna[5], Kalant Jambaladinni[5], Priscilla Kagolla[5], Ashish Gautam[5], Anamika Thakur[6], Abhaypratap Vishwakarma[7], Dayanand Agsar[3], Manish K. Pandey[1] and **Rakesh Kumar**[5,*]

[1] *International Crops Research Institute for the Semi-Arid Tropics (ICRISAT), Hyderabad, India*

[2] *Department of Genetics and Plant Breeding, Ch. Charan Singh University, Meerut, India*

[3] *Department of Microbiology, Gulbarga University (GU), Kalaburagi, India*

[4] *Department of Biosciences & Bioengineering, Indian Institute of Technology Bombay, Mumbai, Maharashtra, India*

[5] *Department of Life Science, Central University of Karnataka (CUK), Kalaburagi, India*

[6] *Department of Biotechnology, Dr. Y.S. Parmar University of Horticulture and Forestry, Nauni, Solan, Himachal Pradesh, India*

[7] *Department of Botany, Deshbandhu College, University of Delhi, New Delhi, India*

Abstract: Now and in the future, meeting the global demand for healthy food for the ever-increasing population is a crucial challenge. In the last seven decades, agricultural practices have shifted to the use of synthetic fertilizers and pesticides to achieve higher yields. Despite the huge contribution of synthetic fertilizers in agronomy, their adverse effects on the environment, natural microbial habitat, and human health cannot be underrated. Besides, synthetic fertilizers are manufactured from non-renewable sources such as earth mining or rock exploitation. In this context, understanding and exploiting soil microbiota appears promising to enhance crop production without jeopardizing the environment and human health. This chapter reviews the historical as well as current research efforts made in identifying the interaction between soil microbes and root exudates for crop improvement. First, microbial consortium *viz.* bacteria, algae, fungi, and protozoa are briefly discussed. Then, the application of bio-stimulants followed by genome editing of microbes for crop improvement is summarized. Finally, the perspectives and opportunities to produce bioenergy and bio-fertilizers are analyzed.

Keywords: Biofertilizer, Crop improvement, Genetic engineering, Microbial consortium, Rhizosphere.

---

* **Corresponding author Rakesh Kumar:** Department of Life Science, Central University of Karnataka (CUK), Kalaburagi, India; E-mail: rakeshgupta.hcu@gmail.com
‡ contributed equally

INTRODUCTION

The world population is constantly increasing and is projected to be 10 billion by 2050. Barea [1] estimated that by 2050, food demand is supposed to increase by 70% in the agricultural area. Although conventional farming (high-yield varieties, irrigation, synthetic pesticides and fertilizers) has shown an increase in food production by 70% from 1970 to 1995 in developing countries, its adverse effects on the environment, plants, humans, and aquatic ecosystem cannot be overlooked [2, 3]. Therefore, it is time to change our trajectory towards advanced microbial agricultural practices to combat pests and provide natural nutrition resources to plants without compromising the sustainable environment [4]. A microbial consortium is set of microorganisms, including bacteria, *Cyanobacteria*, algae, protozoa, yeast, and fungi, that works synergistically for hydrolyzing biomass, there by increasing soil fertility [5]. Soil bacteria are very important for biogeochemical cycle and agriculture. Plant-soil bacteria interaction plays a key role in determining the plants' health and growth. Usually, such beneficial bacteria are termed plant growth promoting *Rhizobacteria* (PGPR), which colonize in rhizosphere [6]. Species of *Rhizobium* (*Allorhizobium*, *Azorhizobium*, *Bradyrhizobium*, *Mesorhizobium*, *Rhizobium*, and *Sinorhizobium*) form symbiotic relationship with legume plants, through flavonoids signals produced by plants. Flavonoids lead to nodule formation by inducing nodulation (nod) genes in Rhizobia [7]. PGPR is being used worldwide to increase crop production [8 - 10]. On the other hand, non-symbiotic PGPR such as *Azospirillum* enhances plant's resistance and ion uptake by producing antibacterial and antifungal compounds, growth regulators and siderophores [11]. Further, *Cyanobacteria* play an important role in raising the oxygen level in the atmosphere and ocean. Oxygenic photosynthesis enabled aquatic and terrestrial environments to undergo diversification and form complex life [12, 13]. *Cyanobacteria Anabaena*, *Calothrix*, *Scytonema*, and *Nostoc* have been widely used in rice cultivation. These *Cyanobacteria* develop specialized cells heterocysts to fix the aerobic nitrogen, particularly when nitrate and ammonia are limited in soil [14]. Recently, a pot experiment study has demonstrated that inoculation of *Nostoc* caused significant increase in root length. However, half dose of recommended chemical fertilizer with *Nostoc* improved the growth and production of rice. Pathum Thani [15]. Rice sheath blight is a serious disease in Asian countries caused by pathogenic fungi *Rhizoctonia solani*. Application of *Nostoc piscinale* (SCAU04) and *Anabaena variabilis* (SCAU26) found to produce bioactive substances to inhibit *R. solani* by 90%, and secrete phytohormones to promote plant growth and development, and induce resistance against disease. Fungi are mostly considered harmful pathogens for both plants and animals, because they produce mycotoxins as secondary metabolites. The major mycotoxins are aflatoxin, ochratoxins, trichothecenes, fumonisins, zearalenone, cyclopiazonic acid, and putulin [16]. In contrast, *Trichoderma*, *Aspergillus*, and *Clonostachys rosea* are beneficial fungi, found to be very effective against mycotoxin producing *Fusarium* and *Aspergillus* [17, 18]. These fungi have special characteristics such as promoting plant growth, producing antibiotics, and parasitizes other fungi (hyperparasitism) [19]. Seed coating with PGPR, rhizobia, arbuscular mycorrhizal fungi, and Trichoderma resulted in higher yield and resistance against pathogens in several plant species, thus can be used as an ideal biocontrol agent instead of chemical fungicide [20, 21]. In addition to nitrogen fixation, ion uptake, growth promotion, and protection from toxins, microbes are being explored for wastewater treatment, biodiesel production, bioelectricity, and biosensing [22 - 24]. In this regard, *Saccharomyces cerevisiae*, *Pichia stipitis*, and *Kluyveromyces fagilis* have been used extensively for ethanol production [25]. Metabolic engineering of *Clostridium acetobutylicum* enhanced butanol yield of 0.71 mol butanol/mol glucose, which was 245% higher compared to wild-type strains [26]. Some Oleaginous yeasts like *Cryptococcus psychrotolerans* (IITRFD) and *Rhodosporidiobolus fluvialis* (DMKU-SP314) are used for the production of biodiesel [27, 28]. Here, we have diSome Oleaginous yeasts likscussed the current scenario of microbial uses in crop improvement by biochemical and genetic engineering approaches.

MICROBIAL CONSORTIA

Rhizosphere microorganism plays an essential role in sustainable agriculture, influencing natural plant communities' composition and productivity (Fig. 1). Bacteria, archaea, fungi, algae, viruses, protozoa, oomycetes and microarthropods are the microbial groups residing in the rhizosphere [29]. The leading population of microbes in the rhizosphere is bacteria, trailed by fungi, actinomycetes and other groups. Bacteria, fungi, algae and protozoa coexist in the rhizosphere and exert multiple strategies to utilize minerals and organic wastes. They act as metal sequestering and growth-promoting bioinoculants for plants in metal-stressed soils [29].

Bacteria

Azospirillum, Azotobacter, Bacillus, Enterobacter, Pseudomonas and *Serratia* are successfully used along with *Rhizobium* for microbial consortia for crop improvement [30]. Microbial consortia under extreme environmental conditions enhance crop production. The production of plant growth hormones and vitamins are significantly increased with the application of *Rhizobium* along with *Azotobacter* as consortia [31]. Rhizobium's microbial consortia with *G. intraradices* and *P. striata* show enhanced plant growth in chickpeas root rot along with improved chlorophyll content [32]. Consortia of *Mezorhizobium sp.* and *P.aeruginosa* increased dry weight and nodule formation in chickpeas [33].

Fox *et al.* [34] co-inoculated *Pseudomonas fluorescens* WSM3457 and *Ensifer (Sinorhizobium) medicae* WSM419, increasing nodule numbers in green gram. A physiological defence response was activated against *Sclerotium rolfsii*, a collar rot pathogen using *P. fluorescens, Trichoderma* and *Rhizobium* consortium [35]. Similarly, improved yield with disease resistance was observed with consortia of *B. subtilis, T. harzianum* and *P. aeruginosa* [36].

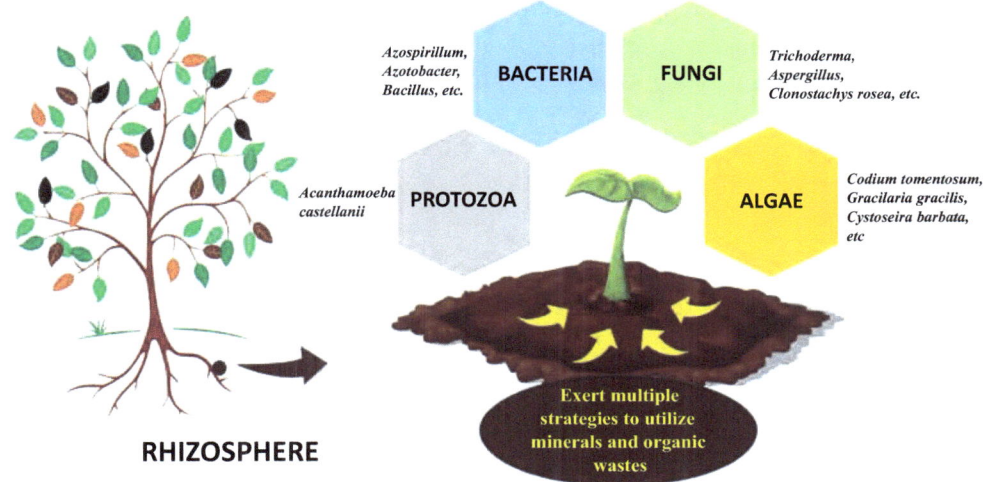

Fig. (1). Schematic representation for rhizosphere microbial diversity and plant strategies to use minerals and organic wastes.

Root length of *Arabidopsis* was significantly enhanced using consortia of *Bacillus, Burkholderia, Pseudomonas, Ralstonia* and *Variovorax* in response to abiotic stress [37]. Jha and Subramanium [38] reported an increase in NPK concentration and reduction in Na and Ca concentration in response to salinity stress in paddy using a consortium of *P. pseudoalcaligenes* with *B. pumilus*. Increased production of flavonoids and lipochitooligosaccharide, along with enhanced nodulation was observed using a combination of *A. brasilense* and *Rhizobium* [39]. Cyanobacterial consortia of *Anabaena – Azotobacter* biofilms and *Anabaena sp.-Providencia sp.* elicited plant defense response in maize hybrids [40].

Fungi

Arbuscular Mycorrhizal fungi (AMF) in combination with plant growth-promoting *Rhizobacteria* (PGPR), such as nitrogen fixing rhizobia, phosphate solubilizing rhizobia and free living bacteria such as *Azospirillum., Bacillus sp.,* and *Pseudomonas sp.* shows synergistic interaction enhancing growth and productivity in various crops [41, 42]. Sharma *et al.* [43], emphasized that Mycorrhizal fungi and PGPR consortia act as biostimulators, biofertilizers and bioprotectants on plant growth and health. The phosphorus use efficiency was increased in common bean (*Phaseolus vulgaris L.*) for symbiotic nitrogen fixation using consortia of *Glomus intraradices*, a phosphate solubilizer and *R. tropici* CIAT899, a nitrogen fixer [44]. Gao *et al.* [45] evaluated the impact of consortia of *Bradyrhizobium sp.* BXYD3 and *G. mossae* in soybean show alteration of pathogen defence-related genes reducing the severity of *Cylindrocladium parasiticum* incidence. Draught resistance was observed in the finger millet plant using consortia of *Pseudomonas fluorescens* (KB-7), *Pseudomonas poae* (KA-5), *Streptomyces flavofuscus* (SA-11) and *Streptomyces labedae* (SB-9), thus increasing plant growth [46]. Several studies were performed using microbial consortia to grow and develop plants (Table **1**).

Table 1. Few important microbial consortia for crop improvement.

Microbial Consortia	Potential Effect (Plant/Crop Variety)	References
Penicillium simplicissimum and *Penicillium janthinelium*	Enhanced rosette leaves (Arabidopsis)	[47, 48]
Penicillium chrysogenum and *Trichoderma harzianum*	Early flowering onset (Tomato)	[49]
Fusarium oxysporum and *Trichoderma viridae*	Improved dry weight for shoot (Tomato, Arabidopsis)	[50, 51]
Bukholderia ambifaria, Bacillus megaterium Enterobacter cloacae Pantoea ananatis, and *Pseudomonas sp.*	Improved yield (Maize)	[52]
Enterobacter, Serratia, Pseudomonas, Microbacterium and *Achromobacter*	Drought and salt stress resistance (Avocado)	[53]
Pseudomonas sp. (2 strains) + Mixed Mycorrhiza	Improved nutritional and industrial features (Tomato)	[54]
Burkholderia sp. (2 strains)	Enhanced grain yield (40%); inhibition of *Fusarium* sp. (Fenugreek)	[55]
R. Tropici, *Bacillus megaterium*	Improved growth in low phosphorus soil (*Phaseolus vulgaris*)	[56]
Pseudomonas, Enterobacter and *Serratia*	No statistically significant improvement in yield (Rapeseed)	[57]
Mesorhizobium cicero, Anabaena	Modulates plant physiological attributes (*Cicer arietinum*)	[58]
Bacillus cereus, Bacillus sp., and *Bacillus subtilis*	Enhanced production in salinity stress (Wheat)	[59]

(Table 1) cont.....

Microbial Consortia	Potential Effect (Plant/Crop Variety)	References
Xanthomonas sp., Stenotrophomonas sp., and *Microbacterium sp.*	Biocontrol agent against *Hyaloperonospora arabisopsidis* (Arabidospis thaliana)	[60]
Acinetobacter sp., Rahnella aquatilis, Ensifer meliloti Glomus (2 strains), *Sclerocystis sp.,* and *Acaulspora sp.*	Improved parameters (Wheat and *Vicia faba*)	[61]
Pseudomonas syringae pv. Syringae and *Pseudomonas tolaasii*	Enhanced root formation with biocontrol activity (Carrot)	[62]
Pseudomonas fluorescens and *Funneliformis mosseae*	Reproductive and vegetative traits enhancement with improved yield, increased N and P uptake and increased maize root colonization. (*Zea mays*)	[63]
Bacillus elkanii	Improved nodulation, nutrient uptake, nitrogen fixation, plant growth and yield (*Glycine max*)	[64]
Bradyrhizobium and *P. aeruginosa*	Improved nodule weight (Soybean)	[65]
Pseudomonas putida and *Bacillus subtilis*	Systemic resistance to Macrophomina phaseolina (Mung bean)	[66]
Gracilaria corticata, Kappaphycus alvarezii, Padina pavonica, Sargassum johnstonii Ulva lactuca and *Ulva reticulate.*	Aged seed germination (*Allium cepa* L., *Brassica oleracea* var. capitata)	[67]
Rhizophagus irregularis and *serendipita indica*	Restricted lead uptake (*Osmium basilicum* (sweet basil)	[68]
Ochrobactrum pseudogrignonense, Pseudomonas sp., and *Bacillus subtilis*	Drought resistance (Black gram and Pea)	[69]
R. legumnosarum and *P. fluorescens*	Improved growth, yield, symbiotic association with 50% phosphorus reduction (Lentils)	[70]
R. Tropici and *Serratia grimesii*	Early nodulation and growth (*Phaseolus vulgaris*)	[71]
Pseudomonas putida, Novoshingobium sp.	Reduction in abscisic and salicylic acid, thus reducing salt stress effect (*Cirus macrophylla*)	[72]
Funneliformis mosseae and *Diversispora versiformis*	Enhanced root and shoot development and nitrogen content in roots; salt stress resistance (*Chrysanthemum morifolium*)	[73]
Rhizophagus irregularis, and *Bacillus amyloliquefaciens*	Improved biomass and photosynthetic efficiency (*Antennaria dioica, Campanula rotunifolia, Fragaria vesca, Geranium sanguineum, Lotus corniculants Thymus serpyllum, Trifolium repents, Viola tricolor*)	[74]
Pseudomonas sp. and *Rhizobium sullae*	Enhanced growth and antioxidant level with reduced Cd accumulation (*Sulla coronaria*)	[75]
Brevibacillus fluminis, Brevibacillus agri, and *Bacillus paralicheniformis*	Salt stress management (Brinjal, potato, tomato and chilli)	[76]

(Table 1) cont.....

Microbial Consortia	Potential Effect (Plant/Crop Variety)	References
Pseudomonas sp., and *Bacillus sp.*	Promoted growth, yield and nutritional status (Tomato)	[77]
Brettanomyces naardensis, Acaulospora bireticulata, Funneliformis sp.	Reduced incidence of diseases such as root rot and charcoal rot (*Helianthus annus*)	[78]
R. Tropici, *Pseudomonas fluorescens*	Promoted nodulation (*Phaseolus vulgaris*)	[79]
Bacillus subtilis, Bacillus megaterium, and *Bacillus sp.*	Increased yield of seeds and essential oil contents in plants (*Cuminum cyminum*)	[80]
Azotobacter vinelandii, and *Rhizophagus irregularis*	Promoted cluster shift at high fertilization levels and limits root growth at low fertilization levels (Wheat)	[81]
Bacillus diazoefficiens, and *Bacillus velezensis*	Promoted nodule growth and nitrogen fixation (*Glycine max*)	[82]
Bacillus megaterium, and *P. agglomerans*	Alleviation of aluminium and draught stress (*Vigna radiata*)	[83]
Trichoderma sp., and *Pichia guilliermondi*	Enhanced biomass and fruit yield with better shoot growth (Tomato)	[84]
Trichoderma atroviride, and *B. amyloliquefaciens*	Resistance against *F. graminearum* (Wheat, Maize)	[85]
Rhizophagus irregularis, Pseudomonas jessenii, and *P. synxantha*	Enhanced activities of alkaline phosphate, dehydrogenase in soil along with colonization of PGPR (*Triticum aestivum*)	[86]
Pantoea alli, Pseudomonas reactans, and *Rhizoglomus irregular*	Increase in K^+ content and decrease in Na^+ content in plant tissues (*Zea mays*)	[87]
Pseudomonas sp. and, *Serratia proteamaculans, Alkaligenes sp.*, and *Bacillus sp.*	Growth promotion in a saline petroleum contaminated soil (Saltgrass)	[88]

Algae

Algae are essential microbes in soil which affect various crops' growth and yield through different mechanisms [89]. A consortium of *Codium tomentosum, Gracilaria gracilis* and *Cystoseira barbata* positively influenced seed germination for tomato, pepper and aubergine [90]. Similarly, salinity stress was alleviated in *Capsicum annuum var.* using algal extracts of *Jania rubens* and *Padina pavonica* [91]. *Nostoc muscorum* and *Ulva lactua,* along with *Rhizobium leguminosarum,* influence the overall growth of faba beans in terms of improved root and shoot length, dry weight of nodules, pods and other growth parameters along with nutritional status of the plant [92]. Enhanced growth parameters, carbohydrate content and seed germination reported for wheat using algal fertilizer of *Gracilaria corticata, Nizimuddinia zunardini* and *Ulva fasciata* [93]. Similarly, consortia of *Stephanoystis crassipes, Neohodamela larix* and *Ahnfeltiopsis flabelliformis* acts as a biofertilizer enhancing the growth of buckwheat [94]. The photosynthetic performance and growth of willow (*Salix viminalis L.*) enhanced using consortium of *Anabaena sp.* PCC 7120, *Microcystis aeruginosa* MKR 0105 and *Chlorella sp.,* under limited fertilizer content [95].

Thus, the combined effect of plant growth-promoting bacteria (PGPR) with other microbe's increases plant/crop biomass and yield, provides abiotic stress resistance, improves nutrient uptake, act as a biocontrol agent and therefore, these consortia needs to be employed commercially for complete benefits package [96].

BIO-STIMULANTS: INTERACTION OF ROOT EXTRACT WITH SOIL MICROBES

Agriculture is facing simultaneous challenges of increasing productivity to feed the growing world population while at the same time reducing the environmental effects on ecosystems and human health. Several groundbreaking technological ideas have been suggested to aid sustainability in agricultural production systems by using a decreased usage of synthetic pesticides and fertilizers. An eco-sustainable and promising innovation is the use of plant biostimulants (PB) that boost the growth of plants, flowering, fruit development, crop yields, and enhance nutrition efficiency, as well as enhance the tolerance to stresses (Biotic & abiotic) [97]. Plant-microbial interaction is one of the primary form of communication that defines the zone below ground (Fig. **2**). Certain substances identified in root exudates play a vital role in root-microbe interactions, including flavonoids found in legume root exudates, which trigger the Rhizobium meliloti genes required for nodulation. Microbial interactions promote plant growth in a number of ways, including biological nitrogen fixation by various classes of proteobacteria, stress tolerance provided by the involvement of endophytic microbes, and direct and indirect benefits conferred by plant growth-promoting *Rhizobacteria* (PGPR) [98]. Additionally, exudation provides a carbon-rich environment, and plant roots also produce signals which initiate cross-talks with the soil microbes (Fig. **2**). Nitrogen-fixing interaction has been observed in tree roots and the filamentous, gram-positive actinobacterium *Frankia*, with 200 angiosperm species belonging to eight families [99]. Various studies have reported nitrogen fixing bacteria can solubilize and mineralize inorganic and organic pools of soil phosphorus, which convert it into plant-available form, resulting in increased uptake of phosphorus in plants [100]. Most fungi have plant growth promoting properties and have possessed the ability to solubilize P and enhance N uptake in host plants [101, 102]. Up to 70–90% of plant P is supplied by arbuscular mycorrhizal fungi (AMF), and their contribution greatly improve plant growth under low P condition [103, 104]. There is also evidence that plant colonization by AMF is related to enhancing N uptake [105] and improving drought tolerance [106].

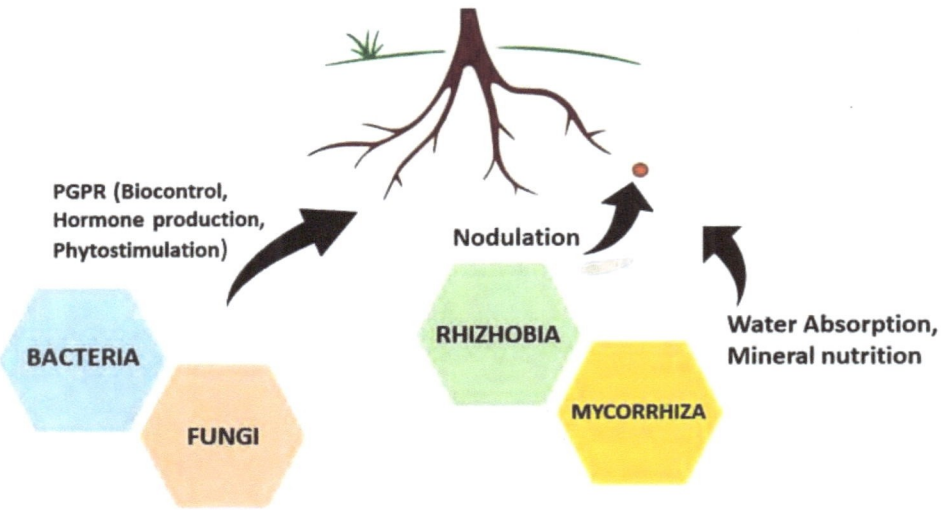

Fig. (2). Microbial interaction with the roots of the plant.

Many asymbiotic relationships have been drawn between microbes and plant roots, such as *Azospirillum* with grass family crops like *Hordeum vulgare, Sorghum bicolor*, and *Triticum aestivum, Acetobacter* associated with *Saccharum officinarum* or *Ipomoea batatas* and *Achromobacter* with *Oryza sativa* [107]. Previous study reported [108] the presence of tryptophan found in the root tip region. Tryptophan is the precursor for indole acetic acid, which suggest that PGPR have been utilizing root exudate pools as a source for promoting plant growth. Several soil bacteria are known to synthesize growth hormones, which have an impact on plant growth. Production of gibberellic acid and cytokinin was observed in *Arthrobacter* [109], *Azospirillum* [110], and *Azotobacter* [111]. These approaches may involve the discovery of new PGP microbes in agricultural fields [112], which help to find the existence of an essential root microbiome that will help a crop better cope with abiotic stress [113].

Recently, in sorghum seedlings, *Streptomyces* isolates showed moderate PGPR activity by enhancing the growth of root [114]. The relative abundance of one sequence variant from the genus *Streptomyces* is positively associated with drought tolerance in plant species [115]. Several other examples exist, including the ability to tolerate drought through higher photosynthesis, evapotranspiration, and stomatal conductance in *Capsicum annuum* that has been inoculated with different root bacteria obtained from naturally drought-tolerant plants [116]. Several species of micro-organisms including; *Pseudomonas* spp., *Acinetobacter* spp., *Azospirillum* ssp., and various AMF have been identified which enhance the uptake of Zn [117], Cu, Mn [118], Ca, and Mg [119].

AGRONOMICALLY IMPORTANT SOIL MICROBES

Sustainable agriculture for global food security is an urgent need for future generations, causing minimum deterioration of the ecosystem [120, 121]. Excessive use of chemical fertilizers and pesticides can be detrimental to soil quality [122, 123]. Thus, beneficial plant-associated microbes could be used for crop improvement in terms of nitrogen fixation, phosphate and potash solubilization, siderophore and phytohormone production, and biotic and abiotic stress tolerance for environmental benefits [124, 125].

Nitrogen-Fixing Bacteria

Nitrogen fixation is a pivotal phenomenon to make Nitrogen (N) available for plant growth and development. Nitrogenase, an oxygen-sensitive enzyme complex, converts atmospheric nitrogen into ammonia utilizing ATP as the energy source [23]. Pathania *et al.* [126] emphasized that the microbial sp. such as *Azospirillum, Azotobacter, Bacillus, Burkholderia, Cyanobacteria*, Enterobacter, Erwinia, Flavobacterium, Gluconacetobacter diazotrophicus, Pseudomonas, Rhizobium and *Stenotrophomonas* can fix atmospheric nitrogen (Table **1**).

Azotobacter chroocochum

Azotobacter is free-living bacteria causing non-symbiotic nitrogen-fixation, although some fix molecular nitrogen from the atmosphere symbiotically [127]. Ammonium ions and nitrate inhibits nitrogen fixation [128].

Azotobacter vinelandii

Azotobacter is a gram-negative diazotroph causing non-symbiotic nitrogen fixation aerobically. They produce phytohormones, vitamins and pyoverdine pigments [129].

Glucanobacter diazotrophicus

Glucanobacter is a nitrogen-fixing bioinoculant associated with sugar-rich plants and also found with other types of plants [130].

Acetobacter xylinum

Acetobacter oxidizes lactate and acetate into carbon dioxide and water. It belongs to the genus of acetic acid bacteria capable of converting ethanol into acetic acid in the presence of oxygen [131].

Azospirillum lipoferum

It is a gram-negative free-living bacteria affecting the growth and yield of many plants by producing phytohormones [132].

Rhizobium sp.

Rhizobium sp. fix atmospheric nitrogen and lives in a symbiotic relationship with legumes such as Peas, Lathyrus, Vicia, Lentils, Berseem, Kidney beans, Lupinus, Ornithopus, Soybean, Melilotus, Lucerne and Fenugreek [133].

Phosphate Solubilizing Microbes

Phosphorus is an essential element for the growth of plants acquired in the form of phosphate ions from the soil [134]. The most cost-effective and sustainable approach is the use of phosphate solubilizing microbes to make phosphorus available to the plant through mineralization and solubilization of inorganic phosphorus [135], and many genera of bacteria and fungi as phosphate solubilizers (Table **2**). They release organic acids such as oxalic acid, succinic acid and malic acid, thus decreasing the surrounding pH and releasing phosphate ions to make them available [136].

Table 2. Some agronomically beneficial microbes.

Microbes	Agronomical Significance (Targets)	References
Nitrogen Fixing Bacteria		
Azotobacter chroocochum, and *Azotobacter vinelandii*	Non symbiotic nitrogen fixation (Wheat, Sugarcane)	[127 - 129]
Azotobacter strain Azo-8	Nitrogen fixation with manure and urea (Wheat)	[157]
Glucanobacter diazotrophicus	Non symbiotic nitrogen fixation (Sugarcane)	[130]
Acetobacter xylinum	Non symbiotic nitrogen fixation (Rice)	[131, 158, 159]
Azospirillum lipoferum	Non symbiotic nitrogen fixation (Sorghum)	[158, 160]
Burkholderia, Gluconabacter, and *Pseudomonas spp.*	Non symbiotic nitrogen fixation (Sugarcane)	[139, 160]
Clostridia, and *Klebsiella*	Free living anaerobic nitrogen fixation (Different plant varieties)	[158]
Rhizobium leguminosarum, Rhizobium Tripoli, Rhizobium phaseoli, Rhizobium lupine, Rhizobium japonicum, and *Rhizobium meliloti*	Symbiotic nitrogen fixation (Peas, *Lathyrus, Vicia, Lentils, Berseem, Kidney beans, Lupinus, Ornithopus, Soybean, Melilotus, Lucerne,* and *fenugreek*	[133]
Anabaena	Symbiotic nitrogen fixation (Leguminous crops)	[158]

(Table 2) cont.....

Microbes	Agronomical Significance (Targets)	References
Phoshphate/Potassium/ Potash Solubilization Microbes		
Bacillus megaterium, Pseudomonas putida, Pseudomonasstriate, Aspergillus awaneorii, and *Bacillus polymyxa*	Phosphate solubilization (Wheat, Soybean)	[123, 161 - 164]
Trichoderma harzianum Rifai 1295-22	Availability of phosphate, iron and manganese (Crack willow (*Salix fragilis*)	[165, 166]
Potassium/ Potash Solubilization Microbes		
Frateuria aurentia	Potash mobilization (Paddy, Sorghum, Groundnut)	[139, 167, 168]
Bacillus edaphicus	Potassium mobilisation (Paddy, Sorghum, Groundnut)	[139, 167, 168]
Plant Growth Promoting Rhizobacteria (PGPR)		
Azospirillum, Methylobacterium, and *Bacillus* spp.	Enhanced production with NPK fertilizer (Cotton)	[169, 170]
Pseudomonas aeruginosa Z5, and *Bacillus fusiformis S10*	High production with reduced fertilizer (Cotton)	[171]
Bacillus subtilis, Bacillus polymyxa, Pseudomonas fluorescene, and *Pseudomonas putida*	Plant growth promotion (Cotton)	[123, 142 - 144]
Azospririllum brasilense, Azospirillum lipoferum, Paenibacillus, Providencia, and *Pseudomonas fluorescens*	Auxin production	[172]
Biological Control Organisms		
Pseudomonas chlororaphis, and *Pseudomonas azotoformans*	Biopesticide (Barley, Oats, Wheat, Pea)	[173]
Bacillus firmus	Controls Nematodes (Corn, Soybean, Cotton, Sorghum)	[174]
Pseudomonas putida, and *Pseudomonas fluorescens*	Siderophore production	[175]
Pseudomonas strains GRP3A and *PRS9*	Siderophore production, Iron chelation (Maiza)	[176]
Pseudomonas fluorescens, and *Pseudomonas sp.*	Control growth of *Pythium ultimum* and *Xanthomonas campestris* through production of hydrogen cyanide (Tomato)	[177 - 179]
Bacillus subtilis	Controls growth of Verticillium dahlia (Over 400 plant species)	[146, 180]
Metarhizium anisopliae	Insecticide (Grasshopper, Termites, Thrips, Caterpillers, and Aphids)	[123, 151]

(Table 2) cont.....

Microbes	Agronomical Significance (Targets)	References
Beauveria bassiana	Insecticide (Termites, whitefly)	
Verticillium lecanii	Insecticide (Aphids, Whiteflies, rust fungi, and Scale insects)	
Paecilomyces lilacinus, and *Arthrobotrys spp.*	Nematicide (Potato)	
Trichoderma viride	Fungicide (Cucumbers, Tomato, Cabbage, Pepper, various Ornamentals, Cereals and Grain Legume crops)	[123]
Trichoderma virens	Controls growth of *Pythium ultimum,* and *Rhizopus oryzae* (Cotton)	[181]
Gliocladium	Controls pathogens growth (Different plant varieties)	[182]
Trichoderma harzianum Rifai 1295-22	Controls growth of plant pathogens (Crack willow (*Salix fragilis*)	[165, 166]
Microbes for Stress Tolerance		
Trichoderma harzianum	Drought and salinity stress (Wheat)	[157]
Mycorrhizal Fungi	Heavy metal immobilisation (Plant roots)	[155]
Rhizobacteria	Drought, Salt and Salinity Stress (Dry lands)	[153]
Pseudomonas putida	Water stress (Sunflower roots)	[156]

Bacillus megaterium

It is a gram-positive, rod-shaped spore-forming bacteria capable of phosphorus solubilization. It is also a cytokinin promoting bacterium capable of plant root overgrowth [137].

Pseudomonas putida

It lives in most soils, associated with plant roots improving plant health through phosphate solubilization. It also produces siderophores, limiting the growth of fungi and other bacteria [123].

Potash Mobilizing Bacteria

Frateuria aurentia

It is a species of proteobacteria that works well in soil with low K content to mobilize available potash near the plant's roots. The availability of potash can be increased by the use of such bacteria in powder form [138].

Plant Growth-Promoting *Rhizobacteria* (PGPR)

PGPR causes phytostimulation, *i.e.*, production of phytohormones such as auxins, cytokinins, gibberellins, indole 3 acetic acid, abscisic acid and ethylene. Phytohormones enhance plant growth by root initiation, cell enlargement, and cell division [139]. *Azotobacter, Azospirillum, Bacillus, Pseudomonas* and *Rhizobium* are the PGPR known to produce phytohormones and may be used with biofertilization [140]. These phytohormones enhance plant growth by altering the endogenous mechanism of the plant (Table **2**).

Bacillus sp.

Bacillus sp., such as *Bacillus subtilis* and *Bacillus polymyxa*, are gram-positive spore-forming bacteria. *Bacillus subtilis* protect the plant throughout the growing season by colonizing the developing root system of plants. On the other hand, *Bacillus polymyxa* produces exopolysaccharides and causes root hairs to undergo physical changes promoting plant growth [141].

Pseudomonas sp.

Pseudomonas fluoroescens is a non-pathogenic saprophyte producing several secondary metabolites that suppress plant diseases and colonize soil, water and plant surface environments. *Pseudomonas putida* shows mutual interaction with *Saccharomyces cerevisiae*, regulating plant health [142, 143].

Biological Control Organisms

Biocontrol agents are the rhizospheric microbes playing a role in protecting plants from various pathogens (Table **1**). Antagonism, competition and induced resistance are some of the common methods for microbial-based pathogen control. *Aeromonas, Alcaligenens, Bacillus, Pseudomonas, Stenotrophomonas maltophilia, Trichoderma* and *Rhizobium* are some of the rhizospheric microbes which release antibiotics, biosurfactants, toxins, chitinase, β-1, 3-glucanase and volatile organic compounds that cause inhibition of growth of plant pathogens [144 - 146]. These microbes can also create competition for nutrients and trace elements required for growth and development. *e.g.* Siderophore production by *Pseudomonas sp* [147]. They are also able to produce ethylene, jasmonic acid and salicylic acid, which helps to defend plants against pathogens by induced systemic resistance [148].

Metarhizium anisopilae

It is an entomopathogenic fungus controlling several insect pests such as Grasshoppers, Termites, Thrips, Caterpillars, Aphids and many more. It causes infection in the insect by attaching to the insect's surface, penetrating the exoskeleton and causing the insect's death by proliferating inside [123].

Beauveria bassiana

It is a naturally occurring entomopathogenic fungus, functioning as an insecticide, controlling termites, whiteflies and many other insects. The spores are sprayed on affected crops causing the killing of an insect in 48-72h [123]. Its use as a mosquito control agent is still under investigation [149].

Verticillium lecanii

It is a biological pesticide producing insecticidal toxins such as bassainolide, dipicolinic acid controlling aphids, whiteflies, rust fungi, thrips, and scale insects [123].

Paecilomyces lilacinus

It is a naturally occurring fungus controlling nematodes attacking plant roots. The mechanism of action as nematicide is by infecting eggs, juvenile and adult females.

Arthrobotrys spp.

Arthobotrys oligospora is a biological indicator of nematodes, potentially used as a nematicide [150].

Trichoderma viride

Trichoderma is an antagonistic fungus acting as a fungicide for different plants preventing various diseases such as root rots, wilts, brown rots and other diseases. *Botritis, Fusarium, and Sclerotinia* are some of the fungal species suppressed by *Trichoderma* [151].

Microbes for Stress Tolerance

Rhizobacteria play a vital role in stress tolerance in plants. Drought, salt, salinity, and heavy metal tolerances are some of the stress conditions tolerated by hormonal modification and exopolysaccharide (EPS) secretion in plants [152]. *Pseudomonas sp., Trichoderma sp.* and *Hebeloma sp.* are some of the microbes involved in stress tolerance [153 - 156].

Pseudomonas putida

It is a gram negative bacterium found abundantly in soils. *P. putida* synthesizes EPS in sunflower roots while maintaining water availability during water stress [155].

Trichoderma harzianum

Trichoderma is a fungus known to combat salinity and drought stress in wheat varieties [156].

Mycorrihizal Fungi

Hebeloma sp., a Mycorrhizal fungus, has shown its benefits in nitrogen and phosphorus limitations in unfavourable soil pH [153]. They can also immobilize metals, thus reducing heavy metal contamination [154].

GENOME EDITING OF MICROBES TO BENEFIT CROP PLANTS

Genetic engineering is commonly seen in bacteria, yeast, and other fungi to develop agriculturally profitable crops. Bacteria are known to generate numerous biochemical and by-products, which assist plant roots in getting nutrients from the soil. By altering the genetic makeup of microbes, the biosynthetic pathway of these biochemicals or bioproducts have been regulated. Previously, for the modification of the genome in microorganisms, various approaches like homologous recombination, Group II retrohoming, and automated multiplex genome engineering has been used [183, 184]. But all of these methods proved to be laborious and time-consuming. In 2013, CRISPR/Cas was explored as a potential genome modification approach in *E. coli* [185]. Afterward, in other *Saccharomyces cerevisiae* and *Streptomyces species*, it was effectively applied [186, 187]. In agriculturally important microbes, genome editing approaches have been extended, *i.e.*, *B. subtilis* and *B. mycoides*, and fungal pathogens, *i.e.*, *Neurospora crassa, Myceliopthora heterotalica, Aspergillus niger*, and *Aspergillus oryzae, etc* [188].

Plant growth associated bacterial species are usually colonized near roots and release siderophore or related biochemical by-products. Potato endospore and rhizophore of grasses are associated with soil-borne bacteria such as *B. mycoides* EC18 and *B. subtilis* HS3. Both of these bacteria have shown antifungal, endophytic, and plant growth promoting function. Traditional methodology of genome editing, on the other hand, makes genetic modification complicated inside the genome. Recent advances in CRISPR/Cas9 based genome editing have been utilized to develop three *B. subtilis* HS3 mutants and two *B. mycoides* EC18 mutants, respectively [189]. *B. subtilis* HS3 release a volatile organic compound 2, 3-Butanediol, which is known to promote growth and development in grass [190]. Through modifying two genes in *B. mycoides*, Yi *et al.* [189] demonstrated that petrobactin is crucial for growth of plants *via* root colonization, respectively. Recently, several studies have reported use of advanced approaches of genome editing enables to modify *E. coli* genome as to our convenience. Heo *et al.* [191], demonstrated CRISPR-Cas9-directed citrate synthase gene modification in the genome of *E.coli* led to an enhancement in the production of n-butanol. In another study, through CRISPR/Cas9, the β-carotene pathway has been integrated into the *E. coli* genome. They modified the methylerythritol-phosphate and metabolic pathways to enhance the production of β-carotene [192]. In the current scenario, the numerous pathogenic fungal species, such as Puccinia, Fusarium, and others like Blumeria, cause severe damage in several crops such as *Triticum aestivum*, *Oryza sativa*, *Zea mays*, and *Sorghum bicolor*. Several techniques have been utilized for controlling losses due to these diseases, such as the use of non-pathogenic fungal antagonists, conventional breeding, and genetic manipulation. In this concern, the most promising approach for developing fungal-resistance crops is genetic engineering. Fungal disease in plants can be managed by inhibiting infection, growth and reproduction using a competing fungal species [193]. Mutant non-pathogenic fungi could be developed by utilizing CRISPR/Cas9 approach, which could be used to form new competitors for the wild type existing pathogen. Only a small number of fungi serve as cell factories, which could be utilized for the biosynthesis of secondary metabolites [188]. Qin *et al.* [194] demonstrated the knockout of the *ura3* gene in *Ganoderma lucidum* 260125 and *Ganoderma lingzhi* using CRISPR/Cas9 approach. These fungi produce anti-tumor and anti-metastatic ganoderic acids. Another study in durum wheat reported a reduction in crown and foot rot disease percentage range from 40 to 80% by altering trichothecene biosynthesis [195]. The genome modification approach provide a toolkit for pathway engineering in microbes and also ways to modify putative genes involved in pathogenicity, which will help to develop disease resistant agriculturally important crops.

TRANSFER OF MICROBIAL GENE INTO PLANT SPECIES

Genetic engineering facilitates the easy transfer of genes, paving the way for crop improvement through enhanced yield, and resistance to abiotic stress, disease, pest and herbicide. To date, many direct and indirect methods have been developed (Table 3), but gene transfer through *Agrobacterium* is the most efficiently utilized method for crop improvement. Tobacco leaf tissues were used to produce the first genetically modified plant with *Agrobacterium tumefaciens* in 1982 [196]. Nearly 120 crop species, such as rice, wheat, maize, soybean, tobacco and cotton, were genetically modified in plant breeding experiments through this method [197 - 200]. Every technique of gene transfer has its pros and cons (Table **1**), but there is a continuous improvement in gene transformation approaches in the last three decades, leading to significant improvement in agricultural production, crop production, and crop improvement [201].

Table 3. Gene transfer methods.

Method (Attributes)	Advantages & Disadvantages	References
Indirect Method		
A. Agrobacterium-mediated gene transfer (Transfer of gene through Ti plasmid containing T-DNA)	**Advantages:** Facilitates *in vitro* gene transfer; stable transformation; high frequency of transformation; **Disadvantages:** Accidental release of modified Agrobacterium into the environment; Host-specific	[202, 203]
B. Non-Agrobacterium based method (Transfer of gene through plasmid of Rhizobia spp)	**Advantages:** Alternate approach; **Disadvantages:** Limited host range, low frequency of transformation	[204]
Viral mediated gene transfer (Integration of gene of interest into the viral genome)	**Advantages:** Wide range of hosts; **Disadvantages:** High copy number per cell, transient transformation	[205]
Direct Method		
A. Physical Method		
1. Particle gun/ bolistic/ ballistic method (Microcarrier coated with DNA shot under high pressure)	**Advantages:** Wide range of hosts; **Disadvantages:** Tissue damage, high copy no. of the transgene in single target	[206, 207]
2. Electroporation (DNA transfer through pores facilitated by an electric field)	**Advantages:** Both transient and stable transformation; **Disadvantages:** Difficulty in the standardization of dose	[208]
3. Microinjection (Surgical technique through the use of micropipette)	**Advantages:** Easy methodology; **Disadvantages:** Time-consuming	[209]
4. Sonoporation (DNA transfer through pores facilitated by ultrasound waves)	**Advantages:** Safe delivery of DNA; **Disadvantages:** The method is still under research work	[210, 211]
5. Hydrodynamic gene transfer (Destabilization of cell membranes through hydrostatic pressure)	**Advantages:** Easy; more readily utilized; **Disadvantages:** Needs more development	[212, 213]
B. Chemical method		
1. Polyethylene glycol (PEG) based (PEG attached to DNA and transfer DNA by penetrating the membrane)	**Advantages:** Simple methodology, Devoid of costly equipment; **Disadvantages:** Difficulty in the regeneration of plant from a protoplast	[214]

(Table 3) cont.....

Method (Attributes)	Advantages & Disadvantages	References
2. Liposome mediated gene delivery (Based on fusion of cationic liposome and cell surface)	**Advantages:** Easy formulation; less toxic; **Disadvantages:** Possible in limited numbers of spp	[215]
3. Fibre mediated gene delivery (Gene transfer facilitated through silicon carbide fiber)	**Advantages:** Simple; **Disadvantages:** Less transformation efficiency	[216]

The most extensively used herbicide used for killing weeds by non-selective mode of action is glyphosate and glufosinate. Glyphosate inhibits explicitly 5–enolpyruvyl shikimate-3 phosphate synthase (*EPSPS*) required for the biosynthesis of amino acid, playing a pivotal role in the shikimate pathway. Globally, the most widely grown herbicide-tolerant plant is Glyphosate-resistant soybean [217]. Glufosinate (also known as phosphinothricin) inhibits glutamine synthetase enzymes competitively [218]. Various herbicide-tolerant transgenic plants were engineered by transferring specific herbicidal genes from microbes into the plant cell (Table **1**). Transgenic plants have been obtained for many crop varieties such as sorghum, soybean, grapes, apricot and many more in the last three decades [219 - 224]. The cultivation of herbicide-resistant crops leads to increased yield and reduced cost due to simplified weed management strategies [225, 226].

Baloglu *et al.* [227] emphasized that agricultural productivity is drastically affected by the pest, providing the basis for developing insect resistance crops through genetic engineering approaches. Transfer of gene coding for crystal toxin (cry) and vegetative insecticidal protein (vip) from *Bacillus thuringiensis* and *Bacillus cereus* in plant cells provides resistance against various insects, as shown in Table **2** [228 - 230]. Cry toxin works by binding specifically to the receptor, inserts into the cell membrane of the insects midgut and forms pores, causing paralysis followed by death. All the functions are carried out by three domains of Cry protein [231]. The first commercially available insect-resistant crop was cotton, showing the incorporation of cry protein and resistance to *Lepidopteron* pest [232]. Alternatively, insecticidal genes from other sources, including plants and mammals, would be introduced into the plant cell to provide insect resistance [227].

Agricultural production worldwide is decreased due to various abiotic stress factors such as drought, heat, cold, flood and salinity [233, 234]. Plants alter their metabolism (activating signalling cascade and regulatory factors such as transcription and heat shock factors) to withstand abiotic stress [235]. Wani *et al.* [236] suggested the utilization of plant biotechnology, genetics and breeding approaches to develop climate resilient crops to overcome the effect of environmental stress. Various microbial genes have been transferred into different plant varieties to mitigate drought, salt and osmotic stress (Table 4).

Table 4. Few recent transgene/transgene products from microbes used in genetic engineering for crop improvement.

Microbes (Transgene/Products)	Target & (Plant/Crop)	References
For Herbicide Resistance		
Streptomyces hygroscopicus (Bar)	Phosphinothricin (Glufosinate), Bialaphos; [*Salvia militiorrhiza, Lotus japonicas, Ipomoeabatatas* ,Tobacco, Tobacco, Tomato, and Sweet potato]	[221, 237 - 240]
Klebsiella ozaenae (Bxn)	Bromoxynil (3,5-dibromo-4-hydroxybezonitrile); [Tobacco, *Trifolium subterraneum* L.]	[241, 242]
Ochrobactrum anthropi (PgrA)	Paraquate [Tobacco]	[243]
Myxococcus xanthus (MxPPO)	Oxyfluorfen, acifluorfen [*Tall fiscue*]	[244]
Arabidopsis thaliana (AtDHAR1)	Methylviologen [Potato]	[245]
Pseudomonas putida (G6/EPSSPS)	Glyphosate [Rice]	[246]
Pseudomonas flouroscens; Bacillus licheniformis (G2/EPSSPS and GT)	Glyphosate [Soybean]	[247, 248]
Bradyrhizobium japonicum (DAAO)	Glyphosate [Arabidopsis]	[249]
Rhizobium sp. RC1 (Dehd)	Monochloroacetic acid [*Nicotiana benthamiana*]	[250]
For Insect Resistance		
Bacllus thuringiensis (Cry1Ac, Cry1Ac – 2A, Cry1Ab & Cry1Ac, Cry1Ac9 & Cry9Aa2, Cry1C, Cry1EC,Cry1Ab & vip3H, Cry1Ah, Cry2A-1Ac-gna,Cry2Ab, Cry2Aa, Cry3B, Cry3a, Cry3A	Lepidoptera, Rice leaf folder, Yellow stem borer, Sap sucking insect, *Phthorimaea operculella, Tryporyza incertulas, Chilo suppressalis, Cnaphalocrocis medicinalis* Guenec, *Spodoptera litura, Sogatella furcifera, Ostrinia furnacalis*, Corn borer, Rice weevil, rice hispa, Lepidoptera, *Phthorimaea operculetta, Helicoverpa armigera*, Fruit borer, Coleoptera, and *Leptinotarsa decemlineata* [Canola, Rice, Indian rice, Potato, Japonica rice, Cotton, Tomato, Chickpea, Brinjal, Alfalfa, and Potato]	[228, 251 - 261]
Bacillus spp, (Bacillus thuringiensis & Bacillus cereus); (vip3A(a), vip3Aa20	NA (Cotton, Maize)	[231, 262]
For Abiotic Stress Resistance		
Escherichia coli (cspA)	Drought tolerance [Maize]	[263]
Bacillus subtilis (cspB)	Drought tolerance [Maize]	[263]

(Table 4) cont.....

Microbes (Transgene/Products)	Target & (Plant/Crop)	References
Escherichia coli (Mannitol -1-phosphate dehydrogenase (mtlD)	Salt and osmotic stress tolerance [Peanut]	[264]
Arthrobacter globiformis (Choline oxidase gene (codA)	Drought tolerance [Sweet potato]	[265]
For Improved Nutritional Values		
Provitamin A biofortified rice		
Erwinia uredovora, Pantoea ananatis & Escherichia coli strain K-12 [Carotene/ Phytoene desaturase gene crtI, crtI & Phosphomannose isomerase (pmi) gene	Vitamin A deficiency (VAD); [Rice cultivar (GR1, GR2, GR2E)]	[266 - 269]
Modified Oil/Fatty Acid Trait		
Lachancea kluyveri (Lackldelta12D), Pichia pastoris (Picpa-omega3D), Micromonas pusilla (Micpu-delta6D), Pyramimonas cordata (Pyrco-delta6E, Pyrco-delta5E)	Conversion of oleic acid to linoleic acid, Conversion of linoleic acid to a - linolenic acid, Conversion of a-linolenic acid to stearidonic acid, Conversion of stearidonic acid to eicosatetraenoic acid, Conversion of eicosapentaenoic acid to docosapentaenoic acid, and increased production of an omega-3 fatty acid, stearidonic acid (SDA); [*Brassica napus* (Argentine Canola)]	[231, 269]
Neurospora crassa (Nc.Fad3)	*Glycine max* L. (Soybean)	
Modified Amino Acid Trait		
Corynebacterium glutamicum (Dihydrodipicolinate synthase cordapA)	Increase free lysine content; [Maize Kernels]	[231, 269]

*NA; Data not available.

Engineering crops for nutritional improvement, such as increasing vitamin content and modified amino and fatty acids, are of utmost importance because of different deficiency and health diseases (Table **2**). The golden rice engineered for carotenoid biosynthesis was a breakthrough study to combat vitamin A deficiency (VAD). There were different golden rice versions to enhance carotenoid accumulation [266 - 269]. Oleic acid, linoleic acid and alpha-linolenic acid are essential amino acids derived from oils. According to WHO, a high proportion of polyunsaturated fatty acid (PUFAs) and low saturated fatty acid content is considered superior for human consumption [270]. Transgenic crop with modified lipid content was engineered, as shown in Table **2**. High levels of eicosapentaenoic acid (EPA) and docosahexaenoic acid (DHA) were produced by engineering *Camelina sativa* with the genes from marine microbes [271, 272]. Similarly, a few transgenic approaches targeted altering the amino acid composition to enhance nutritional value by increasing the amino acid concentration by engineering essential amino acid metabolic pathways [231].

Thus, crop varieties with improved yield, food quality and resistance to abiotic and biotic stresses were contributed through transgenic technology.

USE OF MICROBES FOR THE PRODUCTION OF BIOENERGY FROM AGRICULTURE WASTE

Agricultural crops biomass residue is a good source of bioenergy which is economically feasible and environmentally safe. It helps to meet the rising energy demand in the future and present while curbing greenhouse gas emissions. As demonstrated in various case studies, genetic and metabolic pathway engineering approaches have proven to be important in developing efficient microbial communities which are capable of contributing to bioenergy production [273]. A previous study reported for the production of ethanol, potato wastewater was utilized as a substrate with recombinant *Escherichia coli* strains [274]. Recombinant *E. coli* enables biodiesel production by transferring two enzymes required for the production of ethanol and one enzyme which encodes acyltransferase. Because ethanol was produced indirectly from sugar glycolysis, In future, it is necessary to search for more affordable and cheaper sugar sources necessary for further developments [275]. Similarly, in another study, *S. cerevisiae*, *Fusarium oxysporum*, and *Aspergillus foetidus* have been used in the production of ethanol, and *Malus domestica* pomace has been used as a substrate [276]. Another study on *Malus domestica* reported consistent alcohol production using a psychrophilic *S. cerevesiae* AX1 strain [277]. Wasted rice bran has been found to be a suitable feedstock for bioethanol production due to the presence of low lignin content. But in the case of sugarcane bagasse, prior pre-treatment need to remove high lignin content [278]. Ingale *et al.* [279] reported banana pseudo stem has been a potential substrate for bioethanol production. Various microorganisms such as, *Aspergillus ellipticus*, *Aspergillus fumigatus* and *S. cerevisiae* have been used for biological pre-treatments. In a recent study [280], lignin from barley straw has been utilized as a carbon source for the production of biodiesel using *Rhodococcus* sp. YHY01. The energy density of biodiesel is greater than that of bioethanol, so it can be used in diesel engines. Biodiesel production utilizing oleaginous microbes is a rapidly expanding research field, and several microbes have been identified in the production of biodiesel from distinct carbon sources such as *Cryptococcus curvatus*, *Yarrowia lipolytica*, *Chlorella sp.*, *Rhodococcu ssp.*, *etc* [280]. Similarly, rice straw hydrolysate has been used as substrate for biodiesel production with recombinant *Escherichia coli* strains [281]. Various biotechnological applications utilized rice straw hydrolysate as feedstock for the production of bioethanol and lipid using yeast [282, 283]. Utilizing agricultural waste as a sugar source in *E. coli* mediated production for biodiesel is a promising approach.

CONCLUSION

Plants cannot get nutrition directly from soil organic matter, therefore, largely depend on soil microbes because microbes are capable of decomposing organic residues and recycling soil nutrients. Although applying microbial consortium in agriculture looks promising approach but, organic farming alone cannot meet the global food demand, as there is no such universal microbial consortium that can be used for all types of lands. It was suggested to use both inorganic and organic nutrient sources at a 75:25 ratio. In the last decade, molecular breeding and genome editing techniques have grown enormously with higher precision. Henceforth, it will be interesting to identify and manipulate the genes of metabolic pathways that control the composition of root exudates, which mediates the signals between plant roots and the rhizospheric microbiome. Also, genetic manipulation/identification of a particular strain of soil microbes could enhance plant growth and development remarkably by enhancing the availability of humus, nitrate/ammonia, siderophores, nutrient recycling, and pest control.

ACKNOWLEDGEMENTS

RK would like to thank UGC for providing the UGC-BSR Start-up grant. V.S. acknowledges Chaudhary Charan Singh University (CCSU), Meerut, for collaborating with ICRISAT and the opportunity given as a student to pursue an investigation at ICRISAT. All the authors have read the MS carefully.

REFERENCES

[1]   Barea JM. Future challenges and perspectives for applying microbial biotechnology in sustainable agriculture based on a better understanding of plant-microbiome interactions. J Plant Nutr Soil Sci 2015; 15(2): 261-82.

[2]   Cassman KG, Dobermann A, Walters DT, Yang H. Meeting cereal demand while protecting natural resources and improving environmental quality. Annu Rev Environ Resour 2003; 28(1): 315-58.
[http://dx.doi.org/10.1146/annurev.energy.28.040202.122858]

[3]   Kirchmann H, Kätterer T, Bergström L, Börjesson G, Bolinder MA. Flaws and criteria for design and evaluation of comparative organic and conventional cropping systems. Field Crops Res 2016; 186: 99-106.
[http://dx.doi.org/10.1016/j.fcr.2015.11.006]

[4]   McGuire AM. Agricultural science and organic farming: Time to change our trajectory. Agric Environ Lett 2017; 2(1): 170024.
[http://dx.doi.org/10.2134/ael2017.08.0024]

[5]   Sarma BK, Yadav SK, Singh S, Singh HB. Microbial consortium-mediated plant defense against phytopathogens: Readdressing for enhancing efficacy. Soil Biol Biochem 2015; 87: 25-33.
[http://dx.doi.org/10.1016/j.soilbio.2015.04.001]

[6]   Hayat R, Ali S, Amara U, Khalid R, Ahmed I. Soil beneficial bacteria and their role in plant growth promotion: a review. Ann Microbiol 2010; 60(4): 579-98.
[http://dx.doi.org/10.1007/s13213-010-0117-1]

[7]   Dakora FD. Plant flavonoids: Biological molecules for useful exploitation. Funct Plant Biol 1995;

22(1): 87-99.
[http://dx.doi.org/10.1071/PP9950087]

[8] Viviene NM, Felix DD. Potential use of rhizobial bacteria as promoters of plant growth for increased yield in landraces of African cereal crops. Afr J Biotechnol 2004; 3(1): 1-7.
[http://dx.doi.org/10.5897/AJB2004.000-2002]

[9] Hahn L, Sá ELS, Osório Filho BD, Machado RG, Damasceno RG, Giongo A. Rhizobial inoculation, alone or coinoculated with *Azospirillum brasilense*, promotes growth of wetland rice. Rev Bras Ciênc Solo 2016; 40(0): 40.
[http://dx.doi.org/10.1590/18069657rbcs20160006]

[10] Vargas LK, Volpiano CG, Lisboa BB, Giongo A, Beneduzi A, Passaglia LMP. Potential of rhizobia as plant growth-promoting *Rhizobacteria*. Microbes for legume improvement. Cham: Springer 2017; pp. 153-74.
[http://dx.doi.org/10.1007/978-3-319-59174-2_7]

[11] Pandey A, Kumar S. Potential of *Azotobacters* and *Azospirilla* as biofertilizers for upland agriculture : A review. J Sci Ind Res 1989; 48(3): 134-44.

[12] Lewis LA. Hold the salt: Freshwater origin of primary plastids. Proc Natl Acad Sci USA 2017; 114(37): 9759-60.
[http://dx.doi.org/10.1073/pnas.1712956114] [PMID: 28860199]

[13] Demoulin CF, Lara YJ, Cornet L, *et al.* *Cyanobacteria* evolution: Insight from the fossil record. Free Radic Biol Med 2019; 140: 206-23.
[http://dx.doi.org/10.1016/j.freeradbiomed.2019.05.007] [PMID: 31078731]

[14] Sinha RP, Häder DP. Photobiology and ecophysiology of rice field *Cyanobacteria*. Photochem Photobiol 1996; 64(6): 887-96.
[http://dx.doi.org/10.1111/j.1751-1097.1996.tb01852.x]

[15] Chittapun S, Limbipichai S, Amnuaysin N, Boonkerd R, Charoensook M. Effects of using *Cyanobacteria* and fertilizer on growth and yield of rice, Pathum Thani I: A pot experiment. J Appl Phycol 2018; 30(1): 79-85.
[http://dx.doi.org/10.1007/s10811-017-1138-y]

[16] Kumar V, Basu MS, Rajendran TP. Mycotoxin research and mycoflora in some commercially important agricultural commodities. Crop Prot 2008; 27(6): 891-905.
[http://dx.doi.org/10.1016/j.cropro.2007.12.011]

[17] Kagot V, Okoth S, De Boevre M, De Saeger S. Biocontrol of Aspergillus and Fusarium mycotoxins in africa: Benefits and limitations. Toxins 2019; 11(2): 109.
[http://dx.doi.org/10.3390/toxins11020109] [PMID: 30781776]

[18] Yassein AS, El-Said AHM, El-Dawy EGA. Biocontrol of toxigenic *Aspergillus* strains isolated from baby foods by essential oils. Flavour Fragrance J 2020; 35(2): 182-9.
[http://dx.doi.org/10.1002/ffj.3551]

[19] Vinale F, Sivasithamparam K, Ghisalberti EL, Marra R, Woo SL, Lorito M. Trichoderma–plant–pathogen interactions. Soil Biol Biochem 2008; 40(1): 1-10.
[http://dx.doi.org/10.1016/j.soilbio.2007.07.002]

[20] Rocha I, Ma Y, Souza-Alonso P, Vosátka M, Freitas H, Oliveira RS. Seed coating: A tool for delivering beneficial microbes to agricultural crops. Front Plant Sci 2019; 10: 1357.
[http://dx.doi.org/10.3389/fpls.2019.01357] [PMID: 31781135]

[21] Ma Y. Seed coating with beneficial microorganisms for precision agriculture. Biotechnol Adv 2019; 37(7): 107423.
[http://dx.doi.org/10.1016/j.biotechadv.2019.107423] [PMID: 31398397]

[22] Bardhan P, Gupta K, Mandal M. Microbes as bio-resource for sustainable production of biofuels and other bioenergy products. New and future developments in microbial biotechnology and

bioengineering. Elsevier 2019; pp. 205-22.
[http://dx.doi.org/10.1016/B978-0-444-64191-5.00015-8]

[23] Kumar A, Verma JP. The role of microbes to improve crop productivity and soil health in Ecological wisdom inspired restoration engineering. Singapore: Springer 2019; pp. 249-65.
[http://dx.doi.org/10.1007/978-981-13-0149-0_14]

[24] Kosamia NM, Samavi M, Uprety BK, Rakshit SK. Valorization of biodiesel byproduct crude glycerol for the production of bioenergy and biochemicals. Catalysts 2020; 10(6): 609.
[http://dx.doi.org/10.3390/catal10060609]

[25] Mussatto SI, Machado EMS, Carneiro LM, Teixeira JA. Sugars metabolism and ethanol production by different yeast strains from coffee industry wastes hydrolysates. Appl Energy 2012; 92: 763-8.
[http://dx.doi.org/10.1016/j.apenergy.2011.08.020]

[26] Jang YS, Lee JY, Lee J, *et al.* Enhanced butanol production obtained by reinforcing the direct butanol-forming route in *Clostridium acetobutylicum*. MBio 2012; 3(5): e00314-12.
[http://dx.doi.org/10.1128/mBio.00314-12] [PMID: 23093384]

[27] Deeba F, Pruthi V, Negi YS. Fostering triacylglycerol accumulation in novel oleaginous yeast *Cryptococcus psychrotolerans* IITRFD utilizing groundnut shell for improved biodiesel production. Bioresour Technol 2017; 242: 113-20.
[http://dx.doi.org/10.1016/j.biortech.2017.04.001] [PMID: 28411053]

[28] Poontawee R, Yongmanitchai W, Limtong S. Lipid production from a mixture of sugarcane top hydrolysate and biodiesel-derived crude glycerol by the oleaginous red yeast, *Rhodosporidiobolus fluvialis*. Process Biochem 2018; 66: 150-61.
[http://dx.doi.org/10.1016/j.procbio.2017.11.020]

[29] Buée M, De Boer W, Martin F, van Overbeek L, Jurkevitch E. The rhizosphere zoo: An overview of plant-associated communities of microorganisms, including phages, bacteria, archaea, and fungi, and of some of their structuring factors. Plant Soil 2009; 321(1-2): 189-212.
[http://dx.doi.org/10.1007/s11104-009-9991-3]

[30] Ahemad M, Kibret M. Mechanisms and applications of plant growth promoting *Rhizobacteria*: Current perspective. J King Saud Univ Sci 2014; 26(1): 1-20.
[http://dx.doi.org/10.1016/j.jksus.2013.05.001]

[31] Akhtar N, Qureshi MA, Iqbal A, Ahmad MJ, Khan KH. Influence of *Azotobacter* and IAA on symbiotic performance of Rhizobium and yield parameters of lentil. J Agric Res 2012; 50: 361-72.

[32] Sayeed Akhtar M, Siddiqui ZA. Biocontrol of a root-rot disease complex of chickpea by *Glomus intraradices*, *Rhizobium* sp. and *Pseudomonas straita*. Crop Prot 2008; 27(3-5): 410-7.
[http://dx.doi.org/10.1016/j.cropro.2007.07.009]

[33] Verma JP, Yadav J, Tiwari KN, Kumar A. Effect of indigenous *Mesorhizobium* spp. and plant growth promoting *Rhizobacteria* on yields and nutrients uptake of chickpea (Cicer arietinum L.) under sustainable agriculture. Ecol Eng 2013; 51: 282-6.
[http://dx.doi.org/10.1016/j.ecoleng.2012.12.022]

[34] Fox SL, O'Hara GW, Bräu L. Enhanced nodulation and symbiotic effectiveness of Medicago truncatula when co-inoculated with *Pseudomonas fluorescens* WSM3457 and Ensifer (Sinorhizobium) medicae WSM419. Plant Soil 2011; 348(1-2): 245-54.
[http://dx.doi.org/10.1007/s11104-011-0959-8]

[35] Singh SP, Singh HB, Singh DK. *Trichoderma harzianum* and Pseudomonas sp. mediated management of Sclerotium rolfsii rot in tomato (*Lycopersicon esculentum* Mill.). Bioscan 2013; 8(3): 801-4.

[36] Jain A, Singh A, Singh S, Singh V, Singh HB. Comparative proteomic analysis in pea treated with microbial consortia of beneficial microbes reveals changes in the protein network to enhance resistance against *Sclerotinia sclerotiorum*. J Plant Physiol 2015; 182: 79-94.
[http://dx.doi.org/10.1016/j.jplph.2015.05.004] [PMID: 26067380]

[37] Thijs S, Weyens N, Sillen W, Gkorezis P, Carleer R, Vangronsveld J. Potential for plant growth promotion by a consortium of stress-tolerant 2,4-dinitrotoluene-degrading bacteria: Isolation and characterization of a military soil. Microb Biotechnol 2014; 7(4): 294-306.
[http://dx.doi.org/10.1111/1751-7915.12111] [PMID: 24467368]

[38] Jha Y, Subramanian RB. Paddy plants inoculated with PGPR show better growth physiology and nutrient content under saline conditions. Chil J Agric Res 2013; 73(3): 213-9.
[http://dx.doi.org/10.4067/S0718-58392013000300002]

[39] Smith DL, Praslickova D, Ilangumaran G. Inter-organismal signaling and management of the phytomicrobiome. Front Plant Sci 2015; 6: 722.
[http://dx.doi.org/10.3389/fpls.2015.00722] [PMID: 26442036]

[40] Prasanna R, Bidyarani N, Babu S, Hossain F, Shivay YS, Nain L. Cyanobacterial inoculation elicits plant defense response and enhanced Zn mobilization in maize hybrids. Cogent Food Agric 2015; 1(1): 998507.
[http://dx.doi.org/10.1080/23311932.2014.998507]

[41] Singh R, Parameswaran TN, Prakasa Rao EVS, *et al.* Effect of arbuscular mycorrhizal fungi and *Pseudomonas fluorescens* on root-rot and wilt, growth and yield of *Coleus forskohlii*. Biocontrol Sci Technol 2009; 19(8): 835-41.
[http://dx.doi.org/10.1080/09583150903137601]

[42] Reddy CA, Saravanan RS. Polymicrobial multi-functional approach for enhancement of crop productivity. Adv Appl Microbiol 2013; 82: 53-113.
[http://dx.doi.org/10.1016/B978-0-12-407679-2.00003-X]

[43] Sharma M, Ghosh R. An update on genetic resistance of chickpea to ascochyta blight. Agronomy 2016; 6(1): 18.
[http://dx.doi.org/10.3390/agronomy6010018]

[44] Tajini F, Trabelsi M, Drevon JJ. Combined inoculation with *Glomus intraradices* and *Rhizobium tropici* CIAT899 increases phosphorus use efficiency for symbiotic nitrogen fixation in common bean (*Phaseolus vulgaris* L.). Saudi J Biol Sci 2012; 19(2): 157-63.
[http://dx.doi.org/10.1016/j.sjbs.2011.11.003] [PMID: 23961175]

[45] Gao X, Lu X, Wu M, *et al.* Co-inoculation with rhizobia and AMF inhibited soybean red crown rot: from field study to plant defense-related gene expression analysis. PLoS One 2012; 7(3): e33977.
[http://dx.doi.org/10.1371/journal.pone.0033977] [PMID: 22442737]

[46] Kumar M, Karthikeyan N, Prasanna R. Priming of plant defense and plant growth in disease-challenged crops using microbial consortia. Microbial-mediated Induced Systemic Resistance in Plants. Singapore: Springer 2016; pp. 39-56.
[http://dx.doi.org/10.1007/978-981-10-0388-2_4]

[47] Hossain MM, Sultana F, Kubota M, Koyama H, Hyakumachi M. The plant growth-promoting fungus *Penicillium simplicissimum* GP17-2 induces resistance in *Arabidopsis thaliana* by activation of multiple defense signals. Plant Cell Physiol 2007; 48(12): 1724-36.
[http://dx.doi.org/10.1093/pcp/pcm144] [PMID: 17956859]

[48] Hossain MM, Sultana F, Kubota M, Hyakumachi M. Differential inducible defense mechanisms against bacterial speck pathogen in *Arabidopsis thaliana* by plant-growth-promoting-fungus *Penicillium* sp. GP16-2 and its cell free filtrate. Plant Soil 2008; 304(1-2): 227-39.
[http://dx.doi.org/10.1007/s11104-008-9542-3]

[49] Jogaiah S, Abdelrahman M, Tran LSP, Shin-ichi I. Characterization of rhizosphere fungi that mediate resistance in tomato against bacterial wilt disease. J Exp Bot 2013; 64(12): 3829-42.
[http://dx.doi.org/10.1093/jxb/ert212] [PMID: 23956415]

[50] Bitas V, McCartney N, Li N, *et al. Fusarium oxysporum* volatiles enhance plant growth *via* affecting auxin transport and signaling. Front Microbiol 2015; 6: 1248.

[http://dx.doi.org/10.3389/fmicb.2015.01248] [PMID: 26617587]

[51] Lee S, Yap M, Behringer G, Hung R, Bennett JW. Volatile organic compounds emitted by *Trichoderma* species mediate plant growth. Fungal Biol Biotechnol 2016; 3(1): 7.
[http://dx.doi.org/10.1186/s40694-016-0025-7] [PMID: 28955466]

[52] Kifle MH, Laing MD. Effects of selected diazotrophs on maize growth. Front Plant Sci 2016; 7: 1429.
[http://dx.doi.org/10.3389/fpls.2016.01429] [PMID: 27713756]

[53] Barra PJ, Inostroza NG, Mora ML, Crowley DE, Jorquera MA. Bacterial consortia inoculation mitigates the water shortage and salt stress in an avocado (*Persea americana* Mill.) nursery. Appl Soil Ecol 2017; 111: 39-47.
[http://dx.doi.org/10.1016/j.apsoil.2016.11.012]

[54] Bona E, Cantamessa S, Massa N, *et al*. Arbuscular mycorrhizal fungi and plant growth-promoting pseudomonads improve yield, quality and nutritional value of tomato: A field study. Mycorrhiza 2017; 27(1): 1-11.
[http://dx.doi.org/10.1007/s00572-016-0727-y] [PMID: 27539491]

[55] Kumar H, Dubey RC, Maheshwari DK. Seed-coating fenugreek with *Burkholderia Rhizobacteria* enhances yield in field trials and can combat Fusarium wilt. Rhizosphere 2017; 3: 92-9.
[http://dx.doi.org/10.1016/j.rhisph.2017.01.004]

[56] Korir H, Mungai NW, Thuita M, Hamba Y, Masso C. Co-inoculation effect of rhizobia and plant growth promoting *Rhizobacteria* on common bean growth in a low phosphorus soil. Front Plant Sci 2017; 8: 141.
[http://dx.doi.org/10.3389/fpls.2017.00141] [PMID: 28224000]

[57] Lally RD, Galbally P, Moreira AS, *et al*. Application of endophytic *Pseudomonas fluorescens* and a bacterial consortium to *Brassica napus* can increase plant height and biomass under greenhouse and field conditions. Front Plant Sci 2017; 8: 2193.
[http://dx.doi.org/10.3389/fpls.2017.02193] [PMID: 29312422]

[58] Prasanna R, Ramakrishnan B, Simranjit K, *et al*. Cyanobacterial and rhizobial inoculation modulates the plant physiological attributes and nodule microbial communities of chickpea. Arch Microbiol 2017; 199(9): 1311-23.
[http://dx.doi.org/10.1007/s00203-017-1405-y] [PMID: 28669069]

[59] Shahzad S, Khan MY, Zahir ZA, Asghar HN, Chaudhry UK. Comparative effectiveness of different carriers to improve the efficacy of bacterial consortium for enhancing wheat production under salt affected field conditions. Pak J Bot 2017; 49(4): 1523-30.

[60] Berendsen RL, Vismans G, Yu K, *et al*. Disease-induced assemblage of a plant-beneficial bacterial consortium. ISME J 2018; 12(6): 1496-507.
[http://dx.doi.org/10.1038/s41396-018-0093-1] [PMID: 29520025]

[61] Dal Cortivo C, Barion G, Ferrari M, *et al*. Effects of field inoculation with VAM and bacteria consortia on root growth and nutrients uptake in common wheat. Sustainability 2018; 10(9): 3286.
[http://dx.doi.org/10.3390/su10093286]

[62] Etminani F, Harighi B. Isolation and identification of endophytic bacteria with plant growth promoting activity and biocontrol potential from wild pistachio trees. Plant Pathol J 2018; 34(3): 208-17.
[http://dx.doi.org/10.5423/PPJ.OA.07.2017.0158] [PMID: 29887777]

[63] Ghorchiani M, Etesami H, Alikhani HA. Improvement of growth and yield of maize under water stress by co-inoculating an arbuscular mycorrhizal fungus and a plant growth promoting rhizobacterium together with phosphate fertilizers. Agric Ecosyst Environ 2018; 258: 59-70.
[http://dx.doi.org/10.1016/j.agee.2018.02.016]

[64] Htwe AZ, Moh SM, Moe K, Yamakawa T. Effects of co-inoculation of *Bradyrhizobium japonicum* SAY3-7 and *Streptomyces griseoflavus* P4 on plant growth, nodulation, nitrogen fixation, nutrient uptake, and yield of soybean in a field condition. Soil Sci Plant Nutr 2018; 64(2): 222-9.

[http://dx.doi.org/10.1080/00380768.2017.1421436]

[65] Kumawat KC, Sharma P, Sirari A, *et al.* Synergism of *Pseudomonas aeruginosa* (LSE-2) nodule endophyte with *Bradyrhizobium* sp. (LSBR-3) for improving plant growth, nutrient acquisition and soil health in soybean. World J Microbiol Biotechnol 2019; 35(3): 47.
[http://dx.doi.org/10.1007/s11274-019-2622-0] [PMID: 30834977]

[66] Sharma CK, Vishnoi VK, Dubey RC, Maheshwari DK. A twin rhizospheric bacterial consortium induces systemic resistance to a phytopathogen *Macrophomina phaseolina* in mung bean. Rhizosphere 2018; 5: 71-5.
[http://dx.doi.org/10.1016/j.rhisph.2018.01.001]

[67] Patel RV, Pandya KY, Jasrai RT, Brahmbhatt N. Efficacy of priming treatment on germination, development and enzyme activity of *Allium cepa* L. and *Brassica oleracea* var capitata. RJLBPCS 2018; 4(2): 55.

[68] Sabra M, Aboulnasr A, Franken P, Perreca E, Wright LP, Camehl I. Beneficial root endophytic fungi increase growth and quality parameters of sweet basil in heavy metal contaminated soil. Front Plant Sci 2018; 9: 1726.
[http://dx.doi.org/10.3389/fpls.2018.01726] [PMID: 30538713]

[69] Saikia J, Sarma RK, Dhandia R, *et al.* Alleviation of drought stress in pulse crops with ACC deaminase producing *Rhizobacteria* isolated from acidic soil of Northeast India. Sci Rep 2018; 8(1): 3560.
[http://dx.doi.org/10.1038/s41598-018-21921-w] [PMID: 29311619]

[70] Singh N, Singh G, Aggarwal N, Khanna V. Yield enhancement and phosphorus economy in lentil (*Lens culinaris* Medikus) with integrated use of phosphorus, *Rhizobium* and plant growth promoting *Rhizobacteria*. J Plant Nutr 2018; 41(6): 737-48.
[http://dx.doi.org/10.1080/01904167.2018.1425437]

[71] Tavares MJ, Nascimento FX, Glick BR, Rossi MJ. The expression of an exogenous ACC deaminase by the endophyte *Serratia grimesii* BXF1 promotes the early nodulation and growth of common bean. Lett Appl Microbiol 2018; 66(3): 252-9.
[http://dx.doi.org/10.1111/lam.12847] [PMID: 29327464]

[72] Vives-Peris V, Gómez-Cadenas A, Pérez-Clemente RM. Salt stress alleviation in citrus plants by plant growth-promoting *Rhizobacteria Pseudomonas putida* and *Novosphingobium* sp. Plant Cell Rep 2018; 37(11): 1557-69.
[http://dx.doi.org/10.1007/s00299-018-2328-z] [PMID: 30062625]

[73] Wang Y, Wang M, Li Y, Wu A, Huang J. Effects of arbuscular mycorrhizal fungi on growth and nitrogen uptake of *Chrysanthemum morifolium* under salt stress. PLoS One 2018; 13(4): e0196408.
[http://dx.doi.org/10.1371/journal.pone.0196408] [PMID: 29698448]

[74] Xie L, Lehvävirta S, Timonen S, Kasurinen J, Niemikapee J, Valkonen JPT. Species-specific synergistic effects of two plant growth—promoting microbes on green roof plant biomass and photosynthetic efficiency. PLoS One 2018; 13(12): e0209432.
[http://dx.doi.org/10.1371/journal.pone.0209432] [PMID: 30596699]

[75] Chiboub M, Jebara SH, Abid G, Jebara M. Co-inoculation effects of *Rhizobium sullae* and *Pseudomonas* sp. on growth, antioxidant status, and expression pattern of genes associated with heavy metal tolerance and accumulation of cadmium in *Sulla coronaria*. J Plant Growth Regul 2020; 39(1): 216-28.
[http://dx.doi.org/10.1007/s00344-019-09976-z]

[76] Goswami SK, Kashyap PL, Awasthi S. Deciphering rhizosphere microbiome for the development of novel bacterial consortium and its evaluation for salt stress management in solanaceous crops in India. Indian Phytopathol 2019; 72(3): 479-88.
[http://dx.doi.org/10.1007/s42360-019-00174-1]

[77] He Y, Pantigoso HA, Wu Z, Vivanco JM. Co-inoculation of *Bacillus* sp. and *Pseudomonas putida* at

different development stages acts as a biostimulant to promote growth, yield and nutrient uptake of tomato. J Appl Microbiol 2019; 127(1): 196-207.
[http://dx.doi.org/10.1111/jam.14273] [PMID: 30955229]

[78] Nafady NA, Hashem M, Hassan EA, Ahmed HAM, Alamri SA. The combined effect of arbuscular mycorrhizae and plant-growth-promoting yeast improves sunflower defense against *Macrophomina phaseolina* diseases. Biol Control 2019; 138: 104049.
[http://dx.doi.org/10.1016/j.biocontrol.2019.104049]

[79] Nascimento FX, Tavares MJ, Franck J, Ali S, Glick BR, Rossi MJ. ACC deaminase plays a major role in *Pseudomonas fluorescens* YsS6 ability to promote the nodulation of Alpha- and Betaproteobacteria rhizobial strains. Arch Microbiol 2019; 201(6): 817-22.
[http://dx.doi.org/10.1007/s00203-019-01649-5] [PMID: 30877322]

[80] Mishra BK, Lal G, Sharma YK, Kant K, Saxena SN, Dubey PN. Effect of microbial inoculants on cumin (*Cuminum cyminum* Linn.) growth and yield. Int J Seed Spicees 2019; 53: 53-6.

[81] Raklami A, Bechtaoui N, Tahiri A, Anli M, Meddich A, Oufdou K. Use of *Rhizobacteria* and mycorrhizae consortium in the open field as a strategy for improving crop nutrition, productivity and soil fertility. Front Microbiol 2019; 10: 1106.
[http://dx.doi.org/10.3389/fmicb.2019.01106] [PMID: 31164880]

[82] Sibponkrung S, Kondo T, Tanaka K, *et al.* Co-inoculation of *Bacillus velezensis* strain S141 and *Bradyrhizobium* strains promotes nodule growth and nitrogen fixation. Microorganisms 2020; 8(5): 678.
[http://dx.doi.org/10.3390/microorganisms8050678] [PMID: 32392716]

[83] Silambarasan S, Logeswari P, Cornejo P, Kannan VR. Role of plant growth–promoting *Rhizobacterial* consortium in improving the *Vigna radiata* growth and alleviation of aluminum and drought stresses. Environ Sci Pollut Res Int 2019; 26(27): 27647-59.
[http://dx.doi.org/10.1007/s11356-019-05939-9] [PMID: 31338767]

[84] Xia Y, Sahib MR, Amna A, Opiyo SO, Zhao Z, Gao YG. Culturable endophytic fungal communities associated with plants in organic and conventional farming systems and their effects on plant growth. Sci Rep 2019; 9(1): 1669.
[http://dx.doi.org/10.1038/s41598-018-38230-x] [PMID: 30737459]

[85] Karuppiah V, Li Y, Sun J, Vallikkannu M, Chen J. Vel1 regulates the growth of *Trichoderma atroviride* during co-cultivation with *Bacillus amyloliquefaciens* and is essential for wheat root rot control. Biol Control 2020; 151: 104374.
[http://dx.doi.org/10.1016/j.biocontrol.2020.104374]

[86] Varinderpal S, Sharma S, Kunal , *et al.* Synergistic use of plant growth-promoting *Rhizobacteria*, arbuscular mycorrhizal fungi, and spectral properties for improving nutrient use efficiencies in wheat (*Triticum aestivum* L.). Commun Soil Sci Plant Anal 2020; 51(1): 14-27.
[http://dx.doi.org/10.1080/00103624.2019.1689259]

[87] Moreira H, Pereira SIA, Vega A, Castro PML, Marques APGC. Synergistic effects of arbuscular mycorrhizal fungi and plant growth-promoting bacteria benefit maize growth under increasing soil salinity. J Environ Manage 2020; 257: 109982.
[http://dx.doi.org/10.1016/j.jenvman.2019.109982] [PMID: 31868642]

[88] Xia M, Chakraborty R, Terry N, Singh RP, Fu D. Promotion of saltgrass growth in a saline petroleum hydrocarbons contaminated soil using a plant growth promoting bacterial consortium. Int Biodeterior Biodegradation 2020; 146: 104808.
[http://dx.doi.org/10.1016/j.ibiod.2019.104808]

[89] Mondal S, Halder SK, Yadav AN, Mondal KC. Microbial consortium with multifunctional plant growth-promoting attributes: Future perspective in agriculture. Adv plant microb sustain agricul 2020; 219-58.

[90] Demir N, Dural B, Yildirim K. Effect of seaweed suspensions on seed germination of tomato, pepper

and aubergine AGRIS. FAO 2006.

[91] Rinez I, Saad I, Rinez A, Haouala R. Algal extracts alleviates salinity stress on *Capsicum annuum* var. Baklouti. IJISET 2016; 3(11): 372-82.

[92] El Gamal MA, Massoud ON, Salem OM. The promotive effect of algae and *Rhizobium leguminosarum* on arbuscular mycorrhizal fungi activity and their impact on faba bean plant. J Microbiol 2009; 24.

[93] Shahbazi F, Nejad MS, Salimi A, Gilani A. Effect of seaweed extracts on the growth and biochemical constituents of wheat. IJACS 2015; 8(3): 283-7.

[94] Anisimov MM, Skriptsova AV, Chaikina EL, Klykov AG. Effect of water extracts of seaweeds on the growth of seedling roots of buckwheat (*Fagopyrum esculentum* Moench). IJRRAS 2013; 16(2): 282-7.

[95] Grzesik M, Romanowska-Duda Z, Kalaji HM. Effectiveness of *Cyanobacteria* and green algae in enhancing the photosynthetic performance and growth of willow (*Salix viminalis* L.) plants under limited synthetic fertilizers application. Photosynthetica 2017; 55(3): 510-21.
[http://dx.doi.org/10.1007/s11099-017-0716-1]

[96] Santoyo G, Guzmán-Guzmán P, Parra-Cota FI, Santos-Villalobos S, Orozco-Mosqueda MC, Glick BR. Plant growth stimulation by microbial consortia. Agronomy 2021; 11(2): 219.
[http://dx.doi.org/10.3390/agronomy11020219]

[97] Colla G, Rouphael Y. Biostimulants in horticulture. Sci Hortic 2015; 196: 1-2.
[http://dx.doi.org/10.1016/j.scienta.2015.10.044]

[98] Barea JM, Pozo MJ, Azcón R, Azcón-Aguilar C. Microbial co-operation in the rhizosphere. J Exp Bot 2005; 56(417): 1761-78.
[http://dx.doi.org/10.1093/jxb/eri197] [PMID: 15911555]

[99] Gauthier D, Jaffré T, Prin Y. Abundance of Frankia from *Gymnostoma* spp. in the rhizosphere of Alphitonia neocaledonica, a non-nodulated Rhamnaceae endemicto New Caledonia. Eur J Soil Biol 2000; 36(3-4): 169-75.
[http://dx.doi.org/10.1016/S1164-5563(00)01061-X]

[100] Rodríguez H, Fraga R. Phosphate solubilizing bacteria and their role in plant growth promotion. Biotechnol Adv 1999; 17(4-5): 319-39.
[http://dx.doi.org/10.1016/S0734-9750(99)00014-2] [PMID: 14538133]

[101] Khan MS, Zaidi A, Ahemad M, Oves M, Wani PA. Plant growth promotion by phosphate solubilizing fungi : Current perspective. Arch Agron Soil Sci 2010; 56(1): 73-98.
[http://dx.doi.org/10.1080/03650340902806469]

[102] Vergara C, Araujo KEC, Urquiaga S, *et al.* Dark septate endophytic fungi help tomato to acquire nutrients from ground plant material. Front Microbiol 2017; 8: 2437.
[http://dx.doi.org/10.3389/fmicb.2017.02437] [PMID: 29312163]

[103] Smith SE, Smith FA. Roles of arbuscular mycorrhizas in plant nutrition and growth: New paradigms from cellular to ecosystem scales. Annu Rev Plant Biol 2011; 62(1): 227-50.
[http://dx.doi.org/10.1146/annurev-arplant-042110-103846] [PMID: 21391813]

[104] Heijden MGA, Martin FM, Selosse MA, Sanders IR. Mycorrhizal ecology and evolution: the past, the present, and the future. New Phytol 2015; 205(4): 1406-23.
[http://dx.doi.org/10.1111/nph.13288] [PMID: 25639293]

[105] Hodge A, Storer K. Arbuscular mycorrhiza and nitrogen: Implications for individual plants through to ecosystems. Plant Soil 2015; 386(1-2): 1-19.
[http://dx.doi.org/10.1007/s11104-014-2162-1]

[106] Wu QS, Srivastava AK, Zou YN. AMF-induced tolerance to drought stress in citrus: A review. Sci Hortic 2013; 164: 77-87.
[http://dx.doi.org/10.1016/j.scienta.2013.09.010]

[107] Dobereiner J. Isolation and identification of root associated diazotrophs. Nitrogen fixation with non-legumes. Dordrecht: Springer 1989; pp. 103-8.
[http://dx.doi.org/10.1007/978-94-009-0889-5_13]

[108] Cooke TJ, Poli D, Sztein A, Cohen JD. Evolutionary patterns in auxin action. Auxin Mol Bio 2002; 319-38.

[109] Cacciari I, Lippi D, Pietrosanti T, Pietrosanti W. Phytohormone-like substances produced by single and mixed diazotrophic cultures of Azospirillum and Arthrobacter. Plant Soil 1989; 115(1): 151-3.
[http://dx.doi.org/10.1007/BF02220706]

[110] Umali-Garcia M, Hubbell DH, Gaskins MH, Dazzo FB. Association of azospirillum with grass roots. Appl Environ Microbiol 1980; 39(1): 219-26.
[http://dx.doi.org/10.1128/aem.39.1.219-226.1980] [PMID: 16345490]

[111] Pathak DV. Analogue resistant mutants of A. chroococcum affecting growth parameters in sunflower (*Helianthus annus* L.) under pot culture conditions. Proc Indian National Sci Acad 1995; 18: 203-6.

[112] Schlaeppi K, Bulgarelli D. The plant microbiome at work. Mol Plant Microbe Interact 2015; 28(3): 212-7.
[http://dx.doi.org/10.1094/MPMI-10-14-0334-FI] [PMID: 25514681]

[113] Hartman K, Van der Heijden MG, Wittwer RA, Banerjee S, Walser JC, Schlaeppi K. Cropping practices manipulate abundance patterns of root and soil microbiome members paving the way to smart farming. Microbiome 2018; 6(1): 1-4.
[PMID: 29291746]

[114] Xu L, Naylor D, Dong Z, *et al.* Drought delays development of the sorghum root microbiome and enriches for monoderm bacteria. Proc Natl Acad Sci 2018; 115(18): E4284-93.
[http://dx.doi.org/10.1073/pnas.1717308115] [PMID: 29666229]

[115] Fitzpatrick CR, Copeland J, Wang PW, Guttman DS, Kotanen PM, Johnson MTJ. Assembly and ecological function of the root microbiome across angiosperm plant species. Proc Natl Acad Sci 2018; 115(6): E1157-65.
[http://dx.doi.org/10.1073/pnas.1717617115] [PMID: 29358405]

[116] Marasco R, Rolli E, Ettoumi B, *et al.* A drought resistance-promoting microbiome is selected by root system under desert farming. PLoS One 2012; 7(10): e48479.
[http://dx.doi.org/10.1371/journal.pone.0048479] [PMID: 23119032]

[117] Yazdani M, Pirdashti H. Efficiency of co-inoculation phosphate solubilizer microorganisms (psm) and plant growth promoting *Rhizobacteria* (PGPR) on micronutrients uptake in corn (Zea mays L.). Int Res J Appl Basic Sci 2011; 2: 28-34.

[118] Liu A, Hamel C, Hamilton RI, Ma BL, Smith DL. Acquisition of Cu, Zn, Mn and Fe by mycorrhizal maize (*Zea mays* L.) grown in soil at different P and micronutrient levels. Mycorrhiza 2000; 9(6): 331-6.
[http://dx.doi.org/10.1007/s005720050277]

[119] Khan AG. Role of soil microbes in the rhizospheres of plants growing on trace metal contaminated soils in phytoremediation. J Trace Elem Med Biol 2005; 18(4): 355-64.
[http://dx.doi.org/10.1016/j.jtemb.2005.02.006] [PMID: 16028497]

[120] Kumar Meena R, Kumar Singh R, Pal Singh N, Kumari Meena S, Singh Meena V. Isolation of low temperature surviving plant growth – promoting *Rhizobacteria* (PGPR) from pea (*Pisum sativum* L.) and documentation of their plant growth promoting traits. Biocatal Agric Biotechnol 2015; 4(4): 806-11.
[http://dx.doi.org/10.1016/j.bcab.2015.08.006]

[121] Kumar SS, Kumar V, Kumar R, Malyan SK, Pugazhendhi A. Microbial fuel cells as a sustainable platform technology for bioenergy, biosensing, environmental monitoring, and other low power device applications. Fuel 2019; 255: 115682.

[http://dx.doi.org/10.1016/j.fuel.2019.115682]

[122] Kang Y, Hao Y, Shen M, Zhao Q, Li Q, Hu J. Impacts of supplementing chemical fertilizers with organic fertilizers manufactured using pig manure as a substrate on the spread of tetracycline resistance genes in soil. Ecotoxicol Environ Saf 2016; 130: 279-88.
[http://dx.doi.org/10.1016/j.ecoenv.2016.04.028] [PMID: 27152658]

[123] Singh S, Singh V, Pal K. Importance of microorganisms in agriculture. Retrieved 2019; 7: 2019.

[124] Yan N, Marschner P, Cao W, Zuo C, Qin W. Influence of salinity and water content on soil microorganisms. Int Soil Water Conserv Res 2015; 3(4): 316-23.
[http://dx.doi.org/10.1016/j.iswcr.2015.11.003]

[125] Achal V, Mukherjee A, Zhang Q. Unearthing ecological wisdom from natural habitats and its ramifications on development of biocement and sustainable cities. Landsc Urban Plan 2016; 155: 61-8.
[http://dx.doi.org/10.1016/j.landurbplan.2016.04.013]

[126] Pathania P, Rajta A, Singh PC, Bhatia R. Role of plant growth-promoting bacteria in sustainable agriculture. Biocatal Agric Biotechnol 2020; 30: 101842.
[http://dx.doi.org/10.1016/j.bcab.2020.101842]

[127] Kass DL, Drosdoff M, Alexander M. Nitrogen fixation by *Azotobacter paspali* in association with bahiagrass (Paspalum notatum). Soil Sci Soc Am J 1971; 35(2): 286-9.
[http://dx.doi.org/10.2136/sssaj1971.03615995003500020031x]

[128] Bürgmann H, Widmer F, Sigler WV, Zeyer J. mRNA extraction and reverse transcription-PCR protocol for detection of nifH gene expression by *Azotobacter vinelandii* in soil. Appl Environ Microbiol 2003; 69(4): 1928-35.
[http://dx.doi.org/10.1128/AEM.69.4.1928-1935.2003] [PMID: 12676666]

[129] Menhart N, Thariath A, Viswanatha T. Characterization of the pyoverdines of *Azotobacter vinelandii* ATCC 12 837 with regard to heterogeneity. Biol Met 1991; 4(4): 223-32.
[http://dx.doi.org/10.1007/BF01141185] [PMID: 1838001]

[130] Sevilla M, Kennedy C. Colonization of rice and other cereals by *Acetobacter diazotrophicus*, an endophyte of sugarcane. The quest for nitrogen fixation in rice. IRRI 2000; 151-65.

[131] Cleenwerck I, Vandemeulebroecke K, Janssens D, Swings J. Re-examination of the genus Acetobacter, with descriptions of *Acetobacter cerevisiae* sp. nov. and *Acetobacter malorum* sp. nov. Int J Syst Evol Microbiol 2002; 52(Pt 5): 1551-8.
[PMID: 12361257]

[132] Dobbelaere S, Croonenborghs A, Thys A, *et al.* Responses of agronomically important crops to inoculation with Azospirillum. Funct Plant Biol 2001; 28(9): 1-9.
[http://dx.doi.org/10.1071/PP01074]

[133] Young JPW, Crossman LC, Johnston AWB, *et al.* The genome of *Rhizobium leguminosarum* has recognizable core and accessory components. Genome Biol 2006; 7(4): R34.
[http://dx.doi.org/10.1186/gb-2006-7-4-r34] [PMID: 16640791]

[134] Richardson AE, Simpson RJ. Soil microorganisms mediating phosphorus availability update on microbial phosphorus. Plant Physiol 2011; 156(3): 989-96.
[http://dx.doi.org/10.1104/pp.111.175448] [PMID: 21606316]

[135] Mehta P, Walia A, Chauhan A, Kulshrestha S, Shirkot CK. Phosphate solubilisation and plant growth promoting potential by stress tolerant *Bacillus* sp. isolated from rhizosphere of apple orchards in *trans* Himalayan region of Himachal Pradesh. Ann Appl Biol 2013; 163(3): 430-43.
[http://dx.doi.org/10.1111/aab.12077]

[136] Istina IN, Widiastuti H, Joy B, Antralina M. Phosphate-solubilizing microbe from Saprists peat soil and their potency to enhance oil palm growth and P uptake. Procedia Food Sci 2015; 3: 426-35.
[http://dx.doi.org/10.1016/j.profoo.2015.01.047]

[137] Köberl M, Müller H, Ramadan EM, Berg G. Desert farming benefits from microbial potential in arid soils and promotes diversity and plant health. PLoS One 2011; 6(9): e24452.
[http://dx.doi.org/10.1371/journal.pone.0024452] [PMID: 21912695]

[138] Singh S, Singh V, Pal K. Importance of micro organisms in agriculture. Climate and environmental changes: Impact, challenges and solutions 2017.

[139] Johansen JE, Binnerup SJ, Kroer N, Mølbak L. *Luteibacter rhizovicinus* gen. nov., sp. nov., a yellow-pigmented gammaproteobacterium isolated from the rhizosphere of barley (*Hordeum vulgare* L.). Int J Syst Evol Microbiol 2005; 55(6): 2285-91.
[http://dx.doi.org/10.1099/ijs.0.63497-0] [PMID: 16280484]

[140] Li HB, Singh RK, Singh P, *et al.* Genetic diversity of nitrogen-fixing and plant growth promoting Pseudomonas species isolated from sugarcane rhizosphere. Front Microbiol 2017; 8: 1268.
[http://dx.doi.org/10.3389/fmicb.2017.01268] [PMID: 28769881]

[141] Nihorimbere V, Ongena M, Smargiassi M, Thonart P. Beneficial effect of the rhizosphere microbial community for plant growth and health. Biotechnol Agron Soc Environ 2011; 15(2): 327-37.

[142] Yegorenkova IV, Tregubova KV, Ignatov VV. Paenibacillus polymyxa *Rhizobacteria* and their synthesized exoglycans in interaction with wheat roots: colonization and root hair deformation. Curr Microbiol 2013; 66(5): 481-6.
[http://dx.doi.org/10.1007/s00284-012-0297-y] [PMID: 23314809]

[143] Espinosa-Urgel M, Salido A, Ramos JL. Genetic analysis of functions involved in adhesion of *Pseudomonas putida* to seeds. J Bacteriol 2000; 182(9): 2363-9.
[http://dx.doi.org/10.1128/JB.182.9.2363-2369.2000] [PMID: 10762233]

[144] Romano JD, Kolter R. Pseudomonas-Saccharomyces interactions: Influence of fungal metabolism on bacterial physiology and survival. J Bacteriol 2005; 187(3): 940-8.
[http://dx.doi.org/10.1128/JB.187.3.940-948.2005] [PMID: 15659672]

[145] Lanteigne C, Gadkar VJ, Wallon T, Novinscak A, Filion M. Production of DAPG and HCN by *Pseudomonas* sp. LBUM300 contributes to the biological control of bacterial canker of tomato. Phytopathology 2012; 102(10): 967-73.
[http://dx.doi.org/10.1094/PHYTO-11-11-0312] [PMID: 22713078]

[146] Mansoori M, Heydari A, Hassanzadeh N, Rezaee S, Naraghi L. Evaluation of Pseudomonas and Bacillus bacterial antagonists for biological control of cotton Verticillium wilt disease. J Plant Prot Res 2013; 53(2): 154-7.
[http://dx.doi.org/10.2478/jppr-2013-0023]

[147] Pereg L, McMillan M. Scoping the potential uses of beneficial microorganisms for increasing productivity in cotton cropping systems. Soil Biol Biochem 2015; 80: 349-58.
[http://dx.doi.org/10.1016/j.soilbio.2014.10.020]

[148] Martínez-Viveros O, Jorquera MA, Crowley DE, Gajardo GM, Mora ML. Mechanisms and practical considerations involved in plant growth promotion by *Rhizobacteria*. Soil Sci Plant Nutr 2010; 10(3): 293-319.

[149] Alabouvette C, Olivain C, Steinberg C. Biological control of plant diseases: The European situation. Eur J Plant Pathol 2006; 114(3): 329-41.
[http://dx.doi.org/10.1007/s10658-005-0233-0]

[150] McNeil DG Jr. Fungus fatal to mosquito may aid global war on Malaria. The New York Times 2005; p. 10.

[151] Niu XM, Zhang KQ. *Arthrobotrys oligospora* : A model organism for understanding the interaction between fungi and nematodes. Mycology 2011; 2(2): 59-78.
[http://dx.doi.org/10.1080/21501203.2011.562559]

[152] Chen J, Zhou L, Din IU, *et al.* Antagonistic activity of *Trichoderma* spp. against *Fusarium oxysporum*

in rhizosphere of *Radix pseudostellariae* triggers the expression of host defense genes and improves its growth under long-term monoculture system. Front Microbiol 2021; 12: 579920.
[http://dx.doi.org/10.3389/fmicb.2021.579920] [PMID: 33790872]

[153] Kaushal M, Wani SP. Plant-growth-promoting *Rhizobacteria*: Drought stress alleviators to ameliorate crop production in drylands. Ann Microbiol 2016; 66(1): 35-42.
[http://dx.doi.org/10.1007/s13213-015-1112-3]

[154] Tibbett M, Sanders FE, Cairney JWG. The effect of temperature and inorganic phosphorus supply on growth and acid phosphatase production in arctic and temperate strains of ectomycorrhizal *Hebeloma* spp. in axenic culture. Mycol Res 1998; 102(2): 129-35.
[http://dx.doi.org/10.1017/S0953756297004681]

[155] Fomina MA, Alexander IJ, Colpaert JV, Gadd GM. Solubilization of toxic metal minerals and metal tolerance of mycorrhizal fungi. Soil Biol Biochem 2005; 37(5): 851-66.
[http://dx.doi.org/10.1016/j.soilbio.2004.10.013]

[156] Sandhya VZ, SK Z A, Grover M, *et al*. Alleviation of drought stress effects in sunflower seedlings by the exopolysaccharides producing *Pseudomonas putida* strain GAP-P45. Biol Fertil Soils 2009; 46(1): 17-26.
[http://dx.doi.org/10.1007/s00374-009-0401-z]

[157] Shukla N, Awasthi RP, Rawat L, Kumar J. Seed biopriming with drought tolerant isolates of *Trichoderma harzianum* promote growth and drought tolerance in *Triticum aestivum*. Ann Appl Biol 2015; 166(2): 171-82.
[http://dx.doi.org/10.1111/aab.12160]

[158] Singh NK, Chaudhary FK, Patel DB. Effectiveness of *Azotobacter* bio-inoculant for wheat grown under dryland condition. J Environ Biol 2013; 34(5): 927-32.
[PMID: 24558807]

[159] Pathak DV, Kumar M. Microbial inoculants as biofertilizers and biopesticides. Microbial inoculants in sustainable agricultural productivity. New Delhi: Springer 2016; pp. 197-209.
[http://dx.doi.org/10.1007/978-81-322-2647-5_11]

[160] Sokollek SJ, Hertel C, Hammes WP. Cultivation and preservation of vinegar bacteria. J Biotechnol 1998; 60(3): 195-206.
[http://dx.doi.org/10.1016/S0168-1656(98)00014-5] [PMID: 9729803]

[161] Dobbelaere S, Vanderleyden J, Okon Y. Plant growth-promoting effects of diazotrophs in the rhizosphere. Crit Rev Plant Sci 2003; 22(2): 107-49.
[http://dx.doi.org/10.1080/713610853]

[162] Kalayu G. Phosphate solubilizing microorganisms: Promising approach as biofertilizers. Int J Agron 2019; 2019: 1-7.
[http://dx.doi.org/10.1155/2019/4917256]

[163] Song OR, Lee SJ, Lee YS, Lee SC, Kim KK, Choi YL. Solubilization of insoluble inorganic phosphate by *Burkholderia cepacia* DA23 isolated from cultivated soil. Braz J Microbiol 2008; 39(1): 151-6.
[http://dx.doi.org/10.1590/S1517-83822008000100030] [PMID: 24031195]

[164] Pankaj P. Microbiological and molecular analysis of biodegradation potential of indigenous bacterial cultures against cypermethrin, fipronil and sulfosulfuron pesticides. Pantnagar, India: Department of Microbiology GBPUAT 2015.

[165] Adams P, De-Leij FAAM, Lynch JM. *Trichoderma harzianum* Rifai 1295-22 mediates growth promotion of crack willow (*Salix fragilis*) saplings in both clean and metal-contaminated soil. Microb Ecol 2007; 54(2): 306-13.
[http://dx.doi.org/10.1007/s00248-006-9203-0] [PMID: 17345130]

[166] Altomare C, Norvell WA, Björkman T, Harman GE. Solubilization of phosphates and micronutrients

by the plant-growth-promoting and biocontrol fungus *Trichoderma harzianum* rifai 1295-22. Appl Environ Microbiol 1999; 65(7): 2926-33.
[http://dx.doi.org/10.1128/AEM.65.7.2926-2933.1999] [PMID: 10388685]

[167] Chitra K, Sharavanan P. Studies on potassium solubilizing bacteria from southern indian tea soils. Int J Curr Microbiol Appl Sci 2014; 3(5): 1045-52.

[168] Sheng XF. Growth promotion and increased potassium uptake of cotton and rape by a potassium releasing strain of *Bacillus edaphicus*. Soil Biol Biochem 2005; 37(10): 1918-22.
[http://dx.doi.org/10.1016/j.soilbio.2005.02.026]

[169] Gomathy M, Sathya Prakash D, Thangaraju M, Sundaram SP, Manicka Sundaram P. Impact of biofertigation of Azophosmet on cotton yield under drip irrigation. Res J Agric Biol Sci 2008; 4(6): 695.

[170] Dhale D, Chatte S, Jadhav VT. Response of bioinoculents on growth, yield and fiber quality of cotton under irrigation. Agric Biol J N Am 2011; 2(2): 376-86.
[http://dx.doi.org/10.5251/abjna.2011.2.2.376.386]

[171] Yasmin S, Hafeez FY, Schmid M, Hartmann A. Plant-beneficial *Rhizobacteria* for sustainable increased yield of cotton with reduced level of chemical fertilizers. Pak J Bot 2013; 45(2): 655-62.

[172] Kuss AV, Kuss VV, Lovato T, Flôres ML. Nitrogen fixation and *in vitro* production of indolacetic acid by endophytic diazotrophic bacteria. Pesqui Agropecu Bras 2007; 42(10): 1459-65.
[http://dx.doi.org/10.1590/S0100-204X2007001000013]

[173] Bayer Crop Science. 2017. Available from: https://www.cropscience.bayer.us/products/seedgrowth

[174] Wilson MJ, Jackson TA. Progress in the commercialisation of bionematicides. BioControl 2013; 58(6): 715-22.
[http://dx.doi.org/10.1007/s10526-013-9511-5]

[175] Vansuyt G, Robin A, Briat JF, Curie C, Lemanceau P. Iron acquisition from fe-pyoverdine by *Arabidopsis thaliana*. Mol Plant Microbe Interact 2007; 20(4): 441-7.
[http://dx.doi.org/10.1094/MPMI-20-4-0441] [PMID: 17427814]

[176] Sharma A, Johri BN. Growth promoting influence of siderophore-producing *Pseudomonas* strains GRP3A and PRS9 in maize (*Zea mays* L.) under iron limiting conditions. Microbiol Res 2003; 158(3): 243-8.
[http://dx.doi.org/10.1078/0944-5013-00197] [PMID: 14521234]

[177] Singh N, Siddiqui ZA. Effects of *Bacillus subtilis*, *Pseudomonas fluorescens* and *Aspergillus awamori* on the wilt-leaf spot disease complex of tomato. Phytoparasitica 2015; 43(1): 61-75.
[http://dx.doi.org/10.1007/s12600-014-0427-0]

[178] Haas D, Défago G. Biological control of soil-borne pathogens by fluorescent pseudomonads. Nat Rev Microbiol 2005; 3(4): 307-19.
[http://dx.doi.org/10.1038/nrmicro1129] [PMID: 15759041]

[179] Mondal KK, Dureja P, Prakash Verma J. Management of *Xanthomonas camprestris* pv. malvacearum-induced blight of cotton through phenolics of cotton rhizobacterium. Curr Microbiol 2001; 43(5): 336-9.
[http://dx.doi.org/10.1007/s002840010312] [PMID: 11688797]

[180] Kim HS, Park J, Cho SW, *et al.* Isolation and characterization of Bacillus strains for biological control. J Microbiol 2003; 41(3): 196-201.

[181] Howell CR. Cotton seedling preemergence damping-off incited by *Rhizopus oryzae* and *Pythium* spp. and its biological control with *Trichoderma* spp. Phytopathology 2002; 92(2): 177-80.
[http://dx.doi.org/10.1094/PHYTO.2002.92.2.177] [PMID: 18943091]

[182] Kaewchai S. Mycofungicides and fungal biofertilizers. Fungal Divers 2009; 38: 25-50.

[183] Wang HH, Church GM. Multiplexed genome engineering and genotyping methods: applications for

synthetic biology and metabolic engineering. Methods in enzymology. Academic Press 2011; 498: pp. 409-26.

[184] Enyeart PJ, Chirieleison SM, Dao MN, *et al.* Generalized bacterial genome editing using mobile group II introns and Cre-*lox*. Mol Syst Biol 2013; 9(1): 685.
[http://dx.doi.org/10.1038/msb.2013.41] [PMID: 24002656]

[185] Jiang W, Bikard D, Cox D, Zhang F, Marraffini LA. RNA-guided editing of bacterial genomes using CRISPR-Cas systems. Nat Biotechnol 2013; 31(3): 233-9.
[http://dx.doi.org/10.1038/nbt.2508] [PMID: 23360965]

[186] DiCarlo JE, Norville JE, Mali P, Rios X, Aach J, Church GM. Genome engineering in *Saccharomyces cerevisiae*s using CRISPR-Cas systems. Nucleic Acids Res 2013; 41(7): 4336-43.
[http://dx.doi.org/10.1093/nar/gkt135] [PMID: 23460208]

[187] Cobb RE, Wang Y, Zhao H. High-efficiency multiplex genome editing of Streptomyces species using an engineered CRISPR/Cas system. ACS Synth Biol 2015; 4(6): 723-8.
[http://dx.doi.org/10.1021/sb500351f] [PMID: 25458909]

[188] Muñoz IV, Sarrocco S, Malfatti L, Baroncelli R, Vannacci G. CRISPR-Cas for fungal genome editing: A new tool for the management of plant diseases. Front Plant Sci 2019; 10: 135.
[http://dx.doi.org/10.3389/fpls.2019.00135] [PMID: 30828340]

[189] Yi Y, Li Z, Song C, Kuipers OP. Exploring plant-microbe interactions of the *Rhizobacteria Bacillus subtilis* and *Bacillus mycoides* by use of the CRISPR-Cas9 system. Environ Microbiol 2018; 20(12): 4245-60.
[http://dx.doi.org/10.1111/1462-2920.14305] [PMID: 30051589]

[190] Ryu CM, Farag MA, Hu CH, *et al.* Erratum: Bacterial volatiles promote growth in Arabidopsis (Proceedings of the National Academy of Sciences of the United States of America (April 15, 2003) 100: 8 (4927-4932)). Proc Natl Acad Sci USA 2003; 100(14): 8607.

[191] Heo MJ, Jung HM, Um J, Lee SW, Oh MK. Controlling citrate synthase expression by CRISPR/Cas9 genome editing for n-butanol production in *Escherichia coli*. ACS Synth Biol 2017; 6(2): 182-9.
[http://dx.doi.org/10.1021/acssynbio.6b00134] [PMID: 27700055]

[192] Li Y, Lin Z, Huang C, *et al.* Metabolic engineering of *Escherichia coli* using CRISPR–Cas9 meditated genome editing. Metab Eng 2015; 31: 13-21.
[http://dx.doi.org/10.1016/j.ymben.2015.06.006] [PMID: 26141150]

[193] Shishido M, Miwa C, Usami T, Amemiya Y, Johnson KB. Biological control efficiency of Fusarium wilt of tomato by nonpathogenic *Fusarium oxysporum* Fo-B2 in different environments. Phytopathology 2005; 95(9): 1072-80.
[http://dx.doi.org/10.1094/PHYTO-95-1072] [PMID: 18943305]

[194] Qin H, Xiao H, Zou G, Zhou Z, Zhong JJ. CRISPR-Cas9 assisted gene disruption in the higher fungus *Ganoderma* species. Process Biochem 2017; 56: 57-61.
[http://dx.doi.org/10.1016/j.procbio.2017.02.012]

[195] Scherm B, Orrù M, Balmas V, *et al.* Altered trichothecene biosynthesis in TRI6-silenced transformants of *Fusarium culmorum* influences the severity of crown and foot rot on durum wheat seedlings. Mol Plant Pathol 2011; 12(8): 759-71.
[http://dx.doi.org/10.1111/j.1364-3703.2011.00709.x] [PMID: 21726376]

[196] Fraley RT, Rogers SG, Horsch RB, *et al.* Expression of bacterial genes in plant cells. Proc Natl Acad Sci 1983; 80(15): 4803-7.
[http://dx.doi.org/10.1073/pnas.80.15.4803] [PMID: 6308651]

[197] Nadolska-Orczyk A, Orczyk W, Przetakiewicz A. Agrobacterium-mediated transformation of cereals : From technique development to its application. Acta Physiol Plant 2000; 22(1): 77-88.
[http://dx.doi.org/10.1007/s11738-000-0011-8]

[198] Wilkins TA, Rajasekaran K, Anderson DM. Cotton biotechnology. Crit Rev Plant Sci 2000; 19(6):

511-50.
[http://dx.doi.org/10.1080/07352680091139286]

[199] Abhishek A, Kumari R, Karjagi CG, *et al.* Tissue culture independent *Agrobacterium tumefaciens* mediated in planta transformation method for tropical maize (*Zea mays*. L). PNAS Section B. Biological Sciences 2016; 86(2): 375-84.

[200] Que Q, Chilton MD, Elumalai S, Zhong H, Dong S, Shi L. Repurposing macromolecule delivery tools for plant genetic modification in the era of precision genome engineering. Transgenic Plants. SpringerLink 2019; pp. 3-18.
[http://dx.doi.org/10.1007/978-1-4939-8778-8_1]

[201] Kavipriya C, Yuvaraja A, Senthil K, Menaka C. Genetic transformation methods for crop improvement: A brief review. Agric Rev 2019; 40(4): 281-8.
[http://dx.doi.org/10.18805/ag.R-1904]

[202] Guo M, Ye J, Gao D, Xu N, Yang J. Agrobacterium-mediated horizontal gene transfer: Mechanism, biotechnological application, potential risk and forestalling strategy. Biotechnol Adv 2019; 37(1): 259-70.
[http://dx.doi.org/10.1016/j.biotechadv.2018.12.008] [PMID: 30579929]

[203] Slater SC, Goldman BS, Goodner B, *et al.* Genome sequences of three agrobacterium biovars help elucidate the evolution of multichromosome genomes in bacteria. J Bacteriol 2009; 191(8): 2501-11.
[http://dx.doi.org/10.1128/JB.01779-08] [PMID: 19251847]

[204] Rathore DS, Mullins E. Alternative non-agrobacterium based methods for plant transformation. Annu Plant Review online 2018; 891-908.
[http://dx.doi.org/10.1002/9781119312994.apr0659]

[205] Fiandaca MS, Federoff HJ. Using viral-mediated gene delivery to model Parkinson's disease: Do nonhuman primate investigations expand our understanding? Exp Neurol 2014; 256: 117-25.
[http://dx.doi.org/10.1016/j.expneurol.2013.03.014] [PMID: 23524194]

[206] Rajasekaran K. Biolistic transformation of cotton zygotic embryo meristem. Transgenic Cotton. Totowa, NJ: Humana Press 2013; pp. 47-57.
[http://dx.doi.org/10.1007/978-1-62703-212-4_4]

[207] Liu J, Nannas NJ, Fu F, *et al.* Genome-scale sequence disruption following biolistic transformation in rice and maize. Plant Cell 2019; 31(2): 368-83.
[http://dx.doi.org/10.1105/tpc.18.00613] [PMID: 30651345]

[208] Young JL, Dean DA. Electroporation-mediated gene delivery. Adv Genet 2015; 89: 49-88.
[http://dx.doi.org/10.1016/bs.adgen.2014.10.003] [PMID: 25620008]

[209] Baskaran P, Dasgupta I. Gene delivery using microinjection of agrobacterium to embryonic shoot apical meristem of elite indica rice cultivars. J Plant Biochem Biotechnol 2012; 21(2): 268-74.
[http://dx.doi.org/10.1007/s13562-011-0078-x]

[210] Du X, Wang J, Zhou Q, *et al.* Advanced physical techniques for gene delivery based on membrane perforation. Drug Deliv 2018; 25(1): 1516-25.
[http://dx.doi.org/10.1080/10717544.2018.1480674] [PMID: 29968512]

[211] Cochran M, Wheatley MA. *In vitro* gene delivery with ultrasound-triggered polymer microbubbles. Ultrasound Med Biol 2013; 39(6): 1102-19.
[http://dx.doi.org/10.1016/j.ultrasmedbio.2013.01.013] [PMID: 23562023]

[212] Jinturkar KA, Rathi MN, Misra A. Gene delivery using physical methods. Challenges in delivery of therapeutic genomics and proteomics. Elsevier 2011; pp. 83-126.
[http://dx.doi.org/10.1016/B978-0-12-384964-9.00003-7]

[213] Wang W, Li W, Ma N, Steinhoff G. Non-viral gene delivery methods. Curr Pharm Biotechnol 2013; 14(1): 46-60.
[PMID: 23437936]

[214] Sahab S, Hayden MJ, Mason J, Spangenberg G. Mesophyll protoplasts and PEG-mediated transfections: transient assays and generation of stable transgenic canola plants. Transgenic plants. New York, NY: Humana Press 2019; pp. 131-52.
[http://dx.doi.org/10.1007/978-1-4939-8778-8_10]

[215] Munye MM, Ravi J, Tagalakis AD, McCarthy D, Ryadnov MG, Hart SL. Role of liposome and peptide in the synergistic enhancement of transfection with a lipopolyplex vector. Sci Rep 2015; 5(1): 9292.
[http://dx.doi.org/10.1038/srep09292] [PMID: 25786833]

[216] Simon M, Foroughi-Wehr B. Inhibition of extracellular DNase activity of barley microspores in the presence of polyethylene glycol and silicon carbide fibers. J Plant Physiol 2000; 156(2): 184-9.
[http://dx.doi.org/10.1016/S0176-1617(00)80304-1]

[217] James C, Teng P, Arujanan M, *et al.* Invitational Essays to celebrate the 20th Anniversary of the commercialization of biotech crops (1996 to 2015): Progress and promise. ISAAA Brief. 2015; p. 51.

[218] Lea PJ, Joy KW, Ramos JL, Guerrero MG. The action of 2-amino-4-(methylphosphinyl)-butanoic acid (phosphinothricin) and its 2-oxo-derivative on the metabolism of *Cyanobacteria* and higher plants. Phytochemistry 1984; 23(1): 1-6.
[http://dx.doi.org/10.1016/0031-9422(84)83066-6]

[219] Mishutkina YV, Kamionskaya AM, Skryabin KG. The creation of sugar beet transgenic plants expressing bar gene. Appl Biochem Microbiol 2010; 46(1): 80-6.
[http://dx.doi.org/10.1134/S000368381001014X]

[220] Liu SC, Zhang GC, Yang LF, Mii M, Gai JY, Zhu YL. Bialaphos-resistant transgenic soybeans produced by the Agrobacterium-mediated cotyledonary-node method. J Agric Sci Technol 2014; 16(1): 175-90.

[221] Liu Y, Yang SX, Cheng Y, *et al.* Production of herbicide-resistant medicinal plant *Salvia miltiorrhiza* transformed with the bar gene. Appl Biochem Biotechnol 2015; 177(7): 1456-65.
[http://dx.doi.org/10.1007/s12010-015-1826-5] [PMID: 26364310]

[222] Petri C, Wang H, Burgos L, Sánchez-Navarro J, Alburquerque N. Production of transgenic apricot plants from hypocotyl segments of mature seeds. Sci Hortic 2015; 197: 144-9.
[http://dx.doi.org/10.1016/j.scienta.2015.09.023]

[223] Do PT, Lee H, Mookkan M, Folk WR, Zhang ZJ. Rapid and efficient Agrobacterium-mediated transformation of sorghum (*Sorghum bicolor*) employing standard binary vectors and bar gene as a selectable marker. Plant Cell Rep 2016; 35(10): 2065-76.
[http://dx.doi.org/10.1007/s00299-016-2019-6] [PMID: 27350252]

[224] Metwali EMR, Soliman HIA, Almaghrabi OA, Kaddasa NM. Producing transgenic thompson seedless grape (*vitis vinifera*) plants using agrobacterium tumefaciens. Int J Agric Biol 2016; 18(4): 661-70.
[http://dx.doi.org/10.17957/IJAB/15.0134]

[225] Green JM. The benefits of herbicide-resistant crops. Pest Manag Sci 2012; 68(10): 1323-31.
[http://dx.doi.org/10.1002/ps.3374] [PMID: 22865693]

[226] Brookes G, Barfoot P. Environmental impacts of genetically modified (GM) crop use 1996–2015: Impacts on pesticide use and carbon emissions. GM Crops Food 2017; 8(2): 117-47.
[http://dx.doi.org/10.1080/21645698.2017.1309490] [PMID: 28414252]

[227] Baloglu MC, Kavas M, Gürel S, Gürel E. The use of microorganisms for gene transfer and crop improvement. Crop improvement through microbial biotechnology. Elsevier 2018; pp. 1-25.
[http://dx.doi.org/10.1016/B978-0-444-63987-5.00001-3]

[228] Tohidfar M, Zare N, Jouzani GS, Eftekhari SM. Agrobacterium-mediated transformation of alfalfa (*Medicago sativa*) using a synthetic cry3a gene to enhance resistance against alfalfa weevil. Plant Cell Tissue Organ Cult 2013; 113(2): 227-35.
[http://dx.doi.org/10.1007/s11240-012-0262-2]

[229] Zhao Q, Liu M, Tan M, Gao J, Shen Z. Expression of Cry1Ab and Cry2Ab by a polycistronic transgene with a self-cleavage peptide in rice. PLoS One 2014; 9(10): e110006.
[http://dx.doi.org/10.1371/journal.pone.0110006] [PMID: 25333312]

[230] Rausch MA, Chougule NP, Deist BR, Bonning BC. Modification of Cry4Aa toward improved toxin processing in the gut of the pea aphid, Acyrthosiphon pisum. PLoS One 2016; 11(5): e0155466.
[http://dx.doi.org/10.1371/journal.pone.0155466] [PMID: 27171411]

[231] Kumar K, Gambhir G, Dass A, et al. Genetically modified crops: Current status and future prospects. Planta 2020; 251(4): 91.
[http://dx.doi.org/10.1007/s00425-020-03372-8] [PMID: 32236850]

[232] Perlak FJ, Fuchs RL, Dean DA, McPherson SL, Fischhoff DA. Modification of the coding sequence enhances plant expression of insect control protein genes. Proc Natl Acad Sci 1991; 88(8): 3324-8.
[http://dx.doi.org/10.1073/pnas.88.8.3324] [PMID: 2014252]

[233] Suzuki N, Rivero RM, Shulaev V, Blumwald E, Mittler R. Abiotic and biotic stress combinations. New Phytol 2014; 203(1): 32-43.
[http://dx.doi.org/10.1111/nph.12797] [PMID: 24720847]

[234] Tuteja N, Gill SS, Eds. Climate change and plant abiotic stress tolerance. John Wiley & Sons 2013.
[http://dx.doi.org/10.1002/9783527675265]

[235] Raza A, Razzaq A, Mehmood S, et al. Impact of climate change on crops adaptation and strategies to tackle its outcome: A review. Plants 2019; 8(2): 34.
[http://dx.doi.org/10.3390/plants8020034] [PMID: 30704089]

[236] Wani SH, Sah SK, Hossain MA, Kumar V, Balachandran SM. Transgenic approaches for abiotic stress tolerance in crop plants. Advances in plant breeding strategies: agronomic, abiotic and biotic stress traits. Cham: Springer 2016; pp. 345-96.
[http://dx.doi.org/10.1007/978-3-319-22518-0_10]

[237] Lohar DP, Schuller K, Buzas DM, Gresshoff PM, Stiller J. Transformation of *Lotus japonicus* using the herbicide resistance bar gene as a selectable marker. J Exp Bot 2001; 52(361): 1697-702.
[http://dx.doi.org/10.1093/jexbot/52.361.1697] [PMID: 11479335]

[238] Zang N, Zhai H, Gao S, Chen W, He S, Liu Q. Efficient production of transgenic plants using the bar gene for herbicide resistance in sweetpotato. Sci Hortic 2009; 122(4): 649-53.
[http://dx.doi.org/10.1016/j.scienta.2009.06.023]

[239] De Block M, Botterman J, Vandewiele M, et al. Engineering herbicide resistance in plants by expression of a detoxifying enzyme. EMBO J 1987; 6(9): 2513-8.
[http://dx.doi.org/10.1002/j.1460-2075.1987.tb02537.x] [PMID: 16453789]

[240] Choi HJ, Chandrasekhar T, Lee HY, Kim KM. Production of herbicide-resistant transgenic sweet potato plants through *Agrobacterium tumefaciens* method. Plant Cell Tissue Organ Cult 2007; 91(3): 235-42.
[http://dx.doi.org/10.1007/s11240-007-9289-1]

[241] Stalker DM, McBride KE, Malyj LD. Herbicide resistance in transgenic plants expressing a bacterial detoxification gene. Science 1988; 242(4877): 419-23.
[http://dx.doi.org/10.1126/science.242.4877.419] [PMID: 17789813]

[242] Dear BS, Sandral GA, Spencer D, Khan MRI, Higgins TJV. The tolerance of three transgenic subterranean clover (*Trifolium subterraneum* L.) lines with the bxn gene to herbicides containing bromoxynil. Aust J Agric Res 2003; 54(2): 203-10.
[http://dx.doi.org/10.1071/AR02134]

[243] Jo J, Won SH, Son D, Lee BH. Paraquat resistance of transgenic tobacco plants over-expressing the *Ochrobactrum anthropi* pqrA gene. Biotechnol Lett 2004; 26(18): 1391-6.
[http://dx.doi.org/10.1023/B:BILE.0000045638.82348.7a] [PMID: 15604769]

[244] Lee KW, Ahsan N, Lee SH, *et al.* Responses of MxPPO overexpressing transgenic tall fescue plants to two diphenyl-ether herbicides, oxyfluorfen and acifluorfen. Acta Physiol Plant 2008; 30(5): 745-54.
[http://dx.doi.org/10.1007/s11738-008-0177-z]

[245] Eltayeb AE, Yamamoto S, Habora MEE, Yin L, Tsujimoto H, Tanaka K. Transgenic potato overexpressing Arabidopsis cytosolic AtDHAR1 showed higher tolerance to herbicide, drought and salt stresses. Breed Sci 2011; 61(1): 3-10.
[http://dx.doi.org/10.1270/jsbbs.61.3]

[246] Te ZH, Lin CY, Shen ZC. Development of transgenic glyphosate-resistant rice with G6 gene encoding 5-enolpyruvylshikimate-3-phosphate synthase. ASC 2011; 10(9): 1307-12.

[247] Guo B, Guo Y, Hong H, *et al.* Co-expression of G2-EPSPS and glyphosate acetyltransferase GAT genes conferring high tolerance to glyphosate in soybean. Front Plant Sci 2015; 6: 847.
[http://dx.doi.org/10.3389/fpls.2015.00847] [PMID: 26528311]

[248] Chhapekar S, Raghavendrarao S, Pavan G, *et al.* Transgenic rice expressing a codon-modified synthetic CP4-EPSPS confers tolerance to broad-spectrum herbicide, glyphosate. Plant Cell Rep 2015; 34(5): 721-31.
[http://dx.doi.org/10.1007/s00299-014-1732-2] [PMID: 25537885]

[249] Han H, Zhu B, Fu X, *et al.* Overexpression of d-amino acid oxidase from *Bradyrhizobium japonicum*, enhances resistance to glyphosate in *Arabidopsis thaliana*. Plant Cell Rep 2015; 34(12): 2043-51.
[http://dx.doi.org/10.1007/s00299-015-1850-5] [PMID: 26350405]

[250] Mohamed E, Rahiman FA, Wahab RA, Zain CR, Javed MA, Huyop F. A plant transformation vector containing the gene dehD for the development of cultivars resistant to monochloroacetic acid. J Anim Plant Sci 2016; 26(4).

[251] Halfhill MD, Richards HA, Mabon SA, Stewart CN. Expression of GFP and Bt transgenes in *Brassica napus* and hybridization with *Brassica rapa*. Theor Appl Genet 2001; 103(5): 659-67.
[http://dx.doi.org/10.1007/s001220100613]

[252] Bashir K, Husnain T, Fatima T, Latif Z, Aks Mehdi S, Riazuddin S. Field evaluation and risk assessment of transgenic indica basmati rice. Mol Breed 2004; 13(4): 301-12.
[http://dx.doi.org/10.1023/B:MOLB.0000034078.54872.25]

[253] Ramesh S, Nagadhara D, Reddy VD, Rao KV. Production of transgenic indica rice resistant to yellow stem borer and sap-sucking insects, using super-binary vectors of *Agrobacterium tumefaciens*. Plant Sci 2004; 166(4): 1077-85.
[http://dx.doi.org/10.1016/j.plantsci.2003.12.028]

[254] Meiyalaghan S, Jacobs JME, Butler RC, Wratten SD, Conner AJ. Expression of cry1Ac9 and cry9Aa2 genes under a potato light-inducible Lhca3 promoter in transgenic potatoes for tuber moth resistance. Euphytica 2006; 147(3): 297-309.
[http://dx.doi.org/10.1007/s10681-005-9012-4]

[255] Ye R, Huang H, Yang Z, *et al.* Development of insect-resistant transgenic rice with Cry1C*-free endosperm. Pest Manag Sci 2009; 65(9): 1015-20.
[http://dx.doi.org/10.1002/ps.1788] [PMID: 19479952]

[256] Kumar M, Shukla AK, Singh H, Tuli R. Development of insect resistant transgenic cotton lines expressing cry1EC gene from an insect bite and wound inducible promoter. J Biotechnol 2009; 140(3-4): 143-8.
[http://dx.doi.org/10.1016/j.jbiotec.2009.01.005] [PMID: 19428707]

[257] Lu Z, Han N, Tian J, *et al.* Transgenic cry1Ab/vip3H+ epsps rice with insect and herbicide resistance acted no adverse impacts on the population growth of a non-target herbivore, the white-backed planthopper, under laboratory and field conditions. J Integr Agric 2014; 13(12): 2678-89.
[http://dx.doi.org/10.1016/S2095-3119(13)60687-5]

[258] Li X, Lang Z, Zhang J, He K, Zhu L, Huang D. Acquisition of insect-resistant transgenic maize

harboring a truncated cry1Ah gene *via* Agrobacterium-mediated transformation. J Integr Agric 2014; 13(5): 937-44.
[http://dx.doi.org/10.1016/S2095-3119(13)60531-6]

[259] Sun H, Lang Z, Lu W, *et al.* Developing transgenic maize (*Zea mays* L.) with insect resistance and glyphosate tolerance by fusion gene transformation. J Integr Agric 2015; 14(2): 305-13.
[http://dx.doi.org/10.1016/S2095-3119(14)60855-8]

[260] Maqbool SB, Riazuddin S, Loc NT, Gatehouse AMR, Gatehouse JA, Christou P. Expression of multiple insecticidal genes confers broad resistance against a range of different rice pests. Mol Breed 2001; 7(1): 85-93.
[http://dx.doi.org/10.1023/A:1009644712157]

[261] Iannacone R, Grieco PD, Cellini F. Specific sequence modifications of a cry3B endotoxin gene result in high levels of expression and insect resistance. Plant Mol Biol 1997; 34(3): 485-96.
[http://dx.doi.org/10.1023/A:1005876323398] [PMID: 9225859]

[262] Fang J, Xu X, Wang P, *et al.* Characterization of chimeric *Bacillus thuringiensis* Vip3 toxins. Appl Environ Microbiol 2007; 73(3): 956-61.
[http://dx.doi.org/10.1128/AEM.02079-06] [PMID: 17122403]

[263] Castiglioni P, Warner D, Bensen RJ, *et al.* Bacterial RNA chaperones confer abiotic stress tolerance in plants and improved grain yield in maize under water-limited conditions. Plant Physiol 2008; 147(2): 446-55.
[http://dx.doi.org/10.1104/pp.108.118828] [PMID: 18524876]

[264] Bhauso TD, Thankappan R, Kumar A, Mishra GP, Dobaria JR, Venkat Rajam M. Over-expression of bacterial'mtlD'gene confers enhanced tolerance to salt-stress and water-deficit stress in transgenic peanut ('*Arachis hypogaea*') through accumulation of mannitol. Aust J Crop Sci 2014; 8(3).

[265] Park SC, Kim MD, Kim SH, *et al.* Enhanced drought and oxidative stress tolerance in transgenic sweetpotato expressing a codA gene. J Plant Biotechnol 2015; 42(1): 19-24.
[http://dx.doi.org/10.5010/JPB.2015.42.1.19]

[266] Ye X, Al-Babili S, Klöti A, *et al.* Engineering the provitamin A (β-carotene) biosynthetic pathway into (carotenoid-free) rice endosperm. Science 2000; 287(5451): 303-5.
[http://dx.doi.org/10.1126/science.287.5451.303] [PMID: 10634784]

[267] Al-Babili S, Beyer P. Golden Rice – five years on the road – five years to go? Trends Plant Sci 2005; 10(12): 565-73.
[http://dx.doi.org/10.1016/j.tplants.2005.10.006] [PMID: 16297656]

[268] Paine JA, Shipton CA, Chaggar S, *et al.* Improving the nutritional value of Golden Rice through increased pro-vitamin A content. Nat Biotechnol 2005; 23(4): 482-7.
[http://dx.doi.org/10.1038/nbt1082] [PMID: 15793573]

[269] ISAAA database 2019. Available from: https://www.isaaa .org/gmapp roval datab ase/default.asp

[270] FAO. Fats and fatty acids in human nutrition. 2014. Available from: https://agris.fao.org/agris -search/search.do?recor

[271] Ruiz-Lopez N, Haslam RP, Napier JA, Sayanova O. Successful high-level accumulation of fish oil omega-3 long-chain polyunsaturated fatty acids in a transgenic oilseed crop. Plant J 2014; 77(2): 198-208.
[http://dx.doi.org/10.1111/tpj.12378] [PMID: 24308505]

[272] Usher S, Han L, Haslam RP, *et al.* Tailoring seed oil composition in the real world: optimising omega-3 long chain polyunsaturated fatty acid accumulation in transgenic *Camelina sativa*. Sci Rep 2017; 7(1): 6570.
[http://dx.doi.org/10.1038/s41598-017-06838-0] [PMID: 28747792]

[273] Srivastava RK. Bio-energy production by contribution of effective and suitable microbial system. Mater Sci Technol 2019; 2(2): 308-18.

[274] Abanoz K, Stark BC, Akbas MY. Enhancement of ethanol production from potato-processing wastewater by engineering *Escherichia coli* using *Vitreoscilla* haemoglobin. Lett Appl Microbiol 2012; 55(6): 436-43.
[http://dx.doi.org/10.1111/lam.12000] [PMID: 22994421]

[275] Tsigie YA, Wang CY, Truong CT, Ju YH. Lipid production from *Yarrowia lipolytica* Po1g grown in sugarcane bagasse hydrolysate. Bioresour Technol 2011; 102(19): 9216-22.
[http://dx.doi.org/10.1016/j.biortech.2011.06.047] [PMID: 21757339]

[276] Gulhane PA, Gomashe AV, Kadu K. Apple pomace: A potential substrate for ethanol production. Int J Res Stud Biosci 2015; 3(6): 110-4.

[277] Kourkoutas Y, Koutinas AA, Kanellaki M, Banat IM, Marchant R. Continuous wine fermentation using a psychrophilic yeast immobilized on apple cuts at different temperatures. Food Microbiol 2002; 19(2-3): 127-34.
[http://dx.doi.org/10.1006/fmic.2001.0468]

[278] Sun Y, Cheng J. Hydrolysis of lignocellulosic materials for ethanol production: A review. Bioresour Technol 2002; 83(1): 1-11.
[http://dx.doi.org/10.1016/S0960-8524(01)00212-7] [PMID: 12058826]

[279] Ingale S, Joshi SJ, Gupte A. Production of bioethanol using agricultural waste: Banana pseudo stem. Braz J Microbiol 2014; 45(3): 885-92.
[http://dx.doi.org/10.1590/S1517-83822014000300018] [PMID: 25477922]

[280] Bhatia SK, Gurav R, Choi TR, *et al.* Bioconversion of barley straw lignin into biodiesel using *Rhodococcus* sp. YHY01. Bioresour Technol 2019; 289: 121704.
[http://dx.doi.org/10.1016/j.biortech.2019.121704] [PMID: 31276990]

[281] Wang B, Lin H, Zhan J, Yang Y, Zhou Q, Zhao Y. Biodiesel synthesis by a one-step method in a genetically engineered *Escherichia coli* using rice straw hydrolysate and restaurant oil wastes as raw materials. J Appl Microbiol 2012; 113(3): 531-40.
[http://dx.doi.org/10.1111/j.1365-2672.2012.05357.x] [PMID: 22681508]

[282] Da Cunha-Pereira F, Hickert LR, Sehnem NT, De Souza-Cruz PB, Rosa CA, Ayub MAZ. Conversion of sugars present in rice hull hydrolysates into ethanol by *Spathaspora arborariae*, *Saccharomyces cerevisiae*, and their co-fermentations. Bioresour Technol 2011; 102(5): 4218-25.
[http://dx.doi.org/10.1016/j.biortech.2010.12.060] [PMID: 21220201]

[283] Yu X, Zheng Y, Dorgan KM, Chen S. Oil production by oleaginous yeasts using the hydrolysate from pretreatment of wheat straw with dilute sulfuric acid. Bioresour Technol 2011; 102(10): 6134-40.
[http://dx.doi.org/10.1016/j.biortech.2011.02.081] [PMID: 21463940]

CHAPTER 2

Genome Editing Against Bacterial Plant Pathogens

Ashish Warghane[1,*], Neha G. Paserkar[2] and Sumit Bhose[3]

[1] *School of Applied Sciences and Technology, Gujarat Technological University, Chandkheda, Ahmedabad, Gujarat, India*

[2] *Department of Plant Science, McGill University, Quebec H9X 3V9, Canada*

[3] *Sea6Energy Pvt. Ltd., Bangalore, Karnataka, India*

Abstract: Meeting the crucial demand for sustainable agriculture is an upcoming challenge worldwide, leading to global food security concerns. Approximately 50% of agricultural loss is caused by both biotic and abiotic stresses. As per the estimation of Agrios, 42% of crop loss is characterized by biotic stress alone. Bacteria are the second largest contributor in terms of economic losses caused by various plant diseases. Hence, there is a need to develop elite cultivars in amalgamation with readily available sequenced plant database and progressive genome editing. This has proved to be a groundbreaking/milestone in the field of plant breeding for any desired trait. Until now, many new plant breeding techniques (NPBTs) have been introduced for crop improvement. These techniques include site-specific mutagenesis, cisgenesis, intragenesis, breeding with transgenic inducer lines, *etc*. This book chapter provides a comparative understanding of enrichment in plant genome editing approach about bacterial pathogens aiming for sustainable agriculture development. This chapter also brings a broad aspect of the application, advantages, unsighted aspects of genome editing, and future challenges.

Keywords: Bacterial plant pathogen, NPBTs, Plant genome editing, Sustainable agriculture.

INTRODUCTION

The world's population is rapidly rising and will reach about 9.8 billion by 2050. To fulfill nutrient requirements for the rising population, much more food is needed, but at present, the challenges in the agriculture sector due to various biotic as well as abiotic factors are the biggest concern. Plants are continuously exposed to a large set of pathogens, including bacteria, fungi, oomycetes, and viruses. The world has consistently seen about 20–40% yield loss due to the biotic impacts [1, 2].

---

[*] **Corresponding author Ashish Warghane:** School of Applied Sciences and Technology, Gujarat Technological University, Chandkheda, Ahmedabad, Gujarat, India; E-mail: warghane.ashish@gmail.com

Prakash M. Halami & Aravind Sundararaman (Eds.)
All rights reserved-© 2024 Bentham Science Publishers

Understanding the molecular mechanism between plants and communities of bacteria, fungi, and other microorganisms has been a significant area of investigation in plant pathology for many years. However, we cannot deny that despite decades of research, we have a very limited understanding of the molecular mechanisms of host-pathogen interactions. The pathogen and the host play an endless arms race game between them. When the plant host and pathogen come in contact, the interaction turns into a fight of recognition and evasion. A multilayer defense system including pathogen-associated molecular patterns (PAMPs)-triggered immunity (PTI) and effector-triggered immunity (ETI) has involved plants in battling against interfering pathogens for survival. The PTI gets activated through the recognition of PAMPs by pattern recognition receptors (PRRs) which results in the production of reactive oxygen species (ROS), callose deposition, and transcriptional reprogramming, which usually prevents the invasion of non-adapted pathogens. In contrary, to modulate host cell physiology, pathogens secrete effectors to interrupt PTI and this results in effector-triggered susceptibility (ETS). In resistant plant varieties, these effectors or byproducts can be recognized by intracellular immune receptors and induce ETI (a robust resistance response). This is usually associated with localized plant cell death leading to pathogen arrest. However, as these pathogens have high evolutionary potential, they can overcome the host's ETI response *via* loss and/or modification of ETI-eliciting effectors as well as meta-effector interactions [3 - 5]. Hence, phytopathogens are difficult to control.

The most effective approaches that control plant disease depend on resistant varieties and agrochemicals. However, as explained earlier, many plant pathogens have high evolutionary potential, novel genotypes no longer sensitive to the resistance gene or the phytosanitary product can rapidly emerge *via* mutation or recombination. Hence, the enhancement of plant resistance plays an important role in adjusting crop production to meet global population increases [6, 7]. The major aim of this chapter is to highlight the applications of genome editing against different bacterial plant pathogens.

EXISTING GENOME EDITING TECHNIQUES AND ADVANCEMENT UNTIL NOW

Over the past few years, new plant breeding techniques have been the most useful approach for developing new crop improvement, including pathogen- resistance [8, 9]. New plant breeding techniques include the usage of Meganucleases (MN), also known as homing endonucleases, zinc finger nucleases (ZFNs), transcription activator-like effector nucleases (TALENs), and newly emerged clustered regularly interspaced short palindrome repeats (CRISPR)/CRISPR-associated protein 9 (Cas9), which have revolutionized targeted modifications of genomes and have greatly transformed the researches on plants [10]. The requirement of sophisticated protein engineering rendered MN, ZFN, and TALEN techniques less practicable.

Meganucleases

Meganucleases, also called homing endonucleases, have been used for more than 15 years to induce gene targeting. Although they have been rarely used in crop-related gene editing to date, their scope of applications, especially in gene therapy, has further enhanced due to the recent advances in re-engineering meganuclease specificity [11, 12]. They are divided into five families based on the sequence and structure motifs: LAGLIDADG, GIY-YIG, HNH, His-Cys box, and PD-(D/E)XK. Among these families, the LAGLIDADG proteins have been found in all kingdoms of life and are the most well studied [12]. These proteins generally encode within introns or inteins, but freestanding members also exist. They are highly specific endonucleases capable of recognizing and cleaving the exon-exon junction sequence wherein their intron resides (Fig. 1). Additionally, unlike restriction enzymes, the proteins facilitate lateral mobility of genetic elements within an organism. This process is referred to as "homing" and gives the name to HEs [10].

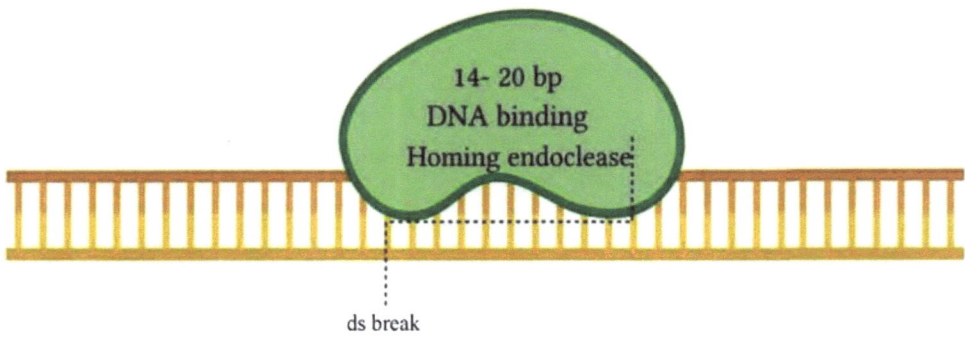

Fig. (1). Schematic representation of Meganuclease components and Meganuclease based genome editing. This figure was created by authors using BioRender.

MN allows insertion, deletion, single-site mutation, and correction in a highly site-specific and controlled fashion. Furthermore, viral vectors are also available for endonuclease delivery as a novel approach to plant engineering; therefore, they are regularly used for applications in medicine, public health, and agronomy. In agriculture, LAGLIDADG Meganucleases I-CreI has been modified for agricultural applications in maize. The endonuclease gene was delivered to immature embryos to generate transgenic plants, with deletions and insertions detected at the HE cut site [13]. To improve prospects of nuclease-mediated improvement of plants, multigene plant transformation vectors have recently been constructed, with a cloning system based on ZFNs and mega nucleases [14, 15].

Zinc-Finger Nucleases (ZFNs)

Zinc finger nucleases are the first genome editing tool to perform targeted genome editing [16]. Zinc finger nucleases (ZFNs) are artificially engineered DNA-binding proteins that are used for the targeted editing of the genome by creating double-strand breaks in DNA at user-specified locations [17]. ZFNs consist of two parts, first Zinc- fingers are the DNA binding domain, and the nucleases are the restriction enzymes (Fig. **2**). For the first time, in 1996, it was shown that protein domains such as "zinc fingers" coupled with enzyme *FokI* endonuclease (naturally found in *Flavobacterium okenakoites*) domains act as site-specific nucleases (zinc finger nucleases) [18]. Zinc-finger nucleases (ZFNs) are artificial restriction enzymes generated by fusing a zinc finger DNA-binding domain to a DNA-cleavage domain (*FokI*). Zinc finger domains can be engineered to target specific desired DNA sequences, and this enables zinc-finger nucleases to target unique sequences within complex genomes [19]. By taking advantage of endogenous DNA repair machinery, these reagents can be used to precisely alter the genomes of higher organisms.

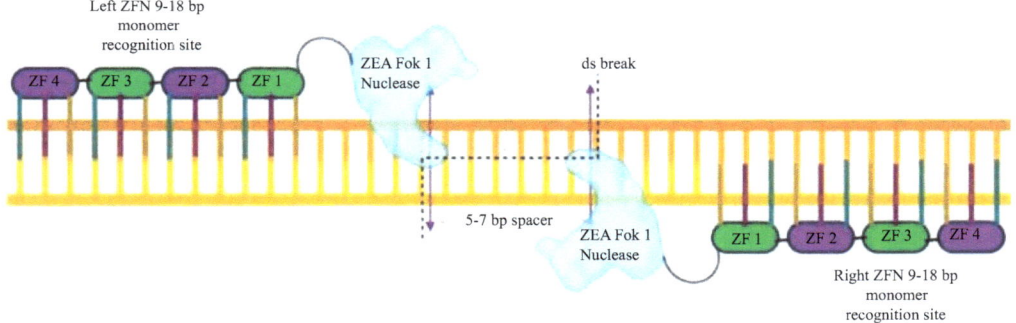

Fig. (2). Schematic representation of ZFN components and ZFN based genome editing. This figure was created by authors using BioRender.

It was revealed that simultaneous expression of ZFNs and delivery of a simple heterologous donor molecule lead to precisely targeted addition of herbicide-tolerance gene at the intended locus in a significant number. Zinc finger proteins have made a significant contribution to antiviral resistance in plants by blocking DNA binding sites of viral replication proteins [20]. In the field of improving crop disease resistance, ZFNs have made little impact by editing host plant genes involved in disease development as they are complex to be engineered and difficult to be multiplexed [21 - 23].

Zhang *et al.*, 2010 reported a competent method for targeted mutagenesis of *Arabidopsis* genes through regulated expression of zinc finger nucleases (ZFNs)—enzymes engineered to create DNA double-strand breaks at specific target loci [24]. The *ADH1* and *TT4* ZFNs were placed under the control of an estrogen-inducible promoter and introduced into *Arabidopsis* plants by the floral-dip transformation. Primary transgenic *Arabidopsis* seedlings induced to express the *ADH1* or *TT4* ZFNs exhibited somatic mutation frequencies of 7% or 16%, respectively. The induced mutations were typically insertions or deletions (1–142 bp) that were localized at the ZFN cleavage site and likely derived from imprecise repair of chromosome breaks by nonhomologous end-joining (Table **1**). This study revealed that it is possible to obtain mutations in any *Arabidopsis* target gene regardless of its mutant phenotype [24, 22].

Table 1. Application of zinc finger nucleases-based genome editing in agriculturally important plants.

Crop/Fruit/Vegetable	Gene	Mutation Type	Editing Efficiency (%)	Aim of Study	Method of Zinc Finger Nuclease System Delivery	Phenotype	Gene Function	References
Arabidopsis thaliana	*ADH1*, *TT4*	Insertion or deletion	33%	Selectable or screenable phenotypes (*adh1*, allyl alcohol resistance; *tt4*, lack of anthocyanins in the seed coat)	Floral dip method	No effect	(*adh1*, Alcohol dehydrogenase1; *tt4*, Transparent testa4)	[24]
Zea mays	*IPK1*	Insertional disruption	-	IPK1 is an enzyme that catalyses the final step in phytate biosynthesis in maize seeds. Phytate reduction is agriculturally important	Plasmid DNA delivery	Herbicide-tolerant phenotype	Inositol-1,3,4,5,6-pentakisphosphate 2-kinase	[20]
Nicotiana benthamiana	*ALS SuRA* and *SuRB*	insertion/deletions (indels)	2%	*Acetolactate synthase* genes	Protoplast transformation	Confers resistance to one or more herbicides (P191A, chlorsulphuron; S647T, imazaquin; W568L, chlorsulphuron and imazaquin)	Confers resistance to imidazolinone and sulphonylurea herbicides	[16]
Arabidopsis thaliana	*ABI4* gene	Deletion and substitution	3%	*ABA-INSENSITIVE4* gene	Floral dip method	ABA and glucose insensitivity	*ABI4* is an ERF/AP2 transcription factor and plays a role in ABA and sugar signaling during the seed development of *Arabidopsis*	[24]

Townsend *et al.* revealed high-frequency ZFN-stimulated gene targeting at endogenous plant genes, namely the tobacco acetohydroxyacid synthase (*SuRA* and *SuRB*) genes, for which specific mutations are known to confer resistance to imidazoline and sulfonylurea herbicides.

Transcription Activator-Like Effector Nuclease Technique (TALEN)

Three renowned engineered nucleases, *i.e.*, ZFN, TALEN, and CRISPR/Cas9 have revolutionized the biological field for genetic engineering. Like ZFN, TALEN is based upon DNA-protein interaction. TALEN (Transcription Activator-Like Effector Nuclease technique) is one of the site-specific mutagenesis techniques restricted to plant pathogens which introduce the same changes in the genome as CRISPR-Cas9. TALEN, along with ZFN, is a meganuclease and is generally considered a first-generation genome editing tool. TALEN is a multicomponent structure consisting of two unique DNA recognition TALE protein repeats separated by spacer forming homodimer (Fig. 3) and a DNA cleavage domain derived from *FokI* [25].

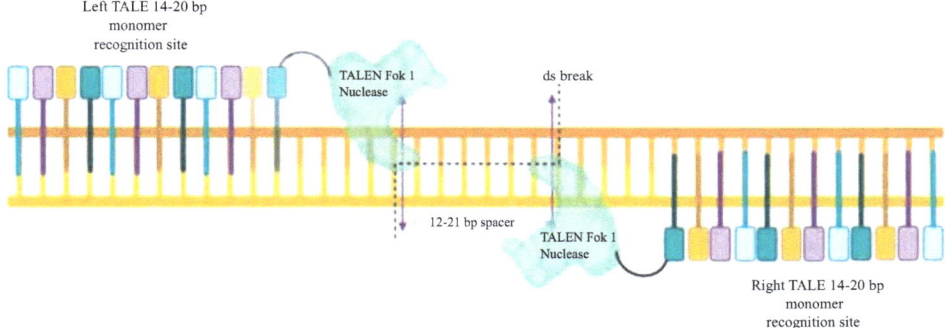

Fig. (3). Schematic representation of TALEN subcomponents and TALEN based genome editing. This figure was created by authors using BioRender.

TALE protein is potent for recognizing and binding target specific DNA sequence, which was first discovered in the year 2007 [26]. Each TALE protein monomer is composed of about 34 amino acid tandem repeats termed central repeat domain (CRD). Specific identification of DNA sequence is provided by the 12th and 13th variable amino acid present in the CRD domain, also known as repeat variable residue (RVD) [27]. Upon target recognition by molecular scissors-like TALE protein, double-stranded breaks (DSBs) are introduced. The crops derived *via* TALEN techniques are considered non-GM. Like CRISPR-Cas9 and ZFN can target multiple sites at the same time, TALEN can carry out multiplexing efficiently [28].

In comparison to other genome editing tools, TALEN is relatively easy to design. TALEN is beneficial over ZFN, as it can target any preferred length of the genome. There are some reports which highlight the importance of length in TALEN, which assists in improving the site-specific DNA cleavage activity [29]. To estimate the true potential of TALEN, it is mandated to engineer this genetic tool for the identification of novel sequences. It's challenging to design and assemble as it requires thymidine at the first position, and is composed of larger size and repetitive sequence [23, 30]. However, TALEN encoding cDNA with a larger size of approximately 3kb as compared to ZFN of 1kb makes it less expressive in the expression vector [31]. Along with this, TALEN is tedious in the application, owing to off-target activity and labor-intensive. TALEN mediated plant genome editing examples are mentioned below Table **2**.

Table 2. Application of TALEN-based genome editing in agriculturally important plants.

Crop/Fruit/ Vegetable	Gene	Mutation Type	Editing Efficiency (%)	Aim of Study	Method of TALEN System Delivery	Phenotype	Gene Function	References
Tomato (*Solanum lycopersicum*)	*PROCERA* (negative regulator of GA signaling *PROCERA* (PRO)	Indels (Insertion and deletion mutation	-	To improve understanding of the mechanisms controlling extension growth and the part played by GA in the growth process	*Agrobacterium*-mediated transformation method	Phenotypes consistent with increased GA response.	The procera mutation affects several aspects of the development of the tomato plant	[32]
Arabidopsis thaliana and Tobacco	*ADH1* gene	Deletion	25%	To carry out targeted mutagenesis in *Arabidopsis* protoplasts	PEG-mediated transformation of protoplast	-	Required for survival and acclimation in hypoxic conditions, especially in roots	[33]
Rice	*Os11N3* (also called *OsSWEET14*) a rice bacterial blight susceptibility gene	Indels (Insertion and deletion mutation	63%	To interfere with the virulence function of *AvrXa7* and *PthXo3*, but not the developmental function of *Os11N3*	*Agrobacterium*-mediated transformation method	Resistance phenotype displayed by two T_2 mutant plants compared with the disease susceptibility phenotype of a nontransgenic wt rice plant.	Rice bacterial blight susceptibility gene provides resistance to bacterial blight (*Xanthomonas oryzae* pv. *oryzae*)	[34]
Arabidopsis thaliana	*ADH1, TT4, MAPKKK1, DSK2B,* and *NATA2*	Indels (Insertion and deletion mutation	1.5–12%	To achieve high levels of mutagenesis in somatic cells	Robust Agrobacterium-mediated floral dip transformation method	Normal	->*ADH1* catalyzes the reduction of acetaldehyde using NADH as a reductant ->*TT4*, biosynthesis of flavonoids ->*DSK2B*, function as shuttle factors ferrying polyubiquitinated substrates to the proteasome for degradation.	[25]

(Table 2) cont.....

Crop/Fruit/Vegetable	Gene	Mutation Type	Editing Efficiency (%)	Aim of Study	Method of TALEN System Delivery	Phenotype	Gene Function	References
Rice and Brachypodium	OsDEP1, OsBADH2, OsCKX2, OsSD1, BdABA1, BdABA1, BdSMC6, BdSPL, BdSBP, BdCOI1, BdRHT and BdHTA1	Small deletions, insertions, and nucleotide substitutions	4% to 14%	To test whether TALENs can induce high-frequency gene knockouts in rice and Brachypodium	Embryonic cells of rice or Brachypodium using Agrobacterium tumefaciens	Normal	->OsDEP1 regulates Carbon-nitrogen metabolism -> OsBADH2 regulates inhibits the synthesis of 2-acetyl-1-pyrroline (2AP) responsible for fragrance -> OsCKX2 reduces yield penalty under salinity stress -> OsSD1 improves drought-, heat-, and salt-tolerance simultaneously	[35]
Potato	Acetolactate synthase gene (ALS)	Indels (Insertion and deletion mutation	Transfection efficiency of protoplasts was 38–39% and the site-directed mutation frequency was 7–8% with a few base deletions as the predominant type of mutation	Establish a pipeline for tetraploid potato targeted gene mutation where TALEN was transiently expressed in protoplasts	Transient expression of TALEN in protoplasts	Normal	Acetolactate synthase gene (ALS) encodes for a critical enzyme involved in the biosynthesis of branched-chain amino acids in plants	[36]
Solanum tuberosum, Ranger Russet variety	Vacuolar invertase gene (VInv), which encodes a protein that breaks down sucrose to glucose and fructose	indel (insertion/deletion)	2.1% to 15.9%	TALEN-mediated gene editing creates a valuable trait in a commercial potato cultivar with high efficiency	Protoplasts were transformed with TALEN-encoding plasmids	-	The new potato lines have significantly lower levels of reducing sugars and acrylamide in heat-processed products	[37]
Rice	Rice bacterial blight susceptibility gene OsSWEET11 and Os11N3 (also called OsSWEET14)	Indel (insertion/deletion)	63%	To interfere with the virulence function of AvrXa7 and PthXo3, but not the developmental function of Os11N3	Agrobacterium-mediated rice transformation	Normal phenotype	Os11N3 gene encodes a member of the SWEET sucrose-efflux transporter family and is hijacked by X. oryzae pv. oryzae, using its endogenous TAL effectors AvrXa7 or PthXo3, to activate the gene and thus divert sugars from the plant cell so as to satisfy the pathogen's nutritional needs and enhance its persistence	[34]

CRISPR/Cas9

"CRISPR" is an acronym for "Clustered Regularly Interspaced Short Palindromic Repeat", which is the RNA-guided programmable genome editing technology derived from the natural defense mechanism of bacteria and archaea against invading viruses. CRISPR/Cas9 assists researchers in altering the DNA sequences and amending gene function, transcriptional regulation, fingerprinting cells, epigenetic tweaks, and genome imaging. The discovery of the CRISPR/Cas9 gene-editing system has revolutionized research in animal and plant biology with its utility in genome editing being first demonstrated in 2012 in mammalian cells. Unlike ZFNs and TALENs, CRISPR genome editing is more straightforward and involves designing a guide RNA (gRNA) of about 20 nucleotides complementary to the DNA stretch within the target gene. The protein Cas9 is an enzyme that acts like a pair of molecular scissors, capable of cutting strands of DNA and commonly written as CRISPR-Cas9 (CRISPR associated 9 endonuclease) [38].

CRISPR/Cas9 hijacks the infecting pathogen DNA and utilizes them as a spacer, thereby acting as a DNA template. The DNA recognition component consists of 20 nucleotide target specific sequences at the 5' end of sgRNA, which direct the Cas9 to the target site to carry out a double-stranded break (Fig. **4**). This break activates repair mechanisms by homologous recombination (HR) or non-homologous end-joining (NHEJ). Protospacer adjacent motif (PAM) is a trinucleotide sequence which is the absolute requirement for the CRISPR system, like Cas9 nuclease requires NGG to recognize the target, whereas Cpf1 requires TTT. Such trinucleotides are abundant in any plant genome, meaning an abundant number of target spacers are available as genome editing targets [38]. Examples of CRISPR mediated plant genome editing are given in Table **3**.

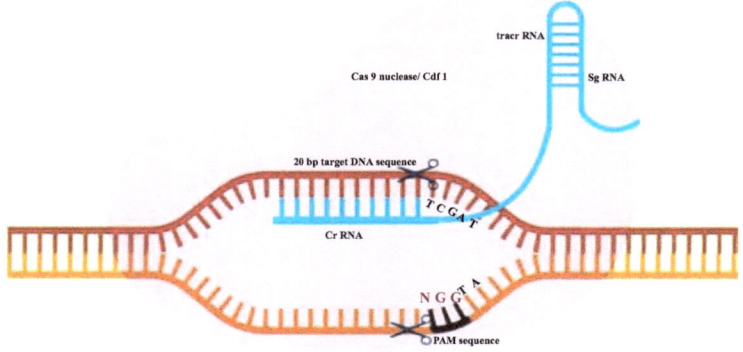

Fig. (4). Schematic representation of CRISPR-Cas9 subcomponents and CRISPR-Cas9 based genome editing CRISPR/Cas9. This figure was created by authors using BioRender.

Table 3. CRISPR mediated plant genome editing.

Crop/fruit/ Vegetable	Gene	Mutation Type	Editing Efficiency (%)	Aim of Study	Method of Cas9 System Delivery	Phenotype	Gene Function	References
Tomato (*Solanum lycopersicum*) and *Arabidopsis*	(*Solanum lycopersicum*) arginine decarboxylase 1 (*SlADC1*) and arginine decarboxylase 2 (*SlADC2*)	Substitution and deletion mutation	-	To investigate the function of Brg11, a T3 bacterial effector from *Ralstonia solanacearum*	*Agrobacterium*-mediated transformation method	-	ADC (arginine decarboxylase) is involved in polyamine biosynthesis that inhibits *Ralstonia solanacearum* niche competitor like *Pseudomonas syringae*, *Xanthomonas euvesicatoria*	[39]
Super Basmati Rice	EBEs of three TALEs (AvrXa7, PthXo3, and TalF) present in the promoter of OsSWEET14	Deletion mutation	9%	To generate resistance against bacterial blight in indigenous varieties against locally prevalent *Xoo* strains.	Bombardment by a gene gun (PDS-1000/Bio-Rad, United States) following optimized protocol for the biolistic transformation of Super Basmati rice	Normal	To provide resistance against bacterial blight *Xanthomonas oryzae* pv. *oryzae* strain	[40]
Apple (*Malus pumila*)	DIPM-1, DIPM-2, and DIPM-4	Insertions and deletions (indels) mutations	0.5–6.9	To increase resistance to fire blight disease	Protoplast transformation, direct delivery of CRISPR RNPs to apple protoplast	–	DIPMs exhibit direct physical interaction with the disease-specific gene of *Erwinia amylovora*, which may act as a susceptible factor	[41]
Wanjincheng orange (*Citrus sinensis* Osbeck)	Gene *CsLOB1* (Lateral Organ Boundaries1) promoter	Insertions, deletions, and substitutions	11.5–64.7	Exhibited enhanced tolerance against citrus canker as compared with wild type	*Agrobacterium*-mediated epicotyl transformation method	Mutant plants showed tolerance against citrus canker (*Xanthomonas axonopodis*)	*CsLOB1* promotes the growth of pathogen and formation of an erumpent pustule	[42]
Duncan grapefruit (*Citrus paradise* Macf.)	*CsLOB1*(*C. sinensis* Lateral Organ Boundaries*) gene	Insertions and short deletions	15.63, 14.29, 54.54, & 81.25	For generating canker-resistant grapefruit	*Agrobacterium*-mediated epicotyl transformation method	Exhibited canker symptoms same as wild type	*CsLOB1* is the disease susceptible gene for bacterial canker disease of citrus	[43]
Kitaake, IR64 and Ciherang-Sub1	Promoters of SWEET 11, SWEET 13 & SWEET 14 genes	insertions, deletions and substitutions	-	To provide broad spectrum and durable resistance against *Xanthomonas oryzae* pv. *oryzae* (*Xoo*)	*Agrobacterium* mediated transformation	Yield was reduced by 10%	SWEET gene function as a sucrose transporter however five of these genes act as susceptibility genes for bacterial blight *Xanthomonas oryzae* pv. *oryzae* (*Xoo*)	[44]
Super Basmati Rice	EBEs of three TALEs (*AvrXa7*, *PthXo3*, and *TalF*) present in the promoter of OsSWEET14	Deletion mutation	9%	To generate resistance against bacterial blight in indigenous varieties, against locally prevalent *Xoo* strains.	Bombardment by a gene gun (PDS-1000/Bio-Rad, United States) following optimized protocol for the biolistic transformation of Super Basmati rice	Normal	To provide resistance against bacterial blight *Xanthomonas oryzae* pv. *oryzae* strain	[40]

(Table 3) cont.....

Crop/fruit/Vegetable	Gene	Mutation Type	Editing Efficiency (%)	Aim of Study	Method of Cas9 System Delivery	Phenotype	Gene Function	References
Tomato	SlJAZ2 gene	Deletion mutation	-	To provide resistance against bacterial speck disease caused by *Pseudomonas syringae* pv. *tomato* (*Pto*) DC3000	*Agrobacterium* mediated transformation	Normal	SlJAZ acts as a coreceptor of COR (coronatine), which stimulates stomatal opening, thereby facilitating the entry of bacterial pathogens in leaf tissues	[45]
Arabidopsis thaliana	AtIAN9 (Immune-Associated Nucleotide-Binding 9 Protein)	Insertion	2.4- 4.8	To provide resistance against *Pseudomonas syringae* (bacterial speck)	*Agrobacterium*-mediated T-DNA insertions into the plant genome	Mutants with clustered leaf trichomes,	IAN9 encodes a plasma membrane-localized protein that genetically behaves as a negative regulator of immunity	[46]
Arabidopsis thaliana	AtErf019 (ethylene-responsive factor 19 gene)	Frame shift mutation	-	*Phytophthora parasitica*	*Agrobacterium tumefaciens*-mediated transformation	Normal	Mediating plant susceptibility to *P. parasitica* through suppression of pathogen-associated molecular pattern-triggered immunity	[47]
Apple	MpDIPM-1/2/4 (DspE-interacting proteins of Malus)	Indel (insertion or deletion)	-0.5 to 6.9%	*Erwinia amylovora* (fire blight)	Direct delivery of CRISPR/Cas9 RNPs to the protoplast system	Normal	To increase resistance to fire blight disease	[48]
Rice	OsXa13	Deletion	-	*Xanthomonas oryzae*	A new rice breeding method: CRISPR/Cas9 system editing of the *Xa13* promoter to cultivate transgene-free bacterial blight-resistant rice	-	Xa13 gene is a pluripotent gene for recessive resistance to bacterial blight that regulates rice bacterial blight resistance and participates in anther development.	[49]
Rice	OSTMS5, OsPi21, OsXa13 (The recessive R gene pi21 imparts broad-spectrum resistance to rice blast)	Homozygous frame-shift mutations	-	*Xanthomonas oryzae*	-	The mutant lines showed only grade 1 lesions	STM develop commercial thermosensitive genic male sterile rice. pi21 imparts broad-spectrum resistance to rice blast.	[50]
Rice	Os8N3 (also known as OsSWEET11)	Insertion	Highly efficient in generating Os8N3 gene editing in rice	*Xanthomonas oryzae*	*Agrobacterium*-mediated transformation	Robust resistant phenotype	Os8N3, conferring disease resistance by expressional loss-of-function in rice, has been considered an essential constituent for pollen development	[51]
Rice	OsCul3a (oscullin3a)	Deletion and substitution	-	*Xanthomonas oryzae*	*Agrobacterium*-mediated transformation method	Opposing effects on the cell-death-related lesion mimic phenotype in the oscul3a mutant	OsCUL3a confers bacterial disease resistance, disrupts cell metabolism, causing enhanced lipid metabolism and suppressed growth-related carbon/nitrogen metabolism	[52]

(Table 3) cont.....

Crop/fruit/Vegetable	Gene	Mutation Type	Editing Efficiency (%)	Aim of Study	Method of Cas9 System Delivery	Phenotype	Gene Function	References
Cacao	*TcNPR3* (cacao Non-Expressor of Pathogenesis-Related 3)	Indel (insertion or deletion)	~27%	*Phytophthora tropicalis*	*Agrobacterium* mediated transformation	Strong phenotypic change	Increased resistance against *Phytophthora tropicalis* and elevated expression of downstream defense genes	[53]
Rice	*OsMPK5* (encodes a stress-responsive rice mitogen-activated protein kinase)	Indels in target, resistance not confirmed	3–8%	Bacterial (*Burkholderia glumae*) pathogens	Protoplast transformation	Normal	*OsMPK5* is a negative regulator of rice defense response, targeted mutation of *OsMPK5* will likely enhance rice disease resistance	[54]
Tomato	*SlDMR6* (Downy mildew resistance 6)	Frameshift deletions and insertions	gRNA efficiency can vary according to the gRNA sequence	*P. syringae*, *Phytophthora capsici*, *Xanthomonas*	*Agrobacterium*-mediated transformation method	Disease resistance phenotype	Resistance to different pathogens, such as the bacterial pathogens *P. syringae* and *Xanthomonas spp.* and the oomycete pathogen *Phytophthora capsici*	[55]

BACTERIAL GENOME EDITING, APPLICATION, AND ITS SIGNIFICANCE

Genome editing by the application of programmable enzymes has revolutionized the biotechnological world. Previously, RNAi-based gene inhibition was used because of easiness of usage but because of transient gene expression, its application was limited [56]. Later the discovery of many techniques like Zinc finger nuclease (ZFN), Transcription Activator-like Effector Nucleases (TALEN), and RNA-guided CRISPR- Cas9 has been found to have broad applications by applying transcriptional activation and repressor domain to express unique phenotypes. Particularly, the discovery of cells and tissue site-specific CRISPR and CRISPR-associated Cas9 proteins have been used to engineer or manipulate not just the cultured primary cells but also the micro-organisms, plants, and animals. This technique has been a quantum leap in both agriculture and synthetic biology.

Here we expand the technology of precise genome editing to multiple bacterial species, as a large variety of micro-organisms have been unexplored to date. To dissect the genetic potential of these tiny microbes there is the need for the ubiquitous genetic tool. The application of genome editing technology has been very limited and ineffective in bacteria. Because bacterial screening is antibiotic marker dependent [57], dearth of potential genome editing tools and poor transformation efficiencies. However, CRISPR-Cas9 system application in bacteria has been evidenced efficient and precise manner. Adaptive immunity is offered to bacteria and archaea by CRISPR loci retaining the genetic memory [58], whereas the resistance is determined by spacer sequence similarity. Table 4 provides the details of genome editing applications in several bacterial species. Currently, the type-II CRISPR-Cas9 system has been used in bacteria which comprises Cas9 nickase mediated and Cas9 mediated genome editing [59]. Strain improvement has been reported in *Bacillus licheniformis* for the production of biochemical products using CRISPR-Cas9 [60]. Further engineered *Pseudomonas* species have been developed by the pnCasPA-BEC system, thereby enhancing the potential of metabolic engineering, and drug target analysis.

Table 4. CRISPR-Cas9 mediated genome engineering targeted against bacterial plant pathogens.

Bacteria Strain Used	Gene	Method of Genome Editing	Mutation Type	Editing Efficiency (%)	Reason for Transformation	Gene Function	References
E. coli	P(3HB-co-4HB)	CRISPRi multiplexing	Insertion	80-90%	To study how the polyhydroxy alkanoate (PHA) biosynthesis pathway fluxes and thereby adjusts PHA composition	Production of poly(3-hydroxybutyrate-co- 4-hydroxybutyrate) [P(3HB-co-4HB)] from glucose	[5]
E. coli	β- carotene	CRISPR– Cas9 based method	Gene deletions, insertions, and replacements	100%	To integrate the β-carotene biosynthetic pathway into the genome and to conduct combinatorial modulation of the MEP pathway and central metabolic pathways to search for improved β-carotene producer	Optimizes the methylerythritol-phosphate (MEP) pathway and central metabolic pathways for β-carotene overproduction optimize the methylerythritol-phosphate (MEP) pathway and central metabolic pathways for β-carotene overproduction	[60]
B. licheniformis	Bacitracin synthase gene cluster *bacABC*, *BL10* and six extracellular protease genes (*mpr, vpr, aprX, epr, bprA, wprA, aprE, bprA, hag, and amyL*)	CRISPR-Cas9n	Single-gene deletion, multiple-gene disruption, large DNA fragment deletion, and single-gene integration	79% 76.5% respectively	To improve nattokinase production by 25.7%	Deletion of *bacABC* may have led to increases in branched-chain amino acid concentrations, resulting in increased fatty acid accumulation.	[60]
P. aeruginosa, P. putida, and Pseudomonas fluorescens	Fts Z depletion (β-galactosidase activity) and pyoverdine production	CRISPRi gene repression using *S. pasteurianus* dCas9 type II CRISPR/Cas system	Deletion	Gene repression of up to 100-fold in β-galactosidase activity in P. aeruginosa and 300-fold in pyoverdine production in *P. putida*	Enables easy depletion of specific proteins in Pseudomonas, accelerating the study and engineering of this widely used model organism	This inducible system enables the study of essential genes, as shown by ftsZ depletions in *P. aeruginosa, P. putida,* and Pseudomonas fluorescens that led to phenotypic changes consistent with depletion of the targeted gene	[61]

(Table 4) cont.....

Bacteria Strain Used	Gene	Method of Genome Editing	Mutation Type	Editing Efficiency (%)	Reason for Transformation	Gene Function	References
Mycobacterium tuberculosis	Optimizes CRISPRi platform for mycobacteria	CRISPR1 Cas9 from *Streptococcus thermophilus* (dCas9$_{Sth1}$)	Deletion	20- to 100-fold knockdown of endogenous gene	To anticipate that the dCas9$_{Sth1}$ CRISPRi system will have broad utility for functional genomics, genetic interaction mapping and drug-target profiling in *M. tuberculosis*	Developing novel drug regimens to shorten the duration of tuberculosis chemotherapy	[62]
Lactobacillus reuteri ATCC PTA 6475	Development and optimization of CRISPR–Cas9 selection in *Lactobacillus reuteri*, a strain with probiotic properties		Deletion	90- 100%	Development of CRISPR-Cas9 genome editing in combination with ssDNA recombineering in *L. reuteri* 6475 that allows identification of mutations at efficiencies up to 100%	CRISPR–Cas9 selection combined with genetic engineering tools such as ssDNA recombineering provides researchers with a toolset that widens the horizon of fine-tuned genome editing in a variety of organisms	[63]
Rhodobacter sphaeroides	Cytosine base editors (CBEs) and adenine base editors (ABEs), generated by fusing endonuclease Cas9 variant to cytosine deaminase PmCDA1 or heterodimer adenine deaminase TadA–TadA	yCRISPR/Cas9 systems	Substitution mutation	97% and 43%	To obtain C-to-T mutation of single and double targets following the first induction step, with the efficiency of up to 97% and 43%; while the second induction step was needed in the case of triple target, with the screening rate of 47%	CBEs and ABEs serve as alternative methods for genetic manipulation in *Rhodobacter*	[64]
Klebsiella pneumoniae	*bla*$_{KPC-2}$ gene	(CRISPR)-Cas9 genome cleavage system and the lambda Red recombination system, pCasKP-pSGKP	Deletion	100%	Simple construction procedures and high efficiency, future applications of the two editing systems should dramatically facilitate a wide variety of investigations, such as gene characterization, drug discovery, and metabolic engineering, in *K. pneumoniae* and relevant *Enterobacteriaceae* species	Major factor that contributed to the carbapenem resistance of a hypermucoviscous carbapenem-resistant *K. pneumoniae* strain	[46]

Genome Editing For Plant Disease Resistance Against Bacterial Pathogens

Among the living bacterial species on the earth, very few hundreds are being involved in crop damage, and this often reveals different symptoms of the disease [65]. Due to various reasons like undetectable asymptomatic infections and the lack of suitable agrochemicals, phytopathogenic bacteria are mostly difficult to control. In general, prevention and exclusion of the pathogen by using genetic resistance, agronomic practices, and biocontrol agents are well-known approaches to control bacterial infections in plants. Phytopathogenic bacteria can be classified as crop-specific; *Clavibacter michiganensis*, which is the causal agent of tomato bacterial ring rot; polyphagous specific, such as *Ralstonia solanacearum*, which causes disease in multiple monocot and dicot species; and "kingdom crosser," such as Dickeya dadantii, an entomo-phytopathogen, which can affect plants and animals.

When any plant gets a bacterial infection, type III effectors are secreted into the plant cell [66]. These effectors mainly disturb the plant's defense pathways and/or activate the S genes for disease development [67]. Thus, to improve plant resistance both S genes and negative regulators of plant innate immune response are good target sites for CRISPR/ Cas9-mediated gene editing. However, few studies have been published on the application of the CRISPR/Cas system to counteract crop bacterial diseases.

CRISPR-Cas9 Mediated Resistance Against *Xanthomonas oryzae*

Bacterial blight disease caused by γ-proteobacterium *Xanthomonas oryzae* pv. *oryzae* is one of the major issues for rice growers. To control losses due to the disease CRISPR/Cas9 mutagenesis of *OsSWEET13* has been performed in recent years [68]. *OsSWEET13* is a susceptible (S) gene which is responsible to encode sucrose transporter involved in plant-pathogen interaction. The bacteria produce an effector protein known as PthXo2 and this induces the expression of *OsSWEET13* in the host and the resultant condition of susceptibility. Earlier to this work a TALEN approach was also used for *OsSWEET14* promoter mutagenesis, where the disruption of this gene rendered the *X. oryzae* effector unable to bind OsSWEET14 and ultimately resulted in disease resistance (69, 34). Similarly, in 2015 a group of researchers obtained a null mutation in *OsSWEET13* to better explore PthXo2-dependent disease susceptibility, and resultant mutants were resistant to bacterial blight [69].

Kim *et al*. (2019) reported a knockout of the endogenous *OsSWEET11* gene, the target of TALEs in rice, resulted in significantly enhanced resistance to *X. oryzae* pv. oryzae (*Xoo*) without altering pollen development. The promoter regions of both *OsSWEET11* and *OsSWEET14* were also targeted by researchers. TALE-binding elements (EBEs) in the promoter region resulted in the rice lines carrying indels conferred robust resistance to most *Xoo* strains [51, 70]. Furthermore, in recent studies, broad-spectrum resistance to *Xoo* is reported in three rice varieties by multiplex genome editing of the promoter regions of *OsSWEET11*, *OsSWEET13*, and *OsSWEET14* [71].

CRISPR Against "*Candidatus liberibacter* spp"/Citrus Greening Bacterium

There are many diseases which affect Citrus trees, among them, Citrus Greening (Huanglongbing, HLB) is considered the most damaging disease of Citrus. HLB kills millions of citrus trees in many citrus-growing areas worldwide [73]. Some of the symptoms shown by HLB disease are yellow shoots, stunted growth, and abnormal pigmentation of fruit and have a bitter taste, fruits unusable for juice production *etc*. This HLB disease has the potential to destroy trees completely after showing the first symptoms of the disease within a few years. It has been estimated that over 60 million citrus trees have been destroyed worldwide by this disease alone [72, 73].

HLB is one of the biggest threats to the citrus industry. Many efforts have been put into understanding the mechanisms of how *Candidatus liberobacter* species causes disease symptoms. But there are large gaps in understanding disease development. CRISPR has one of the chief advantages over all other gene-editing technologies; the modified plant may not contain foreign DNA, so the edited plant can be released to market as commercial products for human consumption. Recently, Irish and Jacob labs at Yale University surveyed the publication related to the Plant molecular gene expression response against HLB infection and identified 28 publications documenting 71 such datasets. These include both transcriptomic and proteomic approaches and describe a diverse range of samples taken from different tissues, stages of HLB progression and Citrus species. Differentially abundant genes and proteins were manually extracted from the publications and collated. Then to make meaningful comparisons possible, phylogenetic orthology inference was used to identify orthologous genes between reference genomes. This allows the straight-forward translation of genes and/or proteins into one common genome annotation, which facilitates meaningful comparison between genomic datasets. This analysis classified 27,855 *C. Sinensis* v2.0 genes into 23,017 orthogroups, and a total of 22,887 unique genes were identified as differentially abundant in at least one dataset (82% of the total genes).

They utilized this analysis as a starting point to select potential HLB susceptibility genes by prioritizing genes which are identified as differentially abundant in multiple studies. These genes are more likely to be biological responses of the

Citrus tree to HLB. For instance, there are 1,029 differentially abundant genes in 10 or more datasets representing 4.65% of the total differentially abundant genes. The subsetted data includes many genes previously implicated in HLB susceptibility. Gene Ontology enrichment analysis identifies carbohydrate metabolism and defence response genes over-represented in the list, consistent with what is understood about the biology of HLB.

CRISPR-Cas9 Mediated Resistance Against Citrus Bacterial Canker

Citrus bacterial canker is the most widespread disease among commercial citrus cultivars and is caused by *Xanthomonas citri* subsp. citri (*Xcc*). Recently two works have reported the application of CRISPR/Cas9 to produce citrus plants resistant to citrus bacterial canker. The canker-resistant mutant was produced by editing the PthA4 effector binding elements in the promoter of the *Lateral Organ Boundaries 1* (*CsLOB1*) gene in Duncan grapefruit. The resultant mutant showed a decrease in typical canker symptoms 4 days post-inoculation with *Xcc*, and no further phenotypic alterations were detectable [74]. Moreover, PCR sequencing confirms that there are no potential off-target mutations in other LOB family genes. In another work, researchers confirmed the link between CsLOB1 promoter activity and CBC disease susceptibility in Wanjincheng orange (*Citrus sinensis* Osbeck). The EBEPthA4 sequence was completely deleted from both *CsLOB1* alleles, resulting in enhancement to citrus bacterial canker. Furthermore, no alteration in plant development was observed after CsLOB1 promoter modification.

CRISPR-Cas9 Mediated Resistance Against *Erwinia amylovora*

Erwinia amylovora is an important pathogen of apple (*Malus* × *domestica*); it is a gram-negative, rod-shaped, facultative anaerobic bacterium. It is found to be an etiological agent of the disease fire blight in apple plants and other related genera. It occurs throughout most apple and pear growing regions of Europe and North America. *E. amylovora* is one of the major threats to apple production worldwide [75]. *E. amylovora* causes its infection *via* the DspA/E effector which interacts with the apple susceptibility protein MdDIPM4. It has been shown that the capacity of the bacterium to induce disease mainly depends on this single delivered effector [76]. The fire blight disease is caused by enterobacterium *Erwinia amylovora* in apple and other commercially important Rosaceae plants [48]. The pathogenicity effector (DspE) of *E. amylovora* interacts with four leucine-rich-repeat, receptor-like serine/threonine kinases produced by DspE-interacting proteins of Malus (DIPM) genes- *DIPM 1, 2, 3, 4* [77, 78] The CRISPR/Cas9 system is used to target *DIPM 1, 2*, and *4* genes in apple protoplast to develop resistance against fire blight disease. The researchers also validated the successful direct delivery of CRISPR/Cas9 ribonucleoproteins (RNPs) (pre-assembled sgRNA/Cas9 complex) into plant protoplasts which enhanced the rapid targeting efficiency, improved on-target, and reduced off-target activity [48, 69].

Significance of Studying Plant-Pathogen Interaction and Application of Crispr-Cas9 for Insight into the Plant-Pathogen Interaction

Plants are subjected to continuous exposure to both harsh environmental conditions and biotic stress. According to the reports of FAO (2020), about 6-20 percent (120 billion US$) of post-harvest loss was caused by insects, fungi, and bacteria. This exposure to biotic stress leads to changes in plant metabolism, ultimately affecting the yield of the crops. CRISPR-Cas9 tools have been applied against viruses and fungus, but there are very few reports of CRISPR-Cas9 mediated genome editing against bacterial plant pathogens, probably because very few bacterial species affect plants. CRISPR-Cas9 based plant genome editing has gained the attention of researchers worldwide due to the ease, convenient design of guide RNA, and its higher efficiency. But to provide immunity against these bacterial pathogens *via* CRISPR-Cas9 mediated site-directed mutagenesis, we need to be well aware of basic plant-pathogen interaction and its invasion into the plants [17].

Normally, when bacterial pathogens encounter plant cells, they inject the effector proteins *via* type III (T3) and tend to become more invasive for the spread of infection. This spread of infection is mediated by the activation of the *S-* gene or interference of host defense mechanisms [58]. This gene has been indeed a good target for designing CRISPR- guide RNA for disease resistance. Multiplexing the guide RNA has made this technology even more robust in providing resistance against a broad range of bacterial plant pathogens. Recently, multiplexing of guide RNA enabled editing all EBEs (Effector binding element), *SWEET 11, SWEET 13,* and *SWEET 14* genes in a single rice line has enabled broad resistance to bacterial blight *Xoo* in Kitaake, IR64, and Ciherang-Sub1 rice cultivars [61]. SWEET genes played an important role in sugar transport, signaling, and developing broad-spectrum resistance against several strains of bacterial blight. This would help us to get into the core mechanism of plant-pathogen interaction, where this strategy would be applicable to provide resistance against other dreadful bacterial pathogens in other crops as well. Below is the table representing the CRISPR-Cas9 mediated genome engineering targeted against bacterial plant pathogens until now [79].

CONCLUDING REMARKS

Biotic and abiotic stresses cause tremendous agricultural loss. But majorly, this loss is contributed by bacterial plant pathogens. With the availability of the plant database, it is easy to retrieve the desired sequence and target them with the available genome editing tools. New plant breeding techniques show the hope of a ray to combat such a problem. As the technology in genome editing advances, we can explore more about the function and its application in synthetic biology as well as important crops. The Lion's share of CRISPR-Cas9 genome editing techniques in the future helps to fight against major plant pathogenic bacteria. These tools are greatly useful in controlling the *Candidatus* bacteria. Genome editing has been successful in many crops. However, targeting multiple genes of different bacterial plant pathogens by multiplexing is still challenging. Our future efforts would be to functionally identify and characterize those genes which are responsible for the surface recognition of pathogens and cross-talk between them without compromising the crop yield. There is a need to evaluate the effect of genome editing techniques on the dawning plant pathogen occupants. Several plant pathogenic bacterias are responsible for serious plant diseases, low yield, and poor food/ fruit quality. This chapter provides basic principles and a wider range of applications of robust genome editing techniques for the management of devastating bacterial pathogens.

LIST OF ABBREVIATIONS

Cas-9	CRISPR Associated Protein 9
CRD	Central Repeat Domain
CRISPR	Clustered Regularly Interspaced Palindromic Repeats
DSBs	Double-Stranded Breaks
ETI	Effector-Triggered Immunity
ETS	Effector-Triggered Susceptibility
MN	Mega Nucleases
NPBTs	New Plant Breeding Techniques
PAMPs	Pathogen Associated Molecular Patterns
PRR	Pattern Recognition Receptor
PTI	PAMP Triggered Immunity
ROS	Reactive Oxygen Species
TALENs	Transcription Activator like Effector Nucleases
ZFNs	Zinc Finger Nucleases

ACKNOWLEDGEMENTS

Dr. Ashish Warghane thanks to the Faculty of Life Science, Mandsaur University, Mandsaur, Madhya Pradesh, India for encouraging to write a book chapter on the

Genome editing against devasting plant bacterial disease. The authors also thanks Dr. Atul Sathe for critically reviewing the draft of the book chapter.

REFERENCES

[1] Borrelli VMG, Brambilla V, Rogowsky P, Marocco A, Lanubile A. The enhancement of plant disease resistance using CRISPR/Cas9 technology. Front Plant Sci 2018; 9: 1245.
[http://dx.doi.org/10.3389/fpls.2018.01245] [PMID: 30197654]

[2] Savary S, Ficke A, Aubertot JN, Hollier C. Crop losses due to diseases and their implications for global food production losses and food security. Food Secur 2012; 4(4): 519-37.
[http://dx.doi.org/10.1007/s12571-012-0200-5]

[3] Chisholm ST, Coaker G, Day B, Staskawicz BJ. Host-microbe interactions: Shaping the evolution of the plant immune response. Cell 2006; 124(4): 803-14.
[http://dx.doi.org/10.1016/j.cell.2006.02.008] [PMID: 16497589]

[4] Laflamme B, Dillon MM, Martel A, Almeida RND, Desveaux D, Guttman DS. The pan-genome effector-triggered immunity landscape of a host-pathogen interaction. Science 2020; 367(6479): 763-8.
[http://dx.doi.org/10.1126/science.aax4079] [PMID: 32054757]

[5] Gosavi G, Yan F, Ren B, et al. Applications of CRISPR technology in studying plant-pathogen interactions: Overview and perspective. Phytopathol Res 2020; 2(1): 21.
[http://dx.doi.org/10.1186/s42483-020-00060-z]

[6] Nejat N, Rookes J, Mantri NL, Cahill DM. Plant–pathogen interactions: Toward development of next-generation disease-resistant plants. Crit Rev Biotechnol 2017; 37(2): 229-37.
[http://dx.doi.org/10.3109/07388551.2015.1134437] [PMID: 26796880]

[7] Dong OX, Ronald PC. Genetic engineering for disease resistance in plants: Recent progress and future perspectives. Plant Physiol 2019; 180(1): 26-38.
[http://dx.doi.org/10.1104/pp.18.01224] [PMID: 30867331]

[8] Lusser M, Davies HV. Comparative regulatory approaches for groups of new plant breeding techniques. N Biotechnol 2013; 30(5): 437-46.
[http://dx.doi.org/10.1016/j.nbt.2013.02.004] [PMID: 23474021]

[9] Nelson R, Wiesner-Hanks T, Wisser R, Balint-Kurti P. Navigating complexity to breed disease-resistant crops. Nat Rev Genet 2018; 19(1): 21-33.
[http://dx.doi.org/10.1038/nrg.2017.82] [PMID: 29109524]

[10] Gaj T, Sirk SJ, Shui SL, Liu J. Genome-editing technologies: Principles and applications. Cold Spring Harb Perspect Biol 2016; : 2016; 8.12: a023754.

[11] Menz J, Modrzejewski D, Hartung F, Wilhelm R, Sprink T. Genome edited crops touch the market: A view on the global development and regulatory environment. Front Plant Sci 2020; 11: 586027.
[http://dx.doi.org/10.3389/fpls.2020.586027]

[12] Silva G, Poirot L, Galetto R, et al. Meganucleases and other tools for targeted genome engineering: Perspectives and challenges for gene therapy. Curr Gene Ther 2011; 11(1): 11-27.
[http://dx.doi.org/10.2174/156652311794520111] [PMID: 21182466]

[13] Gao H, Smith J, Yang M, et al. Heritable targeted mutagenesis in maize using a designed endonuclease. Plant J 2010; 61(1): 176-87.
[http://dx.doi.org/10.1111/j.1365-313X.2009.04041.x] [PMID: 19811621]

[14] Zeevi V, Liang Z, Arieli U, Tzfira T. Zinc finger nuclease and homing endonuclease-mediated assembly of multigene plant transformation vectors. Plant Physiol 2012; 158(1): 132-44.
[http://dx.doi.org/10.1104/pp.111.184374] [PMID: 22082504]

[15] Belfort M, Bonocora RP. Homing endonucleases: From genetic anomalies to programmable genomic clippers. Methods Mol Biol 2014; 1123: 1-26.

[http://dx.doi.org/10.1007/978-1-62703-968-0_1] [PMID: 24510256]

[16] Muntazir M, Aafreen S, Hussain WS, *et al.* Harnessing genome editing techniques to engineer disease resistance in plants. Front in Plant Sci 2019; 10: 1-16.

[17] Carroll D. Genome engineering with zinc-finger nucleases. Genetics 2011; 188(4): 773-82.
[http://dx.doi.org/10.1534/genetics.111.131433] [PMID: 21828278]

[18] Li L, Wu LP, Chandrasegaran S. Functional domains in Fok I restriction endonuclease. Proc Natl Acad Sci 1992; 89(10): 4275-9.
[http://dx.doi.org/10.1073/pnas.89.10.4275] [PMID: 1584761]

[19] Shimizu Y, Bhakta MS, Segal DJ. Restricted spacer tolerance of a zinc finger nuclease with a six amino acid linker. Bioorg Med Chem Lett 2009; 19(14): 3970-2.
[http://dx.doi.org/10.1016/j.bmcl.2009.02.109] [PMID: 19289279]

[20] Shukla VK, Doyon Y, Miller JC, *et al.* Precise genome modification in the crop species Zea mays using zinc-finger nucleases. Nature 2009; 459(7245): 437-41.
[http://dx.doi.org/10.1038/nature07992] [PMID: 19404259]

[21] Khandagale K, Nadaf A. Genome editing for targeted improvement of plants. Plant Biotechnol Rep 2016; 10(6): 327-43.
[http://dx.doi.org/10.1007/s11816-016-0417-4]

[22] Ruiz de Galarreta M, Lujambio A. DNA sensing in senescence. Nat Cell Biol 2017; 19(9): 1008-9.
[http://dx.doi.org/10.1038/ncb3603] [PMID: 28855731]

[23] Jaganathan D, Ramasamy K, Sellamuthu G, Jayabalan S, Venkataraman G. CRISPR for crop improvement: An update review. Front Plant Sci 2018; 9: 985.
[http://dx.doi.org/10.3389/fpls.2018.00985] [PMID: 30065734]

[24] Zhang F, Maeder ML, Unger-Wallace E, *et al.* High frequency targeted mutagenesis in *Arabidopsis thaliana* using zinc finger nucleases. Proc Natl Acad Sci 2010; 107(26): 12028-33.
[http://dx.doi.org/10.1073/pnas.0914991107] [PMID: 20508152]

[25] Christian M, Cermak T, Doyle EL, *et al.* Targeting DNA double-strand breaks with TAL effector nucleases. Genetics 2010; 186(2): 757-61.
[http://dx.doi.org/10.1534/genetics.110.120717] [PMID: 20660643]

[26] Boch J, Scholze H, Schornack S, *et al.* Breaking the code of DNA binding specificity of TAL-type III effectors. Science 2009; 326(5959): 1509-12.
[http://dx.doi.org/10.1126/science.1178811] [PMID: 19933107]

[27] Moscou MJ, Bogdanove AJ. A simple cipher governs DNA recognition by TAL effectors. Science 2009; 326(5959): 1501.
[http://dx.doi.org/10.1126/science.1178817]

[28] Park CY, Kim J, Kweon J, *et al.* Targeted inversion and reversion of the blood coagulation factor 8 gene in human iPS cells using TALENs. Proc Natl Acad Sci 2014; 111(25): 9253-8.
[http://dx.doi.org/10.1073/pnas.1323941111] [PMID: 24927536]

[29] Guilinger JP, Pattanayak V, Reyon D, *et al.* Broad specificity profiling of TALENs results in engineered nucleases with improved DNA-cleavage specificity. Nat Methods 2014; 11(4): 429-35.
[http://dx.doi.org/10.1038/nmeth.2845] [PMID: 24531420]

[30] Stephens J. A gene editing technologies – ZFNs, TALENs, and CRISPR/Cas9. Encycl Appl Plant Sci 2017; 2: 157-61.

[31] Holkers M, Maggio I, Liu J, *et al.* Differential integrity of TALE nuclease genes following adenoviral and lentiviral vector gene transfer into human cells. Nucleic Acids Res 2013; 41(5): e63.
[http://dx.doi.org/10.1093/nar/gks1446]

[32] Lor VS, Starker CG, Voytas DF, Weiss D, Olszewski NE. Targeted mutagenesis of the tomato PROCERA gene using transcription activator-like effector nucleases. Plant Physiol 2014; 166(3):

1288-91.
[http://dx.doi.org/10.1104/pp.114.247593] [PMID: 25217528]

[33] Cermak T, Doyle EL, Christian M, *et al.* Efficient design and assembly of custom TALEN and other TAL effector-based constructs for DNA targeting. Nucleic Acids Res 2011; 39(12): e82.
[http://dx.doi.org/10.1093/nar/gkr218] [PMID: 21493687]

[34] Li T, Liu B, Spalding MH, Weeks DP, Yang B. High-efficiency TALEN-based gene editing produces disease-resistant rice. Nat Biotechnol 2012; 30(5): 390-2.
[http://dx.doi.org/10.1038/nbt.2199] [PMID: 22565958]

[35] Shan Q, Wang Y, Li J, *et al.* Targeted genome modification of crop plants using a CRISPR-Cas system. Nat Biotechnol 2013; 31(8): 686-8.
[http://dx.doi.org/10.1038/nbt.2650] [PMID: 23929338]

[36] Nicolia E, Proux W, Ahman I, *et al.* Targeted gene mutation in tetraploid potato through transient TALEN expression in protoplasts Biotechnol. J Biotechnol 2015; 204: 17-24.

[37] Clasen BM, Stoddard TJ, Luo S, *et al.* Improving cold storage and processing traits in potato through targeted gene knockout. Plant Biotechnol J 2016; 14(1): 169-76.
[http://dx.doi.org/10.1111/pbi.12370] [PMID: 25846201]

[38] Pickar-Oliver A, Gersbach CA. The next generation of CRISPR–Cas technologies and applications. Nat Rev Mol Cell Biol 2019; 20(8): 490-507.
[http://dx.doi.org/10.1038/s41580-019-0131-5] [PMID: 31147612]

[39] Wu D, Von Roepenack-Lahaye E, Buntru M, *et al.* A Plant Pathogen Type III effector protein subverts translational regulation to boost host polyamine levels. Cell Host Microbe 2019; 26(5): 638-649.e5.
[http://dx.doi.org/10.1016/j.chom.2019.09.014] [PMID: 31628081]

[40] Zafar K, Khan MZ, Amin I, *et al.* Precise CRISPR-Cas9 mediated genome editing in super basmati rice for resistance against bacterial blight by targeting the major susceptibility gene. Front Plant Sci 2020; 11: 575.
[http://dx.doi.org/10.3389/fpls.2020.00575] [PMID: 32595655]

[41] Pessina S, Lenzi L, Perazzolli M, *et al.* Knockdown of MLO genes reduces susceptibility to powdery mildew in grapevine. Hortic Res 2016; 3(1): 16016.
[http://dx.doi.org/10.1038/hortres.2016.16] [PMID: 27390621]

[42] Peng A, Chen S, Lei T, *et al.* Engineering canker-resistant plants through CRISPR/Cas9-targeted editing of the susceptibility gene *CsLOB1* promoter in citrus. Plant Biotechnol J 2017; 15(12): 1509-19.
[http://dx.doi.org/10.1111/pbi.12733] [PMID: 28371200]

[43] Jia H, Orbovic V, Jones JB, Wang N. Modification of the PthA4 effector binding elements in Type I CsLOB1 promoter using Cas9/sgRNA to produce transgenic duncan grapefruit alleviating XccΔpthA4:dCsLOB1.3 infection. Plant Biotechnol J 2016; 14(5): 1291-301.
[http://dx.doi.org/10.1111/pbi.12495] [PMID: 27071672]

[44] Oliva R, Ji C, Atienza-Grande G, *et al.* Broad-spectrum resistance to bacterial blight in rice using genome editing. Nat Biotechnol 2019; 37(11): 1344-50.
[http://dx.doi.org/10.1038/s41587-019-0267-z] [PMID: 31659337]

[45] Ortigosa A, Gimenez-Ibanez S, Leonhardt N, Solano R. Design of a bacterial speck resistant tomato by CRISPR/Cas9-mediated editing of *SlJAZ2*. Plant Biotechnol J 2019; 17(3): 665-73.
[http://dx.doi.org/10.1111/pbi.13006] [PMID: 30183125]

[46] Wang X, Tu M, Wang D, *et al.* CRISPR/Cas9-mediated efficient targeted mutagenesis in grape in the first generation. Plant Biotechnol J 2018; 16(4): 844-55.
[http://dx.doi.org/10.1111/pbi.12832] [PMID: 28905515]

[47] Lu W, Deng F, Jia J, *et al.* The *Arabidopsis thaliana* gene *AtERF019* negatively regulates plant resistance to *Phytophthora parasitica* by suppressing PAMP-triggered immunity. Mol Plant Pathol

[48] Malnoy M, Viola R, Jung MH, *et al.* DNA-free genetically edited grapevine and apple protoplast using CRISPR/Cas9 ribonucleoproteins. Front Plant Sci 2016; 7: 1904.
[http://dx.doi.org/10.3389/fpls.2016.01904] [PMID: 28066464]

[49] Li S, Shen L, Hu P, *et al.* Developing disease-resistant thermosensitive male sterile rice by multiplex gene editing. J Integr Plant Biol 2019; 61(12): 1201-5.
[http://dx.doi.org/10.1111/jipb.12774] [PMID: 30623600]

[50] Li X, Yang L, Yin J, Yu N, Ye F. Validation study of the chinese version of addenbrooke's cognitive examination III for diagnosing mild cognitive impairment and mild dementia. J Clin Neurol 2019; 15(3): 313-20.
[http://dx.doi.org/10.3988/jcn.2019.15.3.313] [PMID: 31286702]

[51] Kim YA, Moon H, Park CJ. CRISPR/Cas9-targeted mutagenesis of Os8N3 in rice to confer resistance to Xanthomonas oryzae pv. oryzae. Rice 2019; 12(1): 67.
[http://dx.doi.org/10.1186/s12284-019-0325-7] [PMID: 31446506]

[52] Gao H, Smith J, Yang M, *et al.* Heritable targeted mutagenesis in maize using a designed endonuclease. Plant J 2010; 61(1): 176-87.
[http://dx.doi.org/10.1111/j.1365-313X.2009.04041.x] [PMID: 19811621]

[53] Fister AS, Landherr L, Maximova SN, Guiltinan MJ. Transient expression of CRISPR/Cas9 machinery targeting TcNPR3 enhances defense response in theobroma cacao. Front Plant Sci 2018; 9: 268.
[http://dx.doi.org/10.3389/fpls.2018.00268] [PMID: 29552023]

[54] Xie K, Yang Y. RNA-guided genome editing in plants using a CRISPR-Cas system. Mol Plant 2013; 6(6): 1975-83.
[http://dx.doi.org/10.1093/mp/sst119] [PMID: 23956122]

[55] Daniela P, de TT, Brail Q, Dahlbeck D, Staskawicz B. CRISPR-Cas9 mediated mutagenesis of a DMR6 ortholog in tomato confers broad-spectrum disease resistance. BioRxiv 2016; 064824.

[56] Krueger U, Bergauer T, Kaufmann B, *et al.* Insights into effective RNAi gained from large-scale siRNA validation screening. Oligonucleotides 2007; 17(2): 237-50.
[http://dx.doi.org/10.1089/oli.2006.0065]

[57] Zhu Y, Liu X, Yang ST. Construction and characterization of pta gene-deleted mutant of *Clostridium tyrobutyricum* for enhanced butyric acid fermentation. Biotechnol Bioeng 2005; 90(2): 154-66.
[http://dx.doi.org/10.1002/bit.20354] [PMID: 15759261]

[58] Marraffini LA, Sontheimer EJ. CRISPR interference: RNA-directed adaptive immunity in bacteria and archaea. Nat Rev Genet 2010; 11(3): 181-90.
[http://dx.doi.org/10.1038/nrg2749] [PMID: 20125085]

[59] Xu T, Li Y, Van Nostrand JD, He Z, Zhou J. Cas9-based tools for targeted genome editing and transcriptional control. Appl Environ Microbiol 2014; 80(5): 1544-52.
[http://dx.doi.org/10.1128/AEM.03786-13] [PMID: 24389925]

[60] Li K, Cai D, Wang Z, He Z, Chen S, Kivisaar M. Development of an efficient genome editing tool in *Bacillus licheniformis* Using CRISPR-Cas9 Nickase. Appl Environ Microbiol 2018; 84(6): e02608-17.
[http://dx.doi.org/10.1128/AEM.02608-17] [PMID: 29330178]

[61] Tan SZ, Reisch CR, Prather KLJ. A Robust CRISPR interference gene repression system in Pseudomonas. J Bacteriol 2018; 200(7): e00575-17.
[http://dx.doi.org/10.1128/JB.00575-17] [PMID: 29311279]

[62] Rock JM, Hopkins FF, Chavez A, *et al.* Programmable transcriptional repression in mycobacteria using an orthogonal CRISPR interference platform. Nat Microbiol 2017; 6.2: 16274.
[http://dx.doi.org/10.1038/nmicrobiol.2016.274]

[63] Oh JH, Van Pijkeren JP. CRISPR-Cas9-assisted recombineering in Lactobacillus reuteri. Nucleic acids research 2014; 42.17: e131.

[64] Luo Y, Ge M, Wang B, *et al.* CRISPR/Cas9-deaminase enables robust base editing in *Rhodobacter sphaeroides* 2.4.1. Microb Cell Fact 2020; 19(1): 93.
[http://dx.doi.org/10.1186/s12934-020-01345-w] [PMID: 32334589]

[65] Agrios G. Plant Pathology. 5th. Amsterdam: Elsevier Academic Press 2005; 26-27: pp. 398-401.

[66] Schreiber KJ, Chau-Ly IJ, Lewis JD. What the Wild Things Do: Mechanisms of Plant Host Manipulation by Bacterial Type III-Secreted Effector Proteins. Microorganisms. 2021; 9(5): 1029.

[67] Buttner D. Behind the lines-actions of bacterial type III effector proteins breeding. Crit Rev Plant Sci 2016; 31: 93-123.

[68] Zaidi SSA, Mukhtar MS, Mansoor S. Genome editing: Targeting susceptibility genes for plant disease resistance. Trends Biotechnol 2018; 36(9): 898-906.
[http://dx.doi.org/10.1016/j.tibtech.2018.04.005] [PMID: 29752192]

[69] Zhou H, Liu B, Weeks DP, Spalding MH, Yang B. Large chromosomal deletions and heritable small genetic changes induced by CRISPR/Cas9 in rice. Nucleic Acids Res 2014; 42(17): 10903-14.
[http://dx.doi.org/10.1093/nar/gku806] [PMID: 25200087]

[70] Xu RF, Li H, Qin RY, *et al.* Generation of inheritable and "transgene clean" targeted genome-modified rice in later generations using the CRISPR/Cas9 system. Sci Rep 2015; 5(1): 11491.
[http://dx.doi.org/10.1038/srep11491] [PMID: 26089199]

[71] Oliva R, Ji C, Atienza-Grande G, *et al.* Broad-spectrum resistance to bacterial blight in rice using genome editing. Nat Biotechnol 2019; 37(11): 1344-50.
[http://dx.doi.org/10.1038/s41587-019-0267-z] [PMID: 31659337]

[72] Ghosh D, Bhose S, Mukherjee K, Baranwal VK. Sequence and evolutionary analysis of ribosomal DNA from Huanglongbing (HLB) isolates of Western India. Phytoparasitica 2013; 41(3): 295-305.
[http://dx.doi.org/10.1007/s12600-013-0290-4]

[73] Warghane A, Misra P, Shukla PK, Ghosh DK. Diversity and characterization of Citrus tristeza virus and '*Candidatus* Liberibacter asiaticus' associated with citrus decline in major citrus growing areas of India. Indian Phytopathol 2017; 70: 359-67.
[http://dx.doi.org/10.24838/ip.2017.v70.i3.72495]

[74] Ahlawat YS. Virus diseases of Citrus and Management. 1st., Studium Press (India) Pvt Ltd 2012.

[75] Jia H, Orbovic V, Jones JB, Wang N. Modification of the PthA4 effector binding elements in Type I CsLOB1 promoter using Cas9/sgRNA to produce transgenic Duncan grapefruit alleviating XccΔpthA4:dCsLOB1.3 infection. Plant Biotechnol J 2016; 14(5): 1291-301.
[http://dx.doi.org/10.1111/pbi.12495] [PMID: 27071672]

[76] Winslow CEA, Broadhurst J, Buchanan RE, Krumwiede C Jr, Rogers LA, Smith GH. The families and genera of the bacteria: final report of the committee of the society of American bacteriologists on characterization and classification of bacterial types. J Bacteriol 1920; 5(3): 191-229.
[http://dx.doi.org/10.1128/jb.5.3.191-229.1920] [PMID: 16558872]

[77] Siamer S, Guillas I, Shimobayashi M, Kunz C, Hall MN, Barny MA. Expression of the bacterial type III effector DspA/E in *Saccharomyces cerevisiae* down-regulates the sphingolipid biosynthetic pathway leading to growth arrest. J Biol Chem 2014; 289(26): 18466-77.
[http://dx.doi.org/10.1074/jbc.M114.562769] [PMID: 24828506]

[78] Borejsza-Wysocka EE, Malnoy M, Meng X, *et al.* Silencing of apple proteins that interact with DspE, a pathogenicity effector from *Erwinia amylovora*, as a strategy to increase resistance to fire blight. Acta Hortic 2004; (663): 469-74.
[http://dx.doi.org/10.17660/ActaHortic.2004.663.81]

[79] Pompili V, Dalla Costa L, Piazza S, Pindo M, Malnoy M. Reduced fire blight susceptibility in apple cultivars using a high-efficiency CRISPR/Cas9-FLP/FRT-based gene editing system. Plant Biotechnol J 2020; 18(3): 845-58.
[http://dx.doi.org/10.1111/pbi.13253] [PMID: 31495052]

CHAPTER 3

CRISPR-Cas for Genome Editing - Molecular Scissors for Combating Pathogens

Poornima Devi C. Ramdev[1], Divya K. Shankar[2] and B. Renuka[3,*]

[1] *Department of Microbiology, Yuvaraja's College (Autonomous), Myosre-570005, India*

[2] *Department of Studies in Microbiology, Pooja Bhagavat Memorial Mahajana PG Centre, Myosre-570016, Karnataka, India*

[3] *Promic Svasthya Private Limited, Mysore-570028, Karnataka, India*

Abstract: Clustered Regularly Interspaced Short Palindromic Repeats, abbreviated as CRISPR, is a genome-editing technology that permits the creation of precise knock-out mutants by aiding the modification of gene sequences devoid of the steps involving the insertion of foreign DNA into pathogenic microorganisms. The microorganisms are ubiquitous in nature and harbor in the complex ecosystem of the human being. Cas (acronym for CRISPR-associated) genes are present in many microbial genomes. The variable nature of the microbial genome has been utilized as an integral typing tool in epidemiologic, diagnostic, and evolutionary analyses of the prokaryotic species. The past decade has seen an accumulating growth in the development of gene-editing tools utilizing the CRISPR-Cas system, which essentially is a part of the prokaryotic immune system. The development of these unique gene-editing techniques has empowered researchers to alter and investigate organisms with ease and efficiency as never before. This editing tool can efficiently be programmed and delivered into the bacterial populations to explicitly eliminate members of a targeted micro biome. Manipulation of the gene expression and regulation of the synthesis of metabolites and proteins can be achieved by utilizing an engineered CRISPR-Cas system. Put together, these tools present with the exhilarating opportunity to explore the complex interaction between the individual species of the microbiome and the host organism and thereby reveal novel avenues for the generation of drugs to selectively target the microbiome. CRISPR-Cas technology has been employed to cope with antibiotic resistance in intracellular and extracellular pathogens. The widespread use of antibiotics and the escalation of multidrug-resistant (MDR) bacteria boost the prospect of a post-antibiotic era, which emphasizes the need for novel strategies to target MDR pathogens. The development of permissive synthetic biology techniques offers favorable solutions to carry through safe and efficient antibacterial therapies.

Keywords: CRISPR-Cas, Epigenetic regulation, Genome engineering, Knock-out and -in genes, Synthetic biology.

---

[*] **Corresponding author B. Renuka:** Promic Svasthya Private Limited, Mysore-570028, Karnataka, India; Tel: +91 9035376157; E-mail: renukabs@gmail.com

Prakash M. Halami & Aravind Sundararaman (Eds.)
All rights reserved-© 2024 Bentham Science Publishers

INTRODUCTION

The emanation of disease-causing organisms into multidrug-resistant (MDR) variants has metamorphosed into a critical global health quandary. Essentially, the gradual progression of bacteria to MDR strains is an unbridled phenomenon and requires expeditious remedial efforts to curb the longstanding damage. In accordance with the information bulletin on the global prevalence of antimicrobial resistance published by the World Health Organization (WHO) in 2014, an escalation in the frequency of drug-resistant pathogenic bacteria has rendered the treatment of previously easily curable diseases, like pneumonia, urinary infections and circulatory diseases, worldwide. A global action plan has been accredited by the 68th World Health Assembly in 2015 to contrive the perception and assimilation of antimicrobial resistance (AMR) [1]. The aforementioned proposal suggests developing novel drugs, vaccines, diagnostic technologies, and different interventional strategies to ensure continuous management of bacterial diseases. Therefore, there is a pressing need to discern the drug resistance process in MDR bacteria so as to earmark features (like efflux pump for removal of antibiotics, and site mutation at the activity sites) to intercept the pathogenic activity. Furthermore, discovering pioneering and advanced low molecular weight drugs and metabolically designing the synthesis of such small molecule drugs is critical to mitigating the drug resistance in MDR microorganisms.

The emergence of singular synthetic biology (SB) and eukaryotic genome editing/engineering tools impart encouraging diagnostic and management measures to detect and remedy the prevalent refractory bacterial disorders. Major progress in genetic engineering techniques has successfully helped in focusing and editing pathogenic eukaryotic genomes for superior understanding and the extenuation of the drug resistance processes. The implementation of singular genome engineering and SB techniques, like CRISPR-Cas systems, recombination- mediated genetic engineering, and the eukaryotic intercellular signaling mechanisms, for pathogen targeting has been pivotal in the establishment of antibacterial strategies, specifically, the development of vaccines, new antibiotics, phage therapies, and specialized diagnostics.

Innovative genome engineering techniques created as a result of recent developments in SB have made it easier to manipulate microbial genomes for a variety of therapeutic and scientific purposes [2 - 4]. A novel platform is offered by SB to bring together the concepts of basic research and further aid its application in translational research. This kind of approach opens up great potential for providing revolutionary solutions when dealing with infectious agents. SB is a field with great promise due to its inherent nature of combining biological know-how and engineering principles to devise novel, tunable and modular genetic circuits or products for modulating the existing biological systems. Synthetic biologists have a need and a rising interest in using microbial genome editing technologies to supplement the creation of genetic circuits for specific cellular action or metabolic regulation. Advantageous biological features have been established in the engineered species as a result of a precise modification in the eukaryotic genome. The combination of genome SB and editing tools has also made it possible to use genetically modified bacteria to address a number of significant problems in a variety of fields, from renewable energy to global health. New strategies for preventing bacterial infections have emerged, particularly in light of recent progress in bacterial gene editing techniques that can be directed against a wide variety of bacterial hosts [5]. The generation of new SB tools is proving to be a harbinger for establishing novel strategies to address the imminent peril due to antibiotic resistant bacteria. Implementation of SB in bacteria typically entails the development of brief gene circuits for a chosen pathway or gene function up to the modification of the organism's whole gene pool [6, 7]. Insights into the molecular underpinnings of antibiotic resistance have also been provided by a number of studies utilizing SB. For example, *Escherichia coli* was given a deadly dose of the antibiotic medication triclosan in an effort to identify the potential genes that were involved in resisting triclosan [8]. Identification of the crucial genes in the bacteria (*E. coli*, *Pseudomonas putida*, and *Mycoplasma*) for potential therapeutic use has been possible by synthesizing minimal bacterial genomes using bottom-up and top-down strategies [9]. The potential for growth and the influence of SB in negating the effect of antibiotic resistance are apparent from the aforementioned examples. The application of SB tools and bacterial genome engineering in targeting emergent bacterial pathogens focuses on utilizing SB in introducing pioneering antibacterial therapeutic approaches.

Numerous approaches have been tried to potentially create bacterial gene sets with varying degrees of specificity, efficacy, and application across a wide range of hosts [10]. Bacterial gene editing is more frequently done to introduce mutations into the bacterial genome or to create knock-in and knock-out genes. Although many of these techniques were initially created in *E. coli*, they have recently undergone rapid growth and expansion across a variety of bacterial hosts. It is important that these molecular engineering tools are tractable in various pathogenic bacteria that are paving the avenues for further examination and discernment of these pathogens to impede bacterial infections. The principles and procedures of bacterial genome engineering tools have also been listed in several positive evaluations [11]. Many human diseases are genetically determined; as of now, over 6000 hereditary diseases are known to be accounted for by gene and chromosomal DNA mutations, both by nuclear and mitochondrial [12]. The medicament of hereditary diseases has mostly been symptomatic, until recently gene therapy has emerged as a fundamental new approach which is aimed at eliminating directly the cause of the disease by correcting mutations (genome editing). CRISPR-Cas systems are the newest and most promising genetic tool/genome editing tools, with the 2020 Nobel Prize in Chemistry awarded for their application in genome editing.

CRISPR-Cas System - Discovery and Function

The molecular structure of DNA was determined by J.D. Watson and F.H.C. Crick in 1953 [13]. Ever since then researchers have attempted to establish technologies that could engineer the genetic material of organisms and cells. With the recognition of RNA-guided CRISPR-Cas9 system, a facile and efficient approach for genome editing has now turned out to be a reality. The augmentation of this technology has facilitated scientists to reorganize DNA sequences in a broad range of tissues and organisms. Genomic modifications are erstwhile bottlenecks in genetic experiments now. Currently, CRISPR-Cas9 systems are widely used in biotechnology, basic science, and in the establishment of futuristic therapeutics.

The CRISPR/Cas9 system is a gene-editing technology that can help in inducing double- strand breaks (DSBs), single-strand nicks or guide ribonucleic acids (RNAs) to bind with the protospacer adjacent motif (PAM) sequences [14]. By changing simple sequences of gRNA, Cas9-endonucleases can be introduced into a gene of interest and thereby generate DSBs. The efficiency of Cas9-endonucleases and the proficiency by which the genes can be earmarked led to the developing of libraries of CRISPR-knockout strains for both human and mouse cells, which encompass either the whole genome or sets of specific genes of interest [15]. CRISPR screening helps researchers to establish a systematic and high-throughput genetic modification within the model organisms *in vivo*. This genetic disruption is imperative to completely comprehend the gene function and the associated epigenetic regulation [16]. The added benefit of pooled CRISPR libraries is that multiple genes can be studied at once.

The story of the CRISPR-Cas system began in the year 1987 when Nakata and colleagues reported a set of 29 nucleotide (nt) repeats in *E. coli* during their study period of the *iap* gene [17]. By sequencing many microbial genomes in the next decade, additional repeat elements from the genes of different archaeal and bacterial strains were also reported. Subsequently, this singular family of interspaced repeats was termed clustered repeat elements. The term CRISPR was coined by Mojica and Jansen in the year 2002 [18]. A preeminent breakthrough was recognized in 2005, when the spacer sequences were distanced from the direct repeats implying their extra-chromosomal origins or phage association. The basic functionality and mechanistic basis of the CRISPR-Cas system was known clearly by 2010. This system comprises a non-repetitive genetic locus, spacer sequences and adjoining a few genes that could encode CRISPR-associated (cas) proteins [19]. Several scientists have now proceeded to use CRISPR-Cas technology for biotechnological applications and the development of phage resistant dairy strains [20].

By 2011, it was perspicuous that the CRISPR-Cas systems were ubiquitous in prokaryotes and operated as adaptive immune system to impede the attacking bacteriophages and plasmids. Studies had also showed that the Cas proteins acted at three different levels: (a) integration of the new spacer DNA sequences into CRISPR loci, (b) synthesis of crRNAs and (c) suppression of the invading nucleic acid [21]. The identification of CRISPR-Cas9 to be used as a tool for genomic editing came from the studies of Class-2, Type-II CRISPR-Cas system in *Streptococcus thermophilus* and the related human pathogen *Streptococcus pyogenes*. This system contains four *Cas* genes, out of which three (*Cas1*, *Cas2*, *csn2*) are involved in spacer acquisition, and a fourth, *Cas9* (formerly named *Cas5* and *csn1*), is needed for interference [22]. In support of this view, the inactivation of the *Cas9* gene prevented the cleavage of target DNA. To further study the elements required for immunity, the *S. thermophilus* CRISPR-Cas system was introduced into *E. coli*, which furnished heterologous protection in opposition to the infection with phage and plasmids [23]. Using this as an investigational model, segments of the system were inactivated to elucidate the components essential for protection. The study exemplified that the Cas9 protein singularly was adequate for the CRISPR-encoded intervention, and that the two nuclease domains existing in the protein, namely HNH and RuvC, were both essential for the process [24].

The speculation that CRISPR-Cas systems could provide resistance against intrusive foreign DNA was verified in 2007 [25]. In an appropriate series of experiments, scientists explored a Class 2 system in a strain of *S. thermophilus*, which they exposed to virulent bacteriophages. Next, CRISPR loci of the bacteria that exhibited resistance to infection were analysed. The experiment revealed that resistant bacteria had acquired a new spacer sequence, which matched the sequences within the infecting phage used to select resistance. Deletion of this spacer sequence led to the loss of function, and the phages that were competent enough to grow on the resistant bacteria had amassed sequence variation in the protospacer region in the phage genome. In addition, the deactivation of one of the *Cas* genes (*Cas5*) culminated in the deficit of phage resistance. The data, thus, established a role for *Cas* gene products in CRISPR-Cas–mediated immunity and that the specificity of the mechanism was determined by the spacer sequences [26].

Further, appreciation of the functional aspect of the CRISPR-Cas system came from investigating *E. coli*, which carried a Class 1 CRISPR-Cas complex encoding at least eight different Cas proteins. Out of these, five of the proteins could be purified as a multi-protein complex called CRISPR-associated complex for antiviral defence (abbreviated as Cascade). Cascade was revealed to operate in the pre-cr RNA processing, breaking down the long transcripts in the repeat regions and subsequently producing shorter cr RNA polymers that encompassed the virus-derived sequence [27]. After processing, the mature cr RNA molecule was retained by the Cascade, and was accommodated by a *Cas*-encoded helicase, Cas3, that served as guide molecules which enabled the Cascade to intervene in the phage proliferation. Thus, the results suggested two different steps in CRISPR function: firstly, CRISPR expression and cr RNA maturation, and secondly, the interference step that required the Cas3 protein. The results also provided evidence suggesting that the *E. coli* CRISPR-Cas complex targeted the phage DNA and not the RNA, such that crRNA with complementarity with respect to either of the DNA strands could impede the phage proliferation.

Conclusive evidence that DNA is essentially the target of CRISPR-Cas interference came from very elegant experiments using a strain of *Staphylococci epidermidis* that contained a CRISPR array with a homologous spacer sequence to a gene present in a conjugative plasmid [28]. Transferring of the plasmid into the *Staphylococci* strain ensued only when the spacer sequence was deleted or mutated. A self-splicing noncoding sequence was introduced into target sequence on the plasmid. Thus, the CRISPR spacer was complementary to the RNA and not to the DNA, as it is impeded by the noncoding sequence, but, would still be spliced, re-establishing the sensitive target. Undeniably, the introduction of the self-splicing intron was adequate to overcome the CRISPR-Cas hindrance towards plasmid transfer, strongly indicating the DNA as the primary target [29].

This conclusion was further supported by studying *S. thermophilus*, whose CRISPR-Cas complex was seen to bisect both plasmid DNA and bacteriophage *in vivo*. The CRISPR-Cas complex forms the adaptive immune system of archaea and bacteria. It defends against invaders, including the mobile genetic elements (MGEs) and bacteriophages or phages [30]. The CRISPR-Cas system breakdown the foreign genetic entities into three steps. Spacer acquisition or adaptation [31] is the preliminary stage in which the spacer sequence is integrated into the CRISPR array subsequent to its recognition. The following stage is the synthesis or the expression of CRISPR RNA (crRNA), in which the pre-CRISPR RNA (pre-crRNA) is synthesised by RNA polymerase (RNAP). This pre-crRNA is then cleaved into small cr RNA by distinct endoribonucleases. Based on this cr RNAs activity, these are also accepted as guide RNAs. Interference is the eventual stage in which the cr RNAs recognize and form base pairs specific to foreign DNA or

RNA with nearly perfect complementarity. This results in the degradation of the cr RNA-foreign nucleic acid complex. On the other hand, in the presence of a mismatch between the spacer and the invader's DNA or mutation in PAM, the degradation does not transpire, rendering the host susceptible to infection [32].

CRISPR/Cas Systems for Gene Editing, Specificity and Molecular Mechanism

To surmount the shortcomings of the CRISPR/Cas9 system, diverse of CRISPR systems have been developed for effective and streamlined genome editing. The Cas9 variant Cj Cas9, obtained from *Campylobacter jejuni*, is constituted of 984 amino acid residues and has been utilized for efficient genome editing *in vivo* and *in vitro*. CjCas9 is distinctly specific and cuts exclusively at a narrow range of sites in the genomes of both mouse and human. Introduced *via* adeno-associated virus (AAV), it has been observed to generate mutations at targeted places in high numbers in mouse muscle cells or retinal pigment epithelium (RPE) cells. For instance, it was employed to study by targeting the Hif1a or Vegfa gene in RPE cells that minimized the size of laser-induced choroidal neovascularization, rendering it a novel choice of remedy for age-related macular degeneration [33].

Cas13 is of lately recognized CRISPR effector and CRISPR/Cas13 can be selectively specific to viral RNAs and the in-house RNAs in plants' cells. The Cas13 system has an increased RNA target specificity and efficiency [34]. Cas13 was used to direct ADAR2 deaminase for modification of RNA (changing adenosine to inosine) in human cells for recovery of the functional proteins to halt disease progression. Recently, CRISPR/Cas13a has been contemplated as an altogether novel CRISPR type that comes under class II type VI.

Understanding host-pathogen interactions and assisting in targeted therapies are made possible by tools improvised from bacteriophages [35]. Using the CRISPR/Cas9 system, a phage may specifically destroy pathogenic bacterial strains that present with sequences that are remarkably similar to benign bacterial strains. The CRISPR/Cas9 system particularly targets the DNA sequence that leads to the creation of double strand breaks, killing the bacteria [36]. Sequence specific antimicrobials were shown to be effective against pathogenic bacteria when used as targeted therapeutics. These antimicrobials target minute changes in the pathogenic strain's sequence to overcome the pathogens' extensive high genome sequence similarity. To eradicate the pathogenic population, it is possible to take advantage of the Cas's capacity to target particular sequences and to distinguish between genomic DNA that is not targeted and DNA that is targeted by even a single small mismatch. Selective nucleases demonstrated greater antimicrobial activity than the antibiotic chloramphenicol in the research model of *Galleria mellonella* larvae.

A novel two-phage CRISPR system was created using lytic and temperate phage that were specifically programmed to stimulate and eradicate antibiotic-resistant bacteria. This is yet another functional model of the CRISPR system [37]. To mark the genes for antibiotic resistance and give these cells lytic phage resistance initially, lysogenic phage carrying the CRISPR system was used. By doing this, the cells became resistant to lytic phage but developed susceptibility to antibiotics. To eliminate any remaining antibiotic-resistant cells, lytic phage was used in the subsequent stage. Antibiotics could then be used to treat the cells that were concentrated there and were sensitive to antibiotics. Studies indicate that this method may be useful for treating hospital or medical staff skin surfaces. As the phage population increases during infection, the phage can also serve as a detection and diagnostic tool for infection and does not require multiplication of the host bacterium. When a particular bacterium is infected by a phage, the population's phage genome template grows, enabling the phage to destroy the targeted bacteria. The amplification of phage DNA infecting a particular bacterium can be shown by a quantitative PCR following phage infection [38]. Phage reporter systems for *Mycobacterium tuberculosis* and *Bacillus anthracis*, as well as phage phi A1122 for *Yersinia pestis*, are examples of diagnostic phages that are commonly used for excessively pathogenic bacteria. The CRISPR-Cas system operates in a sequence-specific manner by identifying and cleaving the foreign DNA or RNA. The defence mechanism comprises three stages: (a) spacer acquisition or adaptation, (b) cr RNA biogenesis and (c) target interference.

In the initial phase, a disparate sequence of invading MGE called a protospacer is incorporated into the CRISPR array generating a new spacer. This facilitates the host to memorise the intruder's genetic material and helps in displaying the adaptive disposition of this immune network [39]. Two proteins, *viz.*, Cas1 and Cas2, seem to be prevalently implicated in the spacer procurement process as they could be observed in almost all the CRISPR-Cas types. Type III-C, III-D and IV CRISPR-Cas networks are an exception which harbours no homologous proteins. Furthermore, type V-C shows minimal constitution as it consists of only presumed to be effector proteins C2C3 and a Cas1 homologue. In recent years, major advancement has been achieved in realising the biochemical and genetic underpinnings of CRISPR-Cas immunity. Nevertheless, the exact process of spacer acquisition is not fully understood. The choice of protospacers and their refinement procedures before the integration remains widely ambiguous in many CRISPR-Cas types. Recent research elucidates the biochemistry of the spacer integration process. It has been indicated that Cas1 and Cas2 of the type I-E system of *E. coli* form a composite group that assist the incorporation of new spacers in such a way that it is evocative of viral transposases and integrases [40]. Even though both Cas1 and Cas2 are nucleases, the catalytically active sites of Cas2 are expendable for spacer acquisition. A new spacer is usually integrated at the leader-repeat boundary of the CRISPR assemblage, although the primary repeat of the array is duplicated.

The mechanisms of various CRISPR-Cas groups might be conserved only to a specific extent, as numerous studies have elaborated variations with regard to the prerequisites and targets of the adaptation system. While Cas1 and Cas2 are adequate to promote the spacer acquisition in the much-studied type I CRISPR Cas systems, type I-B additionally requires Cas4 for adaptation. The type I-F CRISPR-Cas system of *Pseudomonas aeruginosa* furthermore requires interference machinery to assist the uptake of new spacers [41]. Similarly, the type II-A system requires Cas9, Csn2, and tra cr RNA (trans activating CRISPR RNA) for acquisition. A separate adaptation mode was identified for a type III-B Cas1 protein that is amalgamated to reverse transcriptase. The acquisition from both DNA and RNA has been described. The selection of a target sequence that is incorporated into the CRISPR locus is not an arbitrary event. It has been established that in type I, II and V CRISPR-Cas systems, a short sequence, entitled the PAM, is situated adjacent to the protospacer and is critical for acquisition and interference [42]. Besides, in the task of priming, interference machinery of various ICRISPR-Cas systems can trigger the increased assimilation of new spacers upon cr RNA- guided adherence to protospacer that was chosen upon a primary infection.

This phenomenon reveals a distinct adaptation mode in comparison to ingenuous spacer acquisition, as this necessarily requires a pre-existing spacer complementing the target. It generally results in greater acquisition frequency from the protospacers that lie in close adjacency to the target site. It is intriguing that primed spacer integration does not depend on the target cleavage as it also operates on degenerated targeted sites that usually would result in impaired interference [43]. The precise mechanism remains difficult to understand but it has been established that the interference complex can mobilize Cas1 and Cas2 during PAM- independent binding to DNA.

To facilitate immune reaction cascade, the CRISPR array is transcribed into long precursor crRNA (pre-crRNA) which is subsequently processed into mature guide cr RNAs encompassing memorized sequences of invaders [44]. In type I and III classes, Cas6 family members perform the processing step producing the intermediate species of cr RNAs that are flanked by a short 50 tag. A peculiarity here is that type I-C systems do not encode Cas6 proteins. In its place, Cas5d polypeptide processes pre-cr RNA, yielding intermediate cr RNAs with an11 nucleotide 5′ tag [45]. Additionally, shortening of the 3′ end of the intermediate cr RNA by an unidentified nuclease can occur and yield mature cr RNA species that comprise a full spacer region and a repeat region, which usually exhibit a hair pin like structure in almost all the type I networks. The progressions of crRNAs in class 2 CRISPR- Cas networks vary significantly. For the type II system, tra cr RNA is necessary for the maturation of the pre-cr RNA. The complementary repeat sequence of this RNA facilitates the configuration of RNA duplex with each repeat of the pre-cr RNA, which is made stable by Cas9 protein. The duplex is identified and cleaved by host RNase III giving rise to an intermediate configuration of the cr RNA that goes through subsequent maturation steps [46].

In the final stage of immune reaction, the processed mature cr RNAs act as guides to particularly hinder the invading nucleic acids. Class 1network employs Cascade like complexes to attain target cleavage, however for the class 2 network, a single effector polypeptide is sufficient for targeting impedance. To circumvent the self-targeting phenomenon, type I, II and V circuits selectively recognize the PAM sequence occurring downstream (type II) or upstream (types I and V) with respect to the protospacer. The distinction between self and non-self is facilitated by the 5′label of mature crRNA, which does not bind to the target to aid the disintegration by the complex for the type III networks [47]. In type I circuits, Cascade restricts the entering DNA in a crRNA-contingent manner and subsequently brings in the nuclease Cas3 for target disintegration. Cas3 produces a nick on invader DNA and later degrades the target DNA. The tra cr RNA:cr RNA duplex, in the type II CRISPR-Cas pathway, recruits Cas9 effector proteins to insert a double-strand break in the target DNA. The intervention by type III network consists of peptide complexes like and Cas10-Cmr (types III-B and III-C) and Cas10-Csm (types III-A and III-D) which later target both RNA and DNA [48]. Intriguingly, it is shown that the impedance of type III-A and type III-B networks is not independent of the expression of the target DNA.

BACTERIAL VIRULENCE AND BIOMOLECULAR TARGETS

The disease imparting traits of bacteria are termed virulence factors, and there exists an increasing interest in treatment strategies that influence such traits. However, in accordance with the ecological theory, disease intensity is complex such that it is context-dependent and multifactorial, which thereby complicates the endeavors to identify the most common critical virulence factors. A recent meta-analysis attempted to quantify the disease consequences that are associated with the most commonly studied virulence factor - pyoverdine, which is an iron-scavenging siderophore which is released by opportunistic pathogens like *P. aeruginosa*. In agreement with ecological theory, research had elaborated that the by-product of pyoverdine, although prevalently causing disease, differed considerably over the distinct infection models [49].

Novel bacterial targets are becoming more increasingly desirable for antimicrobial screening because to the high propensity for pathogenic strains to develop resistance towards current treatments. One of the most problematic organisms is MRSA, reported first in 1961. A modern approach for discovering a new class of compounds with a mode of action different from current therapies is to screen hitherto unexploited molecular targets. Comprehensive genomic sequence comparisons facilitated rapid recognition of open reading frames (ORFs), which are more or less conserved in disease causing microorganisms but are missing in eukaryotes. This information allows researchers to judiciously select targets that are unique for bacteria and thereby result in the establishment of broad-spectrum factors, potentially decreasing the risk of mechanism-dependent toxicity. Recently, novel targets were identified, including membrane transporters, virulence gene products, chorismate biosynthesis, cofactor biosynthesis, and fatty acid biosynthesis. The key criterion for each of the aforementioned targets is that they are crucial for bacterial cell survival *in vitro* or during an infection.

The ascertainment of essentiality *via* large-scale genetic screening when combined with whole genomic sequence juxtapositions, has revealed several essential ORFs which are yet to be functionally characterized. The continued rise in antibiotic resistant bacterial infections has prompted alternative approaches for the target identification and management of infections. Anti-virulence therapies function through the inhibition of *in vivo* requisite virulence factors to disable pathogens rather than directly targeting survival or growth [50]. This approach towards treating bacterial-mediated diseases may be advantageous over traditional antibiotics as it only targets components specific for pathogenesis by probably reducing the preference for resistance and restricting the collateral damage to facultative microbiota. During the course of pathogenesis, invading bacteria need to overcome a variety of obstacles that are raised by the host immune system and also the presence of the resident microbiota. Initiating a productive infection demands that the pathogen is capable of sensing and adapting to the changes in the environment within the host, like secretions from the host (*e.g.*, mucus), reduced oxygen tension, pH changes, physical barriers, and active immune response functioning to prevent the pathogen from colonizing. Further, the ability to exploit the resident microbiota is often required for the active establishment of an infection. However, bacterial pathogens are well equipped with a diverse array of strategies to challenge host defences and cause disease. Therefore, disrupting the pathogens mechanism to obstruct host deterrents may serve as a therapeutic strategy to tackle bacterial infections.

Often the standard drug discovery strategies pursue the identification of compounds with bacteriostatic or bactericidal functions by obstructing the targets necessary for growth, *viz.* DNA replication, protein synthesis, and cell wall integrity [51]. Anti-virulence compounds, also referred to as anti-infective compounds, are a substitute for the standard therapeutic intercession of bacterial infections, which have been actively explored. Anti-virulence compounds disarm the pathogen by targeting virulence factors necessary *in vivo* as opposed to growth cycle or viability mechanisms. By disrupting the function of the factors associated with virulence during infection, the pathogenic bacteria are left in a high susceptibility state to clearance by the reactive immune response or to the amplified susceptibility to antibiotic- mediated killing. The inhibitor does not exercise a great deal of selective pressure as classic antimicrobials, hence, possibly slowing the evolution of resistance [52]. During the broad-spectrum antibiotic therapy, no differentiation between the pathogen-associated targets and beneficial microbes exists, resulting in a case of dysbiosis in the host microbiota, thereby rendering the host susceptible to chronic and acute secondary infections.

Attachment of bacteria to a host surface is often considered the first step in initiating and establishing an infection. Nevertheless, the host maintains several strategies so as to prevent and eliminate bacteria that do not have a specific mechanism to facilitate the adherence, including the collegial movement of cilia in the nasopharynx; peristalsis motion in the gastrointestinal tract; and the existence of resident microbiota blocking access towards invading bacteria [53]. Further, the pathogenic organisms will only attach and interact with a specific subset of cells that expresses the appropriate receptor for the attachment. Additional steps beyond the attachment by bacterial adhesins are often required for the infection to proceed including internalization (*e.g.*, phagocytosis), deeper tissue penetration, biofilm formation and, for some pathogens, systemic distribution. The pili are attachment appendages that play a crucial role in the adherence and virulence of numerous pathogens, inclusive of uropathogenic *E. coli* (UPEC) – the leading causative agent for urinary tract infections. A pilus is an elongated multi-subunit process with specific binding terminal subunits, customarily through FimH or PapG, for a particular receptor on the host cell surface. Multiple pili and pili-like formations have been recognized for numerous bacterial pathogens [54].

It is well documented that the contemporary practices of antibiotic treatment are unsustainable due to the spread of antibiotic-resistant pathogens. Resistance methods are easily accrued by both *de novo* mutation and by horizontal gene transfer from the environmental reservoirs. Viable resistance mechanisms have been showing treatment strategies, such as using vancomycin and cationic antimicrobial peptides, for which the resistance was previously thought unfeasible. On the condition that an antibiotic kills or inhibits a sensitive strain's growth, any resistant strains will be capable of growing in a competitor-free environment, thereby creating a strong selection for antibiotic resistance mechanisms. Although resistance often is initially costly to the pathogen, acquired mutations that mitigate this cost quickly spread so that the prevalence of the resistance does not diminish when antibiotic use is reduced [55].

The transformation of virulence (*i.e.*, pathogen-induced host damage) is a crucial puzzle in evolutionary biology and has created a wide range of feedbacks to the underlying theoretical questions: why impair the source of the host? The dominant hypothesis states that the virulence is an inevitable cost incurred while growing within the host and circulating to the next host and is conserved as a consequence of a trade-off between the levy of the host pathology and the advantages of transmission to the new host [56]. Other hypotheses highlight the significance of selection in non-disease settings, where the auxiliary functions of virulence factors can, in tandem, select for the virulence factor-induced damage to the human hosts. Confirmation that virulence factors are non-essential is irrevocably found in rich *in vitro* growth media and is, in general, the ability of a mutated organism to grow to a density that can further be used in assays, which is greatly different from *in vivo* conditions that govern the selection for the resistance to anti-virulence drugs.

The initial rudimentary scenario is the virulence factor at the infection site (*i.e.*, the site of colonization and disruption in the focal host), which is of no benefit, just as the virulence factors generally have any benefit in *in vitro* rich media. If the virulence factor does not confer any perquisites to a pathogen at the location of the infection, targeting that virulence factor in that area will not force any intra-host selection for resistance. Resistance can even be chosen because treatment enables susceptible bacteria to dodge the metabolic costs of inappropriate expression of the virulence factor and reduce potential transmission from a sick host [57]. Although the rationale is clear, the question of the existence of virulence factors that provide no fitness advantage to the pathogens at the location of infection is still debatable.

The outstanding candidates for unhelpful virulence factors are found in opportunistic pathogens that typically exploit distinct environments (for example, nonhuman environments or commensal parts within a human host), with virulence factors being the product of accidental preference in these environments. P dates of ear bleaching can be found among pathogenic *Escherichia coli* (ExPEC) strains outside the intestine (*viz. Escherichia coli* causing urinary diseases and concomitant meningitis). ExPEC strains are opportunistic pathogens that are often derived from healthy intestinal flora but also result in various diseases at different sites outside the intestine, such as the urinary tract or the brain, which are associated with deficient transmission compared to the intestine. Virulence factors associated with ExPEC, including adhesives (such as Pili) and iron gaining agents (such as yersiniabactin), are linked with persistence as a coexisting microorganism in the gut. The outer membrane protein A (OmpA) and lipopolysaccharide {O} have also been demonstrated to be beneficial for interaction with amoebae in the environment. Although these factors are related to virulence, in the mouse model outside the intestine, they confer no measurable benefits in the presence of distinct biological pressures.

Moreover, genetic analyses indicate that accidental selection among microorganisms commensal in the gut - rather than direct selection of virulence outside the intestine - is accountable for the maintenance of these virulence factors. Biogenesis of P pili is targeted in ExPEC strains by means of 2 dicyclic pyridones; however, the selective results of dicyclic pyridones depend on the conditions under which they are used. If the treatment particularly targets the urinary tract, we would expect that resistance would not be chosen, but that disease would be reduced. Nevertheless, in the case of more likely systemic therapy (including both urinary and intestinal tracts), resistance in the intestine may be chosen, as P filament has been demonstrated to offer a selective advantage [58]. Although it is assumed that the numerous virulence factors have been selected from external sources with respect to the human host, the list of virulence factors that may not be locally beneficial at present is very short. We believe this could be due to a lacuna in the research focusing on outlining the costs and benefits of expressing the virulence factor of pathogens at the infection sites (for exceptions, see the plant pathogen literature). We suggest that pathogen fitness is an amount that is overlooked; Future directions of investigations should be to determine the convenience costs and benefits of expressing virulence factors in the relevant host environments.

A major feature of antiviral compounds is that they convert the pathogenic population to a less virulent state rather than eliminating pathogens directly, which means that prolonged courses of antiviral drugs may be necessary to maintain a low virulence state. However, many antiviral medications have been shown to facilitate cleansing. Furanone quorum-sensing inhibitors increase the immuno- or antibiotic-related clearance of *Pseudomonas aeruginosa*, due to quorum sensing effects on modulating immunity and biofilm formation respectively. Under a broad class of beneficial virulence factors, a distinction must be made betwixt virulence factors that offer an immediate and particular benefit to (focal) bacteria expressing the attribute (for example, adhesives) and collectively beneficial virulence agents that confer assistance to a group or neighborhood of bacteria (*e.g.*, excreted iron carriers, enzymes, and toxins). Many virulence elements come under the second cooperative category, which is distinguished by the secretion of expensive polypeptides that scavenge, digest or release growth-promoting resources [59]. From the perspective of social evolution, these guarantees are described as —public goods - individual contributions charged to an enterprise of collective benefit. The conceptual work has illustrated that collectively targeting advantageous virulence factors can significantly reduce resistance selection since communal goods can be taken advantage of by neighbors.

Quorum sensing or signaling is the cell-cell signaling behavior that has accrued a lot of attention as a probable therapeutic target, as it mostly controls many factors of virulence. We highlight the distinction between signal response inhibitors (*i.e.*, inhibitors that disrupt the ability of individual cells to respond to signal molecules) and signal delivery inhibitors (*i.e.*, inhibitors that disrupt the production and/or persistence of stimuli in the environment), where we expect that these two perspectives will pose different risks for the establishment of resistance. The bacterial quorum sensing (QS) can be activated by self-producing extracellular chemical signals in the medium. QS signaling consists primarily of acyl homoserine lactones (AHLs), self-stimulating peptides (AIPs) and autoinducer-2 (AI-2), all of which play major roles in regulating the pathogenesis of bacteria. For example, studies [60] reported that QS signals are implicated in the production of virulence elements such as lectin, exotoxins A, biocyanin, and elastase in *Pseudomonas aeruginosa* during bacterial growth and infection. The production and secretion of enterotoxins, hemolysins, lipase, protein A, and fibronectin are modulated by the signals QS present in *Staphylococcus aureus*. These virulence factors influenced QS to help the bacteria evade the host's immunity and procure nutrition from the host. Anti-QS agents, which are alternatives to antibiotics due to their ability to reduce bacterial virulence and enhance pathogen clearance in various animal models, have been validated to prevent bacterial infection. The translational application of anti-QS agents is even now immature. This review builds around the discoveries and increasing applications of anti-QS agents from multiple research data in the past decades. Our objective is to demonstrate the potential for QS drugs and signal-based methods to curb bacterial infection without causing pathogen resistance to drugs.

Bacterial signals QS are mainly composed of acyl-homoserine lactones (AHL), self- stimulating peptides (AIP), and autoinducer-2 (AI-2), are involved in a number of physiological processes of bacteria inclusive of the formation of biofilms, plasmid conjugation, motility, and antibiotic resistance *via* which bacteria can transform and survive defects [61]. Both Gram-negative and Gram-positive bacteria have different QS signals for cell-to-cell communication. AHL signaling peptides are mostly produced by Gram negative bacteria, and AIP signaling peptides are synthesized by Gram positive bacteria. Both Gram-negative and Gram-positive bacteria synthesize and sense AI-2 signals. These three groups of QS signals are acquiring progressive attention owing to their regulatory role in bacterial proliferation and infection. The Lux-I synthase circuit AHL was considered to be QS, a signaling product in Gram-negative bacteria. Once AHLs amass in the extracellular matrix and overshoot the threshold level, the signaling peptides will disseminate across the cell membrane and then QS bind to specific transcription regulators, thus enhancing the target gene expression. AIPs signaling molecules are produced by Gram-positive bacteria and released by membrane transporters [62]. When the surrounding concentration of AIPs overshoots the threshold, these AIPs bind to a two-component histidine kinase sensor, whose phosphorylation in turn alters the target gene transcription and triggers the related physiological process. For example, QS signaling in *Staphylococcus aureus* is rigidly regulated by additional gene regulator (ARG) associated with the secretion of AIPs. ARG Genes are implicated in the production of various exogenous degradable toxins and enzymes, which are majorly controlled by P2 and P3 promoters. The genes AGR are also involved in encoding AIPs and transmitting histidine kinase signaling. The bacteria can detect and interpret signals from other species in the environment, known as AI-2 interspecies signals. AI-2 signaling in the majority of bacterial strains is activated by Lux S synthase. Lux S is not only involved in regulating AI-2 signaling but also in the catalysis of methyl cycle and has been revealed to control the expression of another 400 genes associated with bacterial processes of surface adherence, motility and toxin production.

Bacteria are found widely in the natural surroundings, on the surface of equipment in hospitals, and in pathological tissues [63]. Biofilm formation is one of the necessary requirements for bacterial adherence and growth. The formation of biofilms is associated with extracellular polymer production and the adhesion matrix and leads to fundamental changes in bacterial proliferation and gene expression. The development of biofilm significantly decreases the sensitivity of bacteria to antibacterial molecules and radiation and seriously affects overall health. Some massive infections are linked to the development of bacterial biofilms on pathological tissues, and most of the infections caused by the bloodstream and the hospital-acquired urinary tract are due to pathogens coated with biofilms on hospital medical devices. A huge number of studies have demonstrated that bacterial quorum sensing (QS) signals play a critical role in biofilm development. The specified QS signal occlusion is an effective means of preventing the biofilm formation of most pathogens, thus improving the sensitivity of pathogens to antibacterial agents and increasing the antibiotic effect of the bacterium [64]. The synthesis of virulence elements, which can help the bacteria avoid the host's immune reaction and create pathological damage, is critical to causing infection. The virulence factors secreted by different strains vary. For instance, Gram-negative *Pseudomonas aeruginosa* secretes virulence factors, such as biocyanin, elastase, lectin, and exotoxins A, and Gram- positive *Staphylococcus aureus* secretes virulence factors, such as fibronectin-binding protein, hemolysin, protein A, lipase, and enterotoxins. Investigations have reported that the synthesis of these virulence factors is influenced by the bacterial signaling systems QS [65]. Inactivating QS to control synthesis and secretion of virulence factors is one attractive broad- spectrum therapeutic strategy.

Bacterial chromosomal alterations have greatly facilitated a better understanding of bacterial disease and virulence mechanisms. Among the methods of genome engineering, the application of the red recombinase λ system for insertion, deletion, or point mutations of the genome has been widely accepted. This method, pioneered by Murphy [66] and subsequently modified by Datsenko and Wanner, encompasses the insertion of single- or double- stranded regions DNA with homologous regions of chromosomes for recombination. Since its conceptualization, this modification strategy has become more efficient through alterations to the method conceptualised by Wanner [67]. This has also been promptly applied to pathogenic bacterial strains in order to investigate the genetic underpinnings of pathogenesis. *Pseudomonas aeruginosa*, an opportunistic pathogen that is the source of hospital infection and chronic infections of CF, uses two main quorum sensing systems: LasR/LasI and RhlR / RhlI, to regulate the synthesis of virulence factors and biofilm development. To understand their role in the virulence of *P. aeruginosa*, red λ recombination has been successfully applied to generate lasR mutant and double mutant strains of *P. aeruginosa* ΔlasR/ΔrhlR to determine their functions. Using the *Caenorhabditis elegans* 24-hour rapid infection assay, it was established that *C. elegans* was killed more rapidly by the wild-type mutant and lasR compared to the rhlR or lasR / rhlR mutant strains, suggesting that RhlR function is critical to virulence. The authors also distinguished a small molecule known as meta-bromo-thiolactone (mBTL), which is comparable to N-acyl-L-homoserine lactone (AHL), a common cell signaling molecule of *P. aeruginosa*, which has been shown to attenuate biocyanine and biofilm formation in the strain Wild type. To determine whether the RhlR or LasR receptor is the molecular target of mBTLand to validate that RhlR is the relevant target *in vivo* for mBTL, a red λ recombination system was used to createlasR and lasR/rhlR mutations.

One of the major modifications of *Salmonella enterica* serovar *Typhimurium*, the pathogenic organism causing typhoid in humans, is that it is capable of surviving inside the phagocytes of the host. Macrophages express the production of induced nitric oxide (NO) synthase (i NO S) in response to the A-lipids, fimbriae and purines that adorn the salmonella coat. It has been previously demonstrated that the NO that is released as an innate response by macrophages can influence the biosynthesis of amino acids in salmonella by targeting DksA. DksA is RNA a major regulatory protein for polymerase in salmonella, which contributes to bacterial resistance. To determine whether DksA is implicated in the anticorrosive defenses in *Salmonella*, Henard and Vazquez-Torres created the dksA mutant using the -Red recombination [68]. They determined that *Salmonella* dksA mutant strains are hypersensitive to the bacteriostatic consequences of NO and were significantly impaired in infection, as evidenced in a murine model of acute systemic infection.

To facilitate high-throughput genome editing, a recombination method called automated multiple genome engineering (MAGE) was recently developed to replace the traditional recombination-based recombination method based on λ-Red recombination (Fig. 1). This method uses the λ-Red system recombinase and a group of oligos to rapidly insert simultaneous modifications into the *Escherichia coli* genome within a span of days. As this method was further refined, the mapped pORTMAGE (ambulatory MAGE) was performed and expanded to include translationally relevant strains such as *Salmonella enterica*. Using pORTMAGE, ten antibiotic resistance modifications were simultaneously inserted into the genomes of *S. enterica* and *E. coli* to investigate the extent to which the molecular processes of antibiotic resistance were preserved among these bacteria. For bacteria that were resistant to standard engineering methods like the red recombination system, a set of II mobile introns was used to edit the site's genome. The predicate group introns II are the bacterial reverse transposon containing the intron RNA and the intron-coded reverse transcript. The group of mobile introns II are ribozymes that can be inserted into specific targets through the retransmission process [69]. Using predictive algorithms, intron RNA can be re-engineered to form a ―targetron‖, so that the location of the DNA chosen target can be modified. This procedure has been applied to several pathogenic strains to understand the mechanisms of virulence. For the many medically relevant types of Clostridium that are opposed to recombination, a goal-based method called the ClosTron technique has been established for successful genome editing. In one example, ClosTron's technique for site-guided mutagenesis of a germination protease called CspC was used in *Clostridium difficile*, a causative agent of food-borne infection and diarrhea. This study aided in determining the functional effect of CspC in the identification of bile acid antibacterial *in vivo* and disease establishment. Yet another example of the application of the targetron technique was the investigation of the virulence mechanism of *Pasteurella multocida*, an animal pathogen causing avian cholera in poultry and wild birds, atrophic rhinitis in pigs, and hemorrhagic septicemia in ungulates.

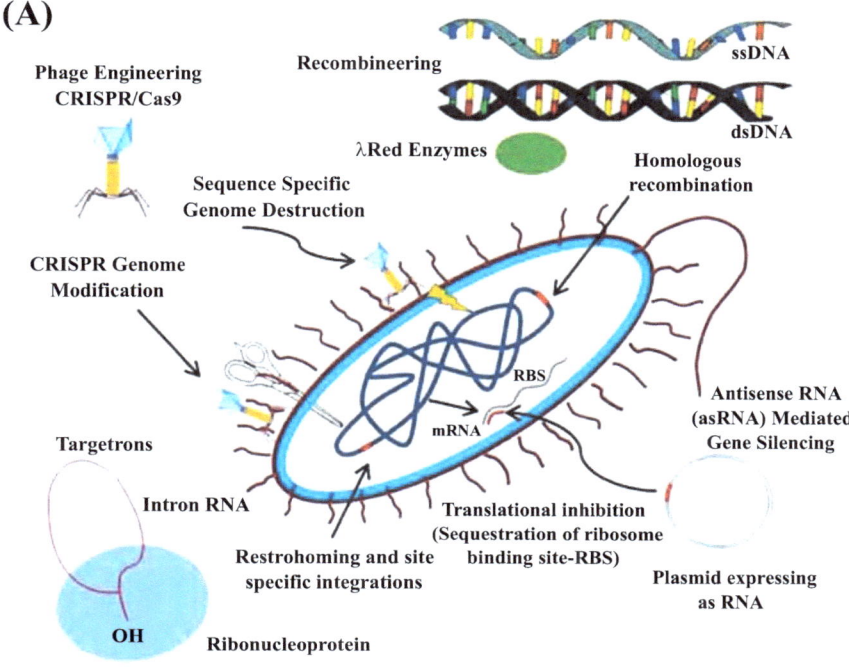

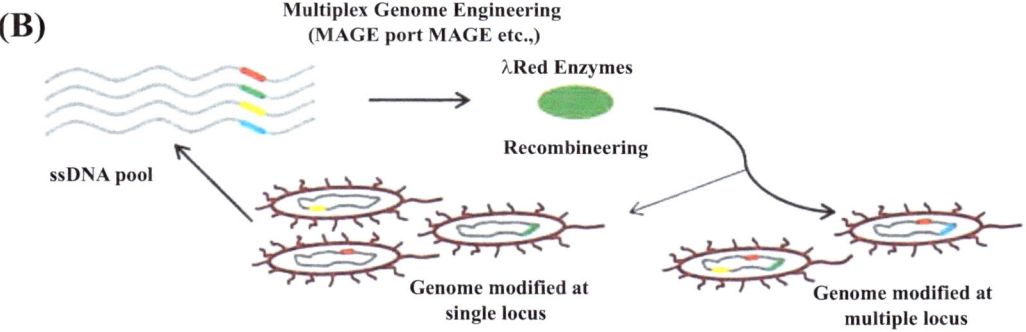

Fig. (1). (a) Schematic of genome engineering tools developed in *E. coli* that have been expanded to broad bacterial hosts (b) Multiplexed Automated Genomic Engineering (MAGE) for modifying bacteria at multiple genomic loci.

The polysaccharide capsule consisting of hyaluronic acid is the main virulence agent. To explore the mechanism of capsule development and to determine the role of the global transcription regulator Fis in capsule development, Steen and colleagues used the targetron technique in *Pasteurella* to create Fis mutations [70]. They demonstrated that the functional Fis protein is not only necessary for capsule formation but is also for the regulation of numerous virulence genes.

APPLICATION OF SYNTHETIC BIOLOGY

Major advances in microbial genome engineering have led to the creation of a proven highly valuable toolbox for altering microbial genome architecture for useful applications. The bacterial genome editing approaches discussed previously clearly demonstrate the efficacy of these tools in synthetic biology avenues to target infectious diseases. Aside from gene deletions, insertions, or mutations to modify genomes using editing tools, one of the primary goals of SB is to construct and integrate gene circuits that process signals *in vivo* to obtain the desired output. Genetic circuits have been introduced into microbes using biological components or functional units for various translational applications [71]. Modular biological components have been used to create networks on the basis of electrical engineering principles, which can then be connected to input and output responses in the form of analog or digital data. Using this engineering substructure, SB has potential utility in the synthesis of industrial chemicals or natural product substitutes, biofuel production, biomedical applications or the understanding and control of bacterial infection [72]. In the next section, we highlight some examples of biological regulatory elements that have been used in biomedical applications focusing on bacterial infections.

CRISPR Naturally occurring systems defend bacteria and archaea against DNA invading elements such as plasmids and viruses [73]. A component of the foreign object DNA has been merged into a tandem array at the host cell locus CRISPR, which is subsequently copied and processed into CRISPR RNA (cr RNA). In a system of type II CRISPR, crRNA forms a complex using CRISPR RNA (tracrRNA) and CRISPR-bound protein 9 (Cas9). The crRNA then directs the Cas9-RNA complex to the target DNA, where the Cas9 nuclease induces the double-stranded discontinuity, thus cleaving the invasive DNA and protecting the host. *In vitro* analysis of Cas9 from *Streptococcus pyogenes* found that cr RNA and tracr RNA can be replaced by a single RNA molecule, the evidence RNA (gRNA) [74].

Bacteria effectively detect and respond to external cues as part of their inherent survival and proliferation tactics. SB utilise these processes to sense and react to clinically relevant signals. The generation and engineering of bacterial circuits that rely on a small-molecule signaling mechanism to target bacterial pathogens have also been used as a futuristic approach in circuit design.

A chief example of this is when Doan and Mars demonstrated that feeding infant mice with genetically altered organisms modified for *E. coli* that decreased regulation increased mice survival rate from *V. cholerae* infection (92% with 8h pretreatment) [75] (Fig. 2). Applying a gene circuit, Saeidi *et al.* engineered *E.*

coli with the *P. aeruginosa* LasR receptor AHL to sense the signal of *P. aeruginosa* AHL N-3-oxodeconoyl-L-HSL and auto regulate the activation of killer and gene products (Pyocin S5 and E7 lysis protein) targeting *P. aeruginosa* [76].

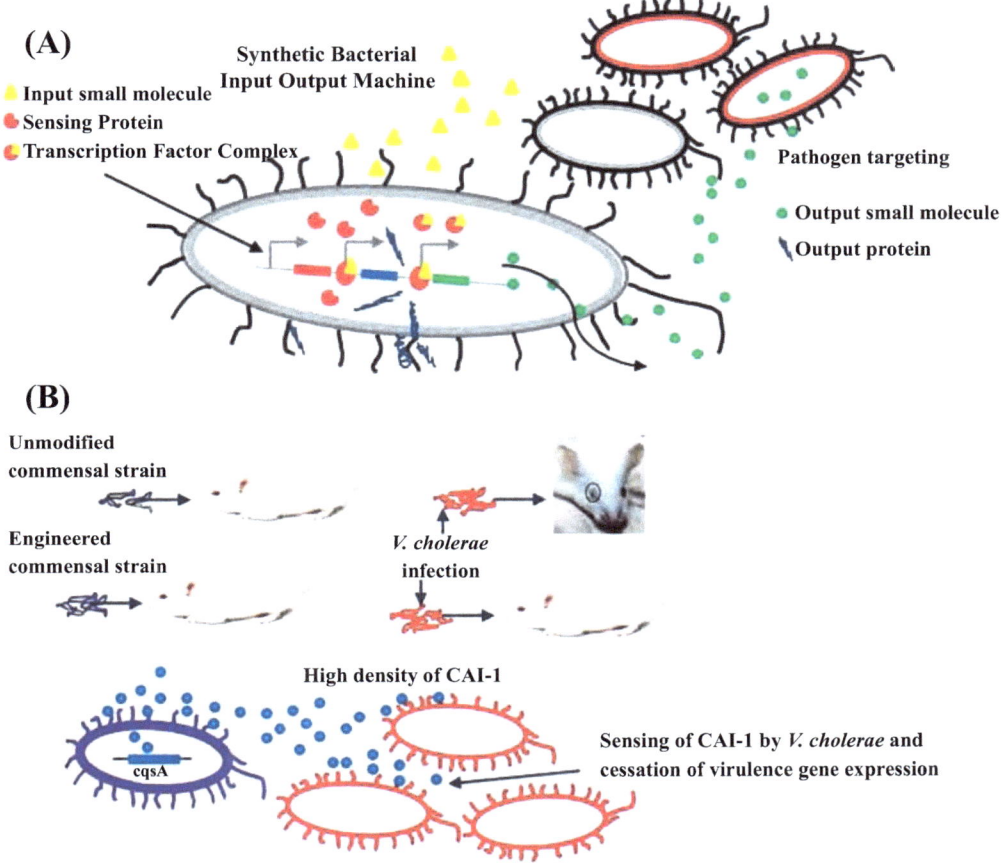

Fig. (2). (a) Synthetic biology circuits can be engineered with desired input and output signals to kill bacterial pathogens; (b) Engineering probiotic bacteria that express QS molecule Cholerae autoinducer-1 (CAI-1) to target *Vibrio cholerae* infection in a mouse model.

In concurrence with this design, QS other circuits were designed to target pathogens based on identification, destruction, and excretion modules to become beneficial bacteria. For instance, in a proof-of-concept investigation, the elimination of the pathogen *P. aeruginosa* was demonstrated using the programmed *E. coli* sensor *P. aeruginosa* QS signaling N-3- oxodeconoyl-L-HSL and activating the synthesis and release of a chimeric killer protein Copy [77]. Furthermore, the application of engineering standard bacterial circuits QS was exhibited by programming *E. coli* to search for and destroy *P. aeruginosa*. This system used CheZ, a motility-enhancing protein and two engineered proteins that secrete and kill DNase I and Microcin S proteins that assisted biofilm disruption and lethality. The majority of these QS circuits are primarily targeted at Gram-negative bacterial circuits, and specifically AHL class QS molecules only. There is a great heterogeneity of QS molecules (other than AHL classes) and circuits that remain unexplored for targeting and can be used effectively to control pathogens. RNA Circuits based on RNA biological elements called riboregulators are also receiving the limelight in SB circuit design due to their configurable and modular nature. These hairpin tools RNA are designed to isolate the ribosome binding site (RBS) at the start of the mRNA initiation site to encode the gene in order to inhibit translation [78]. In one of the applications of these tools, RNA regulatory switches called "toehold" switches were developed by Pardee *et al.* [79].

These have been used for *in vitro* diagnostic applications, cell-free, and paper-based innovation for the recognition of M RNA Ebola and hundreds RNA of antibiotic resistance genes. The RNA-based diagnostic network comprises a reporter gene network (gfp, lacZ), in which RBS is isolated upstream by an attached key. The genetic network and associated cell-free transcription/translation system are frozen onto a porous material or paper and can subsequently be activated by rehydration with the test sample consisting of the mRNA of interest that is to be detected. The antibiotic resistance gene RNA sensors, upon detection of the target gene, exhibited significant stimulation of the reporter gene, thus making this tool reassuring and cost-effective in detecting and distinguishing bacterial infections in clinical samples. Aside from its utility in diagnostics, the ability to fine-tune gene expression and the normative nature of the regulator make them exceptional candidates for chromosomal integration using current genome engineering tools for modifying the biosynthetic pathway in bacterial hosts. It is apparent from the aforementioned examples that genome engineering and synthetic biology techniques in bacteria can have a conspicuous impact on many of the applications for targeting bacterial infection. Resistance and meager success in discovering new antibiotics call for the detection and validation of targets for new bacterial drugs. SB guided genome editing technologies offer new ways to continue the exploration of bacterial target identification and present the potential for developing new antimicrobial therapies. The accessibility to high-throughput bacterial genome editing tools combined with advances in DNA synthesis techniques provides novel avenues for the metabolic engineering of massive gene sets in microbial hosts. Through the rational integration of genetic components, biosynthesis can be reprogrammed to generate new small peptides for therapeutic uses [80].

This is also facilitated by full-genome *in-silico* mining and algorithms that project gene combinations that can be used to engineer biosynthetic pathways to produce new antibiotics. This provides enormous potential for combinational biogenesis of natural product substitutes for the discovery of new antibiotics, as in the case of the antibiotic daptomycin elaborated in the previous section. Besides antibiotics, the establishment of alternative treatment options using phage and the bacterial engineering of probiotics to target and destroy pathogens is essential for fighting infections, including drug-resistant bacterial pathogens. Advances in computational genetic circuit development combined with improvements in the large-scale DNA synthesis herald a new generation of SB based therapeutics. Currently, it is possible to design entire bacterial genomes from complex components. By looking at the genetic code as a counterpart to computer programs, the artificial genome can then be 'turned on' in a compatible cellular domain [81]. In addition, the computational designs of the synthetic circuits are useful for predicting the most favorable combination of the prosthetic parts for the desired cellular function (Fig. **3**). Recently, Voigt and colleagues prepared the Cello software, which permits the user to program the required circuit operations in *E. coli* and assemble the code into DNA synthesis sequences. High-throughput DNA synthesis, computer-aided design (CAD), and advanced genome editing techniques have proved to be of high value for developing reprogrammed bacterial species for therapeutic uses. A merger of advanced computational schemes for predictive systems SB, rapid advancements in techniques for efficient large-scale synthesis DNA in addition to high throughput, automated genome engineering such as MAGE, pORTMAGE implies that SB has tremendous potential presenting new control solutions on the pathogen. Transgenic bacteria present great hope for finding new pointers towards the detection and treatment of infections. Nonetheless, it is also essential to periodically assess the biosafety of these entities to avoid the accidental release of the synthetic bacteria created by several of these applications.

To tackle this concern, Collins and colleagues [82] have developed two engineering protection systems called —Deadman and —Passcode. These switches rely on networks that need distinct inputs (inputs) of small molecules for cell survival. In the presence or absence of specific molecules, the toxin gene transcription is activated, resulting in cell death. Lock switch circuit designs can be incorporated into a wide variety of bacterial hosts to guarantee the safe handling of these modified organisms. It is discernible from the previous examples that SB bridges the interstice between basic research and translated research. It is anticipated that with the continuous technological advancements in this discipline and with the development of novel programmable biological tools, SB has tremendous potential for developing biomedical treatments to intercept and treat diseases caused by bacterial infection.

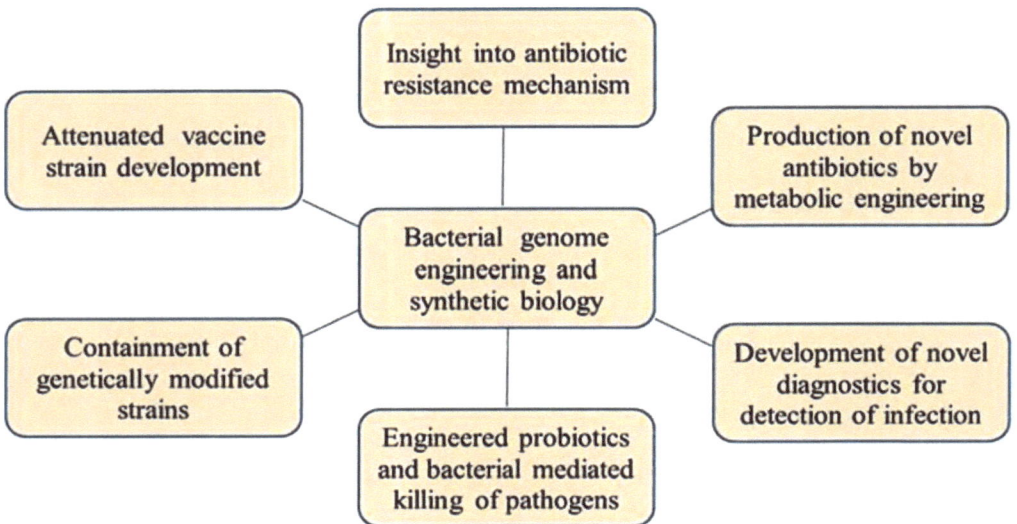

Fig. (3). Multipronged use of SB and genome engineering tools to counter bacterial infections.

Controlling Gene Expression with CRISPR

The field of genome editing began with the discovery of massive nucleases (*e.g.*, the LAGLIDADG family of carrier nucleases) in yeast. After the unearthing of the transcription activator-like effector nuclease and the zinc finger nuclease, the recently discovered short alternating upregulated (CRISPR)/CRISPR (Cas) proteins (Cas) repeats were discovered which opened a new window of applications in the domain of gene editing. Traditional approaches to studying the *in vivo* functions of regulatory elements that control gene expression involve the nucleation of the promoter/enhancer sequences associated with a reporter gene (for example, lacZ). These experiments are restricted to the context outside the genome in which the sequences are deliberated and necessitate the generation of multiple cell lines from mice to address problems related to the copy number and position-dependent effects on reporter gene functionality as well as the undesirable consequences of DNA foreign sequences in the genomic perspective that could disturb endogenous gene expression [83]. The emergence of chromosomal transgenic mice represents an important step towards studying gene expression in the appropriate genetic context; however, it is difficult to work with these cloning vectors of large capacity and still requires generating multiple independent strains of mice, which can take a long period to complete. The developments of tools to activate the expression of arbitrary genes have been transformations in biotechnology and biological research. In metabolic engineering, the regulation of the timing and expression levels of complex polygenic pathways is critical to reduce cellular burden and improve the production of valuable metabolites. To facilitate these goals, lately, a novel CRISPR-Cas transcriptional activation system (CRISPRa) was developed that is efficacious in *Escherichia coli*. This system, in combination with CRISPRi gene suppression, can programmatically and flexibly target various genes for simultaneous activation or suppression. While this system can be operated with heterologous genes, a prominent difficulty is to recognize the rules that establish effective target sites in random promoters in the genome.

Large-scale targeted gene regulation of the genome is a powerful strategy for questioning, disturbing and engineering cellular systems. Here, a method to control gene expression based on Cas9, an RNA endonuclease DNA directed from a system of type II CRISPR was developed. It was demonstrated that catalytically dead Cas9 that lacks endonuclease activity, when expressed with an RNA clue, generates a DNA identification complex that can particularly interfere with RNA polymerase binding, transcription elongation, or transcription factor binding. This complex, termed as CRISPR interference (CRISPRi), can effectively suppress the expression of target genes in *E. coli* without any off-target effects that could be detected. CRISPRi can be used to concurrently suppress multiple target genes, and their effects can be reversed. The aforementioned details provide evidence that the system can be adapted to suppress genes in mammalian cells. The RNA guided recognition platform DNA provides a simple method for selective denaturation of genome-wide gene expression. Systematic genome interrogation and genetic reprogramming of cells require methods to accurately and predictably target gene sets for expression or suppression. Intervention RNAi (RNAi) and engineered DNA-related proteins like zinc finger or transcription-like effector proteins (TALE) have surfaced as powerful techniques for regulating target genes [84, 85]. RNAi can be employed in a relatively direct way to eliminate the expression of target genes. However, RNAi is restricted to specific organisms that have an appropriate host mechanism and can occasionally exhibit notable off-target effects and toxicities. Additionally, custom DNA-binding proteins, *viz.* zinc finger or TALE proteins, remain difficult and somewhat expensive to design, develop, and experimentally within a cellular context. As a result, it remains difficult to employ DNA-binding proteins for the simultaneous modification of multiple genes and the execution of large-scale genetic programs.

Bacteria are attractive targets for a number of engineering applications. Bacterial strains which are capable of using carbon sources like carbon dioxide, methane or lignocellulose, and alternative energy sources such as light or H_2 can provide the basis for a cost-effective and environmentally friendly industrial biosynthesis. Microbial communities, such as those in the human gut, play an important role in human health and disease, and tools for engineering these bacteria have great potential for diagnosis and treatment. To exploit, modulate and modify the functions of these and other bacteria, there is an urgent necessity to construct genetic tools to influence gene expression and to execute complex, multi-gene regulatory programs. The ideal scenario is to construct circuits that can regulate multiple genes simultaneously, respond dynamically to external input or the internal state of a cell, and can be easily reprogrammed to investigate different functional structures. While the possibilities for genome editing and modification are rapidly increasing, our capacity to encode a precisely defined gene expression program that is dynamically responsive to regulatory sequences at the level of DNA remains challenging. Thus, it was sought to develop synthetic transcription factors in bacteria, which could be combined with DNA binding domains that are programmable and regulated by inducible catalysts to engineer complex and dynamically responsive multigenic expression programs. Recently, synthetic control of gene expression has become more pronounced with the advent of programmable transcription factors utilizing the CRISPR-Cas system. The inactive Cas9 protein (dCas9) can be used as a catalyst to target DNA specific sequences with an index RNA (g RNA s) recognizing their targets based on a predictable Watson-Crick rule coupling. This strategy can be used to suppress genes by physical blocking of RNA polymerase (CRISPR or CRISPRESS interference) [86]. For gene activation (CRISPR activation or CRISPRa), the CRISPR complex can be attached to the transcription activator *via* directly binding to dCas9 or *via* domains of recruitment on our gRNA. In bacteria, there is a small number of transcriptional activation domains that have been detailed to be effective when combined with standard DNA binding domains. The two- hybrid bacterial systems were generated using pairs of interacting candidate proteins that were separately incorporated into the subunits of RNA polymerase and DNA binding proteins. It is also feasible to combine RNA polymerase subunits directly with DNA binding domains to initiate transcription. One of the polymerase subunits RNA, RpoZ, was paired with the system CRISPR to initiate gene expression. In juxtaposition, it can be seen that in eukaryotic systems, there are multiple effective steroids, and CRISPRa has been used extensively in diverse applications. The scarcity of CRISPRa reports in bacteria indicates that RpoZ may not be efficacious as a general transcription activator, or that there is a dearth incomplete understanding of the design schemes to predictably activate gene expression in bacteria.

To develop a refined toolkit for gene activation in bacteria, a broad range of candidate proteins were examined for transcription activity in *E. coli*. Several proteins were identified that could constructively activate gene expression when recruited across CRISPRC as a system. By using the most potent activator, SoxS, a single target gene can be activated while simultaneously suppressing a different target gene with CRISPRi, control over the entire multi-gene expression program is possible by using the inducible promoters driving components of the CRISPR-Cas complex. It was found that gene activation in *Escherichia coli* is extremely sensitive to the position of the target site g RNA, consistent with previous findings, and suggests a probable explanation for why synthetic steroids are difficult to develop in bacteria [87]. Lastly, it was demonstrated that bacterial CRISPRa can be employed to elevate the yield of the heterocyclic ethanol biosynthesis pathway. These findings provide a substructure for the application of CRISPRa in bacteria with a diversity of potential applications. Moreover, since SoxSis associated with an extremely conserved site on RNA polymerase, the aforementioned CRISPRa bacterial toolkit may be convenient for application across a range of bacterial species.

CRISPR in the Treatment of Infection

Infectious diseases continue to be a global threat, contributing to increasing morbidity and death rates annually, with the potential to disrupt epidemics. An improved understanding of the etiology of bacteria, viruses, fungi, and parasites, along with rapid diagnosis and treatment of human infection, is essential for improving infectious disease outcomes worldwide. Associated with CRISPR, the prokaryotes' adaptive immune system protects them from foreign intruders. Nowadays, CRISPR-Cas9 technology is routinely employed for competent gene editing, which contributes to the progress achieved in biomedical sciences. The past decade has seen improvement in the comprehension of other diverse CRISPR-Cas systems and has helped expand the applications of CRISPR, especially in the area of infectious diseases [88]. This review has attempted to summarize the biology of CRISPR-Cas systems and discuss the current and emerging applications for assessing host-pathogen interaction processes, for developing precise and portable diagnostic tests, and for promoting prevention and treatment of infectious diseases. The distinguishing feature of the CRISPR-Cas systems is that prokaryotes provide heritable adaptive immunity against foreign genetic elements. The CRISPR genomic site functions as a memory storage unit in which the DNA spacing sequences derived from the invading genetic elements are isolated; The sequences are subsequently called up to direct Cas proteins for the targeted elimination of the foreign invaders.

CRISPR technology provides the tools that promise to elucidate basic host and microbial interactions, to help develop rapid and accurate diagnostics and to facilitate the prevention and medicament of infectious diseases. Understanding the mechanisms by which viruses, bacteria, fungi, and parasites induce human disease is essential to guide optimal clinical care and rational design of targeted therapies and vaccines. CRISPRCas9-dependent gene editing across diverse pathogens has been used to inform the contributions of genes and proteins to molecular pathogenesis [89].

CRISPR-Cas-based "guided RNA nucleases" have been used to target genes encoding antibiotic resistance, virulence factors, and other clinically relevant regions of interest. Thus this technique represents a new configuration of antimicrobial therapy and an approach through which bacterial populations can be manipulated [90]. Recent studies indicate an association between CRISPR-Cas locus intervention and the acquiring of antibiotic resistance. This system protects bacteria from intruding foreign DNA, such as phages, transposons, and plasmids.

This regimen has been shown to represent a potent, selective pressure for gaining antibiotic resistance and virulence components in bacterial pathogens. Treatments based on CRISPR - a Cas3 gene-modifying technique provided by engineered phages can be applied to destroy DNA target pathogens. Cas3 is further destructive than the well-known Cas9 [91]. Research indicates that CRISPR is an efficient way to reduce the recurrence of multiple herpes viruses and was able to eliminate virus DNA in the case of Epstein-Barr virus (EBV).

The anti-herpes virus CRISPR has encouraging applications such as extracting carcinogenic EBV from cancer cells, helping to rid the donated organs of immunocompromised patients from viral intruders, or preventing cold sore eruptions and recurrent eye infections by preventing HSV-1 reactivation.

Exposure to antibiotics leads to tremendous selective pressures leading to the development and spread of antibiotic resistance in pathogenic and commensal bacteria. The time consuming exercise of developing new antibiotics makes this approach unreasonable to combat the accelerated emergence of new antibiotic-resistant pathogens. Hence, alternative approaches such as the establishment of antibacterial therapies based on nucleic acids, bacteriocins, antibacterial peptides, anti-virulence compounds and phage therapies should be made the most of to deal with infections caused by resistant germs [92].

CRISPR in Health and Industry

Cancer, a group of diseases involving the growth of uncommon cells that have the potential to spread to other parts of the body, is a leading life-threatening disease worldwide. Cancer incidence will increase by 50%, resulting in about 15 million deaths annually. Cancer is characterized by the accumulation of several genetic and epigenetic changes in the genome of a cancer cell, which leads to the emergence and development of cancer diseases that disturb cellular signals and result in a cancerous and malignant transformation. Cancer is one of the most dangerous diseases that challenge human life and public health. Although important advances have been made in cancer treatment, including surgery, radiotherapy and chemotherapy, a high tendency to relapse and initial or acquired chemotherapy resistance often leads to a poor prognosis [93]. Therefore, the ability to correct or disable one or more parts of the genome of a cancer cell, for example, restoring the function of tumor suppressor genes, may provide an interesting approach to cancer treatment, which can be done by genome editing. CRISPR-Cas9 facilitated researchers to edit the eukaryotic genomes more accurately and efficiently compared to others (ZFN & TALEN).

The development of a CRISPR-Cas9 mediated genome-editing tool has been bewildering the field of gene therapy, which not only holds broad application potential for the therapeutic treatment of cancer genomes but can also be used for combating neoplastic infections, modifying gene expression, and exploring anticancer drugs. To directly target the genomes of cancer cells, the CRISPR-Cas9 system can also be exploited to fine-engineer immune cells and tumorigenic viruses for cancer immunotherapy applications [94]. The first clinical trial entailing CRISPR began in 2016. The study included removing immune cells from people with lung cancer, utilizing CRISPR to modify a gene expressed in PD-1, and then returning the modified cells to the same individual. Another 20 trials were in progress or nearly ready-made, most of them in China, as of 2017. In 2016, the FDA approved a CRISPR clinical trial that will be used to modify T cells derived from people with various types of cancer and then return the engineered T cells to the same individual.

Application of CRISPR/Cas9-based multiplexed genome editing with two or three genomic targets, has increased galactarate production in *A. niger*. CRISPR/Cas9 induces mutation in the transcriptional activator Gaa R in *A. niger*, which leads to increased production of pectinases. In *A. nidulans,* non-ribosomal peptide synthetase-like (NRPS-like) gene mica was targeted to enhance the production of microperfuranone [95]. The hydrolytic enzyme trehalose production was enhanced in *Myceliophthora thermophila* by a knock-in expression strategy mediated by CRISPR, which enhanced the ethanol production rate during the fermentation of ethanol. Thermophilic species *M. thermophila* and *M. heterothallica* were subjected to genome editing by CRISPR/Cas9 system and fivefold higher lignocellulase production strains were successfully obtained, and it is applied in bio-based fuels and chemicals production [96].

FUTURE PROSPECTS

Genome editing (GE) has transformed biological research with the unique ability to precisely modify the genomes of both eukaryotes and prokaryotes. In recent years, many GE tools for simple and complex genome editing have been explored. CRISPR / Cas9 system has been widely exploited in GE due to its high efficiency, utility, and accuracy. It can be used to add beneficial alleles and remove unwanted alleles concomitantly in a single event. GE includes various techniques, such as the use of zinc finger nuclease (ZFNs), transcriptional activator-like effector nuclease (TALENs), and the latest techniques of short intermodal intercurrent interconnector nuclease (CRISPR)/nuclease associated with CRISPR System 9 (Cas9). To overcome the limitations of the CRISPR/Cas9 system, a variety of efficient CRISPR GE systems were created. The variant CjCas9 Cas9, derived from *Campylobacter jejuni,* consists of 984 amino acid residues (2.95 kbps) and has been used for GE efficiency both *in vitro* and *in vivo*. CjCas9 is very specific and cuts off only a limited number of sites in either mouse or human genomes. Delivered *via* adenovirus (AAV), it has been reported to induce targeted mutations at high frequencies in retinal pigment epithelial cells (RPE) or mouse muscle cells. For example, it has been used to target the Vegfa or Hif1a gene in RPE cells, resulting in laser-induced placental vascular volume reduction, providing a novel option for treating age- related macular degeneration [97]. Cas13 is a recently identified CRISPR responder, and CRISPR/Cas13 can target specific viruses' RNA and endophytes RNA in plant cells. The Cas13 system has high RNA target specificity and efficiency. Cas13 was used to direct ADAR2 deaminase to modulate RNA (changing adenosine into inosine) in human cells to restore functional proteins to halt disease progression.

Antimicrobial resistance (AMR) is a serious peril to modern medicine and may make common infections untreatable. The discovery of relatively new antibiotics has stopped over the past decade [98], and the establishment of new approaches to address the spread of AMR genes will require significant efforts in the coming years. Numerous research groups have independently established how CRISPR-Cas (regularly clustered between symmetric short homologs - linked to CRISPR technology), a bacterial immune system now widely used for genome editing, can specifically remove AMR genes from clusters. Examining the feasibility of combining this molecular therapeutic strategy with radiation, conventional surgery, or chemotherapy may be useful for increasing anti-cancer effectiveness. Radiation therapy is one of the most important methods used in treating cancer patients.

Nevertheless, it has been revealed that tumors containing some genetic mutations, such as p53 and p21 mutations, have poor sensitivity to radiation, which often leads to radiotherapy failure. There are still many hurdles to overcome before using the CRISPR-Cas technique to target AMR natural microbial communities. Determining the appropriate delivery method will be essential to take full advantage of the possibilities of this technology to limit the clinical and environmental spread of AMR by MGEs. A simple modification of CRISPR-Cas formulations to target specific genes of interest will greatly amplify the efficiency with which this can be accomplished. Such advances may have implications for the management of resistance reservoirs and potentially aid in the maintenance or restoration of the antimicrobial activity of the antibiotic. Future research is required to evaluate and improve CRISPR-Cas proliferation in more realistic microbial colonies and to elucidate the risks associated with this technology. Additionally, the social and legislative obstacles associated with the extensive use of this gene editing technology require active participation with communities and the development of explicit guidelines to regulate its responsible and safe use. While the use of naturally occurring CRISPR-Cas systems on plasmids may circumvent some of the problems associated with the release of GMOs, realizing the consequences associated with the large-scale release of any DNA components will be key to the sustainability and risk-free application of this technology.

CONCLUSION

Several years ago, it was primarily hypothesized that a synthetic CRISPR-Cas system could be used as an antimicrobial to destroy specific bacterial genotypes. Recent studies confirmed the ability of CRISPR-Cas to precisely eliminate gene-carrying bacterial strains, including those that determine drug resistance, from the population and re-sensitize bacteria to antibiotic drugs by selectively removing plasmids encoding for AMR. Through the impressive enhancements of CRISPR-Cas9 technology, we can imagine many other potential guidelines for implementing the system in cancer treatment.

ACKNOWLEDGEMENTS

The authors would like to acknowledge Lekhani Writing and Consulting Services Private Limited for the similarity index report.

REFERENCES

[1] Knott GJ, Doudna JA. CRISPR-Cas guides the future of genetic engineering. Science 2018; 361(6405): 866-9.
[http://dx.doi.org/10.1126/science.aat5011] [PMID: 30166482]

[2] Makarova KS, Wolf YI, Iranzo J, et al. Evolutionary classification of CRISPR–Cas systems: A burst of class 2 and derived variants. Nat Rev Microbiol 2020; 18(2): 67-83.
[http://dx.doi.org/10.1038/s41579-019-0299-x] [PMID: 31857715]

[3] Wang J, Li J, Zhao H, et al. Structural and mechanistic basis of PAM-dependent spacer acquisition in CRISPR-Cas systems. Cell 2015; 163(4): 840-53.

[http://dx.doi.org/10.1016/j.cell.2015.10.008] [PMID: 26478180]

[4] Hille F, Richter H, Wong SP, Bratovič M, Ressel S, Charpentier E. The biology of CRISPR- Cas: Backward and forward. Cell 2018; 172(6): 1239-59.
[http://dx.doi.org/10.1016/j.cell.2017.11.032] [PMID: 29522745]

[5] Gong T, Zeng J, Tang B, Zhou X, Li Y. CRISPR-Cas systems in oral microbiome: From immune defense to physiological regulation. Mol Oral Microbiol 2020; 35(2): 41-8.
[http://dx.doi.org/10.1111/omi.12279] [PMID: 31995666]

[6] Tang B, Gong T, Zhou X, *et al*. Deletion of Cas3 gene in *Streptococcus mutans* affects biofilm formation and increases fluoride sensitivity. Arch Oral Biol 2019; 99: 190-7.
[http://dx.doi.org/10.1016/j.archoralbio.2019.01.016] [PMID: 30731369]

[7] Abudayyeh OO, Gootenberg JS, Franklin B, *et al*. A cytosine deaminase for programmable single-base RNA editing. Science 2019; 365(6451): 382-6.
[http://dx.doi.org/10.1126/science.aax7063] [PMID: 31296651]

[8] Zetsche B, Gootenberg JS, Abudayyeh OO, *et al*. Cpf1 is a single RNA-guided endonuclease of a class 2 CRISPR-Cas system. Cell 2015; 163(3): 759-71.
[http://dx.doi.org/10.1016/j.cell.2015.09.038] [PMID: 26422227]

[9] Vickers NJ. Animal communication: When i'm calling you, will you answer too? Curr Biol 2017; 27(14): R713-5.
[http://dx.doi.org/10.1016/j.cub.2017.05.064] [PMID: 28743020]

[10] Wei Y, Terns RM, Terns MP. Cas9 function and host genome sampling in Type II-A CRISPR–Cas adaptation. Genes Dev 2015; 29(4): 356-61.
[http://dx.doi.org/10.1101/gad.257550.114] [PMID: 25691466]

[11] Lebeaux D, Ghigo JM, Beloin C. Biofilm-related infections: bridging the gap between clinical management and fundamental aspects of recalcitrance toward antibiotics. Microbiol Mol Biol Rev 2014; 78(3): 510-43.
[http://dx.doi.org/10.1128/MMBR.00013-14] [PMID: 25184564]

[12] Kazmierczak BI, Schniederberend M, Jain R. Cross-regulation of Pseudomonas motility systems: The intimate relationship between flagella, pili and virulence. Curr Opin Microbiol 2015; 28: 78-82.
[http://dx.doi.org/10.1016/j.mib.2015.07.017] [PMID: 26476804]

[13] Dubern JF, Cigana C, De Simone M, *et al*. Integrated whole-genome screening for *Pseudomonas aeruginosa* virulence genes using multiple disease models reveals that pathogenicity is host specific. Environ Microbiol 2015; 17(11): 4379-93.
[http://dx.doi.org/10.1111/1462-2920.12863] [PMID: 25845292]

[14] Akram FE, El-Tayeb T, Abou-Aisha K, El-Azizi M. A combination of silver nanoparticles and visible blue light enhances the antibacterial efficacy of ineffective antibiotics against methicillin-resistant *Staphylococcus aureus* (MRSA). Ann Clin Microbiol Antimicrob 2016; 15(1): 48.
[http://dx.doi.org/10.1186/s12941-016-0164-y] [PMID: 27530257]

[15] Barton LA, Simon MW. Prophylactic antibiotics. Clin Pediatr 2014; 53(8): 813-3.
[http://dx.doi.org/10.1177/0009922814527508] [PMID: 24647700]

[16] Stiefelhagen P. Suspected severe pneumonia. Antibiotics are ineffective : What now? MMW Fortschr Med 2015; 157(17): 28.
[http://dx.doi.org/10.1007/s15006-015-3568-2] [PMID: 26759867]

[17] (a) Li Z, Knetsch M. Antibacterial strategies for wound dressing: preventing infection and stimulating healing. Current Pharmaceutical Design 2018; 24(8): 936-51.
(b) Fishbain J T, Viscount H B. Surveillance for methicillin-resistant *Staphylococcus aureus* in Battambang, Cambodia. Hawaii Med J 2002; 61(10).

[18] Buysse J. The role of genomics in antibacterial target discovery. Curr Med Chem 2001; 8(14): 1713-26.

[http://dx.doi.org/10.2174/0929867013371699] [PMID: 11562290]

[19] Volker C, Brown JR. Bioinformatics and the discovery of novel anti-microbial targets. Curr Drug Targets Infect Disord 2002; 2(4): 279-90.
[http://dx.doi.org/10.2174/1568005023342326] [PMID: 12570736]

[20] US Department of Health and Human Services. CDC Antibiotic Resistance Threats in the United States, 2013. Atlanta, GA, USA: US Department of Health and Human Services, CDC 2013.

[21] Silver LL. Challenges of antibacterial discovery. Clin Microbiol Rev 2011; 24(1): 71-109.
[http://dx.doi.org/10.1128/CMR.00030-10] [PMID: 21233508]

[22] Dickey SW, Cheung GYC, Otto M. Different drugs for bad bugs: Antivirulence strategies in the age of antibiotic resistance. Nat Rev Drug Discov 2017; 16(7): 457-71.
[http://dx.doi.org/10.1038/nrd.2017.23] [PMID: 28337021]

[23] Maura D, Ballok AE, Rahme LG. Considerations and caveats in anti-virulence drug development. Curr Opin Microbiol 2016; 33: 41-6.
[http://dx.doi.org/10.1016/j.mib.2016.06.001] [PMID: 27318551]

[24] Davies J, Davies D. Origins and evolution of antibiotic resistance. Microbiol Mol Biol Rev 2010; 74(3): 417-33.
[http://dx.doi.org/10.1128/MMBR.00016-10] [PMID: 20805405]

[25] Toprak E, Veres A, Michel JB, Chait R, Hartl DL, Kishony R. Evolutionary paths to antibiotic resistance under dynamically sustained drug selection. Nat Genet 2012; 44(1): 101-5.
[http://dx.doi.org/10.1038/ng.1034] [PMID: 22179135]

[26] D'Costa VM, King CE, Kalan L, et al. Antibiotic resistance is ancient. Nature 2011; 477(7365): 457-61.
[http://dx.doi.org/10.1038/nature10388] [PMID: 21881561]

[27] Habets MGJL, Brockhurst MA. Therapeutic antimicrobial peptides may compromise natural immunity. Biol Lett 2012; 8(3): 416-8.
[http://dx.doi.org/10.1098/rsbl.2011.1203] [PMID: 22279153]

[28] Andersson DI, Hughes D. Antibiotic resistance and its cost: Is it possible to reverse resistance? Nat Rev Microbiol 2010; 8(4): 260-71.
[http://dx.doi.org/10.1038/nrmicro2319] [PMID: 20208551]

[29] Keasling JD. Synthetic biology and the development of tools for metabolic engineering. Metab Eng 2012; 14(3): 189-95.
[http://dx.doi.org/10.1016/j.ymben.2012.01.004] [PMID: 22314049]

[30] Yadav VG, De Mey M, Giaw Lim C, Kumaran Ajikumar P, Stephanopoulos G. The future of metabolic engineering and synthetic biology: Towards a systematic practice. Metab Eng 2012; 14(3): 233-41.
[http://dx.doi.org/10.1016/j.ymben.2012.02.001] [PMID: 22629571]

[31] Burbelo PD, Ching KH, Han BL, Klimavicz CM, Iadarola MJ. Synthetic biology for translational research. Am J Transl Res 2010; 2(4): 381-9.
[PMID: 20733948]

[32] Ruder WC, Lu T, Collins JJ. Synthetic biology moving into the clinic. Science 2011; 333(6047): 1248-52.
[http://dx.doi.org/10.1126/science.1206843] [PMID: 21885773]

[33] Krishnamurthy M, Moore RT, Rajamani S, Panchal RG. Bacterial genome engineering and synthetic biology: Combating pathogens. BMC Microbiol 2016; 16(1): 258.
[http://dx.doi.org/10.1186/s12866-016-0876-3] [PMID: 27814687]

[34] Gibson DG, Glass JI, Lartigue C, et al. Creation of a bacterial cell controlled by a chemically synthesized genome. Science 2010; 329(5987): 52-6.

[http://dx.doi.org/10.1126/science.1190719] [PMID: 20488990]

[35] Jinek M, Chylinski K, Fonfara I, Hauer M, Doudna JA, Charpentier E. A programmable dual-RN--guided DNA endonuclease in adaptive bacterial immunity. Science 2012; 337(6096): 816-21.
[http://dx.doi.org/10.1126/science.1225829] [PMID: 22745249]

[36] Mitchell LA, Wang A, Stracquadanio G, *et al.* Synthesis, debugging, and effects of synthetic chromosome consolidation: SynVI and beyond. Science 2017; 355(6329): eaaf4831.
[http://dx.doi.org/10.1126/science.aaf4831] [PMID: 28280154]

[37] Shen Y, Wang Y, Chen T, *et al.* Deep functional analysis of synII, a 770-kilobase synthetic yeast chromosome. Science 2017; 355(6329): eaaf4791.
[http://dx.doi.org/10.1126/science.aaf4791] [PMID: 28280153]

[38] Wong AC, Levy M. New approaches to microbiome-based therapies. mSystems 2019; 4(3): e00122-19.
[http://dx.doi.org/10.1128/mSystems.00122-19] [PMID: 31164406]

[39] Williams NT. Probiotics. Am J Health Syst Pharm 2010; 67(6): 449-58.
[http://dx.doi.org/10.2146/ajhp090168] [PMID: 20208051]

[40] Kim KO, Gluck M. Fecal microbiota transplantation: An update on clinical practice. Clin Endosc 2019; 52(2): 137-43.
[http://dx.doi.org/10.5946/ce.2019.009] [PMID: 30909689]

[41] Arkin AP, Schaffer DV. Network news: Innovations in 21st century systems biology. Cell 2011; 144(6): 844-9.
[http://dx.doi.org/10.1016/j.cell.2011.03.008] [PMID: 21414475]

[42] Smolke CD, Silver PA. Informing biological design by integration of systems and synthetic biology. Cell 2011; 144(6): 855-9.
[http://dx.doi.org/10.1016/j.cell.2011.02.020] [PMID: 21414477]

[43] Jasny BR, Zahn LM. Genome-sequencing anniversary. A celebration of the genome, part I. Science 2011; 331(6017): 546-6.
[http://dx.doi.org/10.1126/science.331.6017.546-a] [PMID: 21292964]

[44] Nielsen R, Paul JS, Albrechtsen A, Song YS. Genotype and SNP calling from next-generation sequencing data. Nat Rev Genet 2011; 12(6): 443-51.
[http://dx.doi.org/10.1038/nrg2986] [PMID: 21587300]

[45] Kosuri S, Eroshenko N, LeProust EM, *et al.* Scalable gene synthesis by selective amplification of DNA pools from high-fidelity microchips. Nat Biotechnol 2010; 28(12): 1295-9.
[http://dx.doi.org/10.1038/nbt.1716] [PMID: 21113165]

[46] Alcaine SD, Tilton L, Serrano MAC, Wang M, Vachet RW, Nugen SR. Phage-protease-peptide: A novel trifecta enabling multiplex detection of viable bacterial pathogens. Appl Microbiol Biotechnol 2015; 99(19): 8177-85.
[http://dx.doi.org/10.1007/s00253-015-6867-8] [PMID: 26245682]

[47] Dellinger RP, Levy MM, Rhodes A, *et al.* Surviving Sepsis Campaign: International guidelines for management of severe sepsis and septic shock, 2012. Intensive Care Med 2013; 39(2): 165-228.
[http://dx.doi.org/10.1007/s00134-012-2769-8] [PMID: 23361625]

[48] Bernacka-Wojcik I, Lopes P, Catarina Vaz A, *et al.* Bio-microfluidic platform for gold nanoprobe based DNA detection—application to *Mycobacterium tuberculosis*. Biosens Bioelectron 2013; 48: 87-93.
[http://dx.doi.org/10.1016/j.bios.2013.03.079] [PMID: 23660340]

[49] Bhowmick T, Mirrett S, Reller LB, *et al.* Controlled multicenter evaluation of a bacteriophage-based method for rapid detection of *Staphylococcus aureus* in positive blood cultures. J Clin Microbiol 2013; 51(4): 1226-30.
[http://dx.doi.org/10.1128/JCM.02967-12] [PMID: 23390282]

[50] Lukacik P, Barnard TJ, Keller PW, *et al.* Structural engineering of a phage lysin that targets Gram-negative pathogens. Proc Natl Acad Sci 2012; 109(25): 9857-62.
[http://dx.doi.org/10.1073/pnas.1203472109] [PMID: 22679291]

[51] Citorik RJ, Mimee M, Lu TK. Bacteriophage-based synthetic biology for the study of infectious diseases. Curr Opin Microbiol 2014; 19: 59-69.
[http://dx.doi.org/10.1016/j.mib.2014.05.022] [PMID: 24997401]

[52] Smanski MJ, Zhou H, Claesen J, Shen B, Fischbach MA, Voigt CA. Synthetic biology to access and expand nature's chemical diversity. Nat Rev Microbiol 2016; 14(3): 135-49.
[http://dx.doi.org/10.1038/nrmicro.2015.24] [PMID: 26876034]

[53] Brophy JAN, Voigt CA. Principles of genetic circuit design. Nat Methods 2014; 11(5): 508-20.
[http://dx.doi.org/10.1038/nmeth.2926] [PMID: 24781324]

[54] Qi LS, Larson MH, Gilbert LA, *et al.* Repurposing CRISPR as an RNA-guided platform for sequence-specific control of gene expression. Cell 2013; 152(5): 1173-83.
[http://dx.doi.org/10.1016/j.cell.2013.02.022] [PMID: 23452860]

[55] Wang H, La Russa M, Qi LS. CRISPR/Cas9 in genome editing and beyond. Annu Rev Biochem 2016; 85(1): 227-64.
[http://dx.doi.org/10.1146/annurev-biochem-060815-014607] [PMID: 27145843]

[56] Keseler IM, Mackie A, Santos-Zavaleta A, *et al.* The EcoCyc database: Reflecting new knowledge about *Escherichia coli* K-12. Nucleic Acids Res 2017; 45(D1): D543-50.
[http://dx.doi.org/10.1093/nar/gkw1003] [PMID: 27899573]

[57] Hwang WY, Fu Y, Reyon D, *et al.* Efficient *in vivo* genome editing using RNA-guided nucleases. Nat Biotechnol 2013; 31(3): 227.
[http://dx.doi.org/10.1038/nbt.2501] [PMID: 23360964]

[58] Li W, Teng F, Li T, Zhou Q. Simultaneous generation and germline transmission of multiple gene mutations in rat using CRISPR-Cas systems. Nat Biotechnol 2013; 31(8): 684-6.
[http://dx.doi.org/10.1038/nbt.2652] [PMID: 23929337]

[59] Wang H, Yang H, Shivalila CS, *et al.* One-step generation of mice carrying mutations in multiple genes by CRISPR/Cas-mediated genome engineering. Cell 2013; 153(4): 910-8.
[http://dx.doi.org/10.1016/j.cell.2013.04.025] [PMID: 23643243]

[60] Niu Y, Shen B, Cui Y, *et al.* Generation of gene-modified cynomolgus monkey *via* Cas9/RNA-mediated gene targeting in one-cell embryos. Cell 2014; 156(4): 836-43.
[http://dx.doi.org/10.1016/j.cell.2014.01.027] [PMID: 24486104]

[61] Boch J, Bonas U. *Xanthomonas* AvrBs3 family-type III effectors: Discovery and function. Annu Rev Phytopathol 2010; 48(1): 419-36.
[http://dx.doi.org/10.1146/annurev-phyto-080508-081936] [PMID: 19400638]

[62] Li T, Huang S, Jiang WZ, *et al.* TAL nucleases (TALNs): Hybrid proteins composed of TAL effectors and FokI DNA-cleavage domain. Nucleic Acids Res 2011; 39(1): 359-72.
[http://dx.doi.org/10.1093/nar/gkq704] [PMID: 20699274]

[63] Jinek M, Chylinski K, Fonfara I, Hauer M, Doudna JA, Charpentier E. A programmable dual-RN--guided DNA endonuclease in adaptive bacterial immunity. Science 2012; 337(6096): 816-21.
[http://dx.doi.org/10.1126/science.1225829] [PMID: 22745249]

[64] Kleinstiver BP, Pattanayak V, Prew MS, *et al.* High-fidelity CRISPR–Cas9 nucleases with no detectable genome-wide off-target effects. Nature 2016; 529(7587): 490-5.
[http://dx.doi.org/10.1038/nature16526] [PMID: 26735016]

[65] Kleinstiver BP, Prew MS, Tsai SQ, *et al.* Engineered CRISPR-Cas9 nucleases with altered PAM specificities. Nature 2015; 523(7561): 481-5.
[http://dx.doi.org/10.1038/nature14592] [PMID: 26098369]

[66] Cho SW, Kim S, Kim JM, Kim JS. Targeted genome engineering in human cells with the Cas9 RNA-guided endonuclease. Nat Biotechnol 2013; 31(3): 230-2.
[http://dx.doi.org/10.1038/nbt.2507] [PMID: 23360966]

[67] Churchman LS, Weissman JS. Nascent transcript sequencing visualizes transcription at nucleotide resolution. Nature 2011; 469(7330): 368-73.
[http://dx.doi.org/10.1038/nature09652] [PMID: 21248844]

[68] Cong L, Ran FA, Cox D, et al. Multiplex genome engineering using CRISPR/Cas systems. Science 2013; 339(6121): 819-23.
[http://dx.doi.org/10.1126/science.1231143] [PMID: 23287718]

[69] Deltcheva E, Chylinski K, Sharma CM, et al. CRISPR RNA maturation by trans-encoded small RNA and host factor RNase III. Nature 2011; 471(7340): 602-7.
[http://dx.doi.org/10.1038/nature09886] [PMID: 21455174]

[70] Gasiunas G, Barrangou R, Horvath P, Siksnys V. Cas9–crRNA ribonucleoprotein complex mediates specific DNA cleavage for adaptive immunity in bacteria. Proc Natl Acad Sci 2012; 109(39): E2579-86.
[http://dx.doi.org/10.1073/pnas.1208507109] [PMID: 22949671]

[71] Horvath P, Barrangou R. CRISPR/Cas, the immune system of bacteria and archaea. Science 2010; 327(5962): 167-70.
[http://dx.doi.org/10.1126/science.1179555] [PMID: 20056882]

[72] Deltcheva E, Chylinski K, Sharma CM, et al. CRISPR RNA maturation by trans-encoded small RNA and host factor RNase III. Nature 2011; 471(7340): 602-7.
[http://dx.doi.org/10.1038/nature09886] [PMID: 21455174]

[73] Jinek M, Chylinski K, Fonfara I, Hauer M, Doudna JA, Charpentier E. A programmable dual-RN--guided DNA endonuclease in adaptive bacterial immunity. Science 2012; 337(6096): 816-21.
[http://dx.doi.org/10.1126/science.1225829] [PMID: 22745249]

[74] Veits J, Weber S, Stech O, et al. Avian influenza virus hemagglutinins H2, H4, H8, and H14 support a highly pathogenic phenotype. Proc Natl Acad Sci 2012; 109(7): 2579-84.
[http://dx.doi.org/10.1073/pnas.1109397109] [PMID: 22308331]

[75] Saeidi N, Wong CK, Lo TM, et al. Engineering microbes to sense and eradicate *Pseudomonas aeruginosa*, a human pathogen. Mol Syst Biol 2011; 7(1): 521.
[http://dx.doi.org/10.1038/msb.2011.55] [PMID: 21847113]

[76] Hsu PD, Scott DA, Weinstein JA, et al. DNA targeting specificity of RNA-guided Cas9 nucleases. Nat Biotechnol 2013; 31(9): 827-32.
[http://dx.doi.org/10.1038/nbt.2647] [PMID: 23873081]

[77] Cong L, Ran FA, Cox D, et al. Multiplex genome engineering using CRISPR/Cas systems. Science 2013; 339(6121): 819-23.
[http://dx.doi.org/10.1126/science.1231143] [PMID: 23287718]

[78] Pardee K, Green AA, Ferrante T, et al. Paper-based synthetic gene networks. Cell 2014; 159(4): 940-54.
[http://dx.doi.org/10.1016/j.cell.2014.10.004] [PMID: 25417167]

[79] Brophy JAN, Voigt CA. Principles of genetic circuit design. Nat Methods 2014; 11(5): 508-20.
[http://dx.doi.org/10.1038/nmeth.2926] [PMID: 24781324]

[80] Liu CC, Qi L, Lucks JB, et al. An adaptor from translational to transcriptional control enables predictable assembly of complex regulation. Nat Methods 2012; 9(11): 1088-94.
[http://dx.doi.org/10.1038/nmeth.2184] [PMID: 23023598]

[81] Lohmueller JJ, Armel TZ, Silver PA. A tunable zinc finger-based framework for Boolean logic computation in mammalian cells. Nucleic Acids Res 2012; 40(11): 5180-7.

[http://dx.doi.org/10.1093/nar/gks142] [PMID: 22323524]

[82] Moon TS, Lou C, Tamsir A, Stanton BC, Voigt CA. Genetic programs constructed from layered logic gates in single cells. Nature 2012; 491(7423): 249-53.
[http://dx.doi.org/10.1038/nature11516] [PMID: 23041931]

[83] Siuti P, Yazbek J, Lu TK. Synthetic circuits integrating logic and memory in living cells. Nat Biotechnol 2013; 31(5): 448-52.
[http://dx.doi.org/10.1038/nbt.2510] [PMID: 23396014]

[84] Zhou J, Zhu T, Cai Z, Li Y. From cyanochemicals to cyanofactories: A review and perspective. Microb Cell Fact 2016; 15(1): 2.
[http://dx.doi.org/10.1186/s12934-015-0405-3] [PMID: 26743222]

[85] Shen CR, Liao JC. Photosynthetic production of 2-methyl-1-butanol from CO_2 in cyanobacterium *Synechococcus elongatus* PCC7942 and characterization of the native acetohydroxyacid synthase. Energy Environ Sci 2012; 5(11): 9574-83.
[http://dx.doi.org/10.1039/c2ee23148d]

[86] Ramey CJ, Barón-Sola Á, Aucoin HR, Boyle NR. Genome engineering in *Cyanobacteria*: Where we are and where we need to go. ACS Synth Biol 2015; 4(11): 1186-96.
[http://dx.doi.org/10.1021/acssynbio.5b00043] [PMID: 25985322]

[87] Gordon GC, Korosh TC, Cameron JC, Markley AL, Begemann MB, Pfleger BF. CRISPR interference as a titratable, trans-acting regulatory tool for metabolic engineering in the cyanobacterium *Synechococcus sp.* strain PCC 7002. Metab Eng 2016; 38: 170-9.
[http://dx.doi.org/10.1016/j.ymben.2016.07.007] [PMID: 27481676]

[88] Lan EI, Liao JC. Microbial synthesis of n-butanol, isobutanol, and other higher alcohols from diverse resources. Bioresour Technol 2013; 135: 339-49.
[http://dx.doi.org/10.1016/j.biortech.2012.09.104] [PMID: 23186690]

[89] Mimee M, Tucker AC, Voigt CA, Lu TK. Programming a human commensal bacterium, *Bacteroides thetaiotaomicron*, to sense and respond to stimuli in the murine gut microbiota. Cell Syst 2015; 1(1): 62-71.
[http://dx.doi.org/10.1016/j.cels.2015.06.001] [PMID: 26918244]

[90] Whitaker WR, Shepherd ES, Sonnenburg JL. Tunable expression tools enable single-cell strain distinction in the gut microbiome. Cell 2017; 169(3): 538-546.e12.
[http://dx.doi.org/10.1016/j.cell.2017.03.041] [PMID: 28431251]

[91] Qi LS, Larson MH, Gilbert LA, *et al.* Repurposing CRISPR as an RNA-guided platform for sequence-specific control of gene expression. Cell 2013; 152(5): 1173-83.
[http://dx.doi.org/10.1016/j.cell.2013.02.022] [PMID: 23452860]

[92] Choi KR, Jang WD, Yang D, Cho JS, Park D, Lee SY. Systems metabolic engineering strategies: Integrating systems and synthetic biology with metabolic engineering. Trends Biotechnol 2019; 37(8): 817-37.
[http://dx.doi.org/10.1016/j.tibtech.2019.01.003] [PMID: 30737009]

[93] Zhang S, Guo F, Yan W, *et al.* Recent advances of CRISPR/Cas9-based genetic engineering and transcriptional regulation in industrial biology. Front Bioeng Biotechnol 2020; 7: 459.
[http://dx.doi.org/10.3389/fbioe.2019.00459] [PMID: 32047743]

[94] Jin FJ, Wang BT, Wang ZD, Jin L, Han P. CRISPR/Cas9-Based genome editing and its application in *Aspergillus* species. J Fungi 2022; 8(5): 467.
[http://dx.doi.org/10.3390/jof8050467] [PMID: 35628723]

[95] Liu Q, Gao R, Li J, *et al.* Development of a genome-editing CRISPR/Cas9 system in thermophilic fungal *Myceliophthora* species and its application to hyper-cellulase production strain engineering. Biotechnol Biofuels 2017; 10(1): 1-14.
[http://dx.doi.org/10.1186/s13068-016-0693-9] [PMID: 28053662]

[96] Cui L, Bikard D. Consequences of Cas9 cleavage in the chromosome of *Escherichia coli*. Nucleic Acids Res 2016; 44(9): 4243-51.
[http://dx.doi.org/10.1093/nar/gkw223] [PMID: 27060147]

[97] Molla KA, Yang Y. CRISPR/Cas-mediated base editing: Technical considerations and practical applications. Trends Biotechnol 2019; 37(10): 1121-42.
[http://dx.doi.org/10.1016/j.tibtech.2019.03.008] [PMID: 30995964]

[98] Rees HA, Liu DR. Base editing: Precision chemistry on the genome and transcriptome of living cells. Nat Rev Genet 2018; 19(12): 770-88.
[http://dx.doi.org/10.1038/s41576-018-0059-1] [PMID: 30323312]

CHAPTER 4

Genome Editing of Plant Growth-Promoting Microbes (PGPM) Towards Developing Smart Bio-Formulations for Sustainable Agriculture: Current Trends and Perspectives

Sugitha Thankappan[1,*], Asish K. Binodh[2], P. Ramesh Kumar[1], Sajan Kurien[1], Shobana Narayanasamy[3], Jeberlin. B. Prabina[4] and Sivakumar Uthandi[3]

[1] *School of Agricultural Sciences, Karunya Institute of Technology and Sciences (Deemed to be University), Coimbatore-641114, Tamil Nadu, India*

[2] *Centre for Plant Breeding and Genetics, Tamil Nadu Agricultural University, Coimbatore-641003, Tamil Nadu, India*

[3] *Department of Microbiology, Tamil Nadu Agricultural University, Coimbatore-641003, Tamil Nadu, India*

[4] *Department of Soil Science, Agricultural College and Research Institute, Killikulam, Vallanad Post, Tuticorin Dt-628252, Tamil Nadu, India*

Abstract: Plant-associated microbes, referred to as plant microbiomes, are an integral part of the plant system. The multifaceted role of plant microbiota in combating both abiotic and biotic stresses is well documented in different crop species. However, understanding the co-evolution of plant growth- promoting microbes (PGPM) and PGP traits at genetic and molecular levels requires robust molecular tools to unravel the functional gene orthologues involved in plant-microbe interaction. The advent of Clustered Regularly Interspaced Short Palindromic Repeats (CRISPR)/Cas9 (CRISPR-associated protein 9) is of paramount importance in deciphering the plant-microbe interaction and addressing the challenges of unraveling endophytic microbes and their benefits thereof. Our knowledge of plant microbiome composition, signaling cues, secondary metabolites, microbial volatiles, and other driving factors in plant microbiome has been enlightened. In recent years, scientists have focused more on below-ground dialogue in recruiting efficient microbiome/engineered rhizosphere. More recently, base editing techniques using endo-nucleolytic ally deactivated dCas9 protein and sgRNAs (CRISPR interference or CRISPRi) have emerged as a useful approach to study the gene functions and have potential merits in exploring plant-microbe interactions and the signaling cues involved. A systemic understanding of the signaling events and the respective metabolic pathways will enable the application of

---

[*] **Corresponding author Sugitha Thankappan:** School of Agricultural Sciences, Karunya Institute of Technology and Sciences (Deemed to be University), Coimbatore-641114, Tamil Nadu, India; E-mails: akbinodh@gmail.com; kumarisugitha@karunya.edu

Prakash M. Halami & Aravind Sundararaman (Eds.)
All rights reserved-© 2024 Bentham Science Publishers

genome editing tools to enhance the capacity of microbes to produce more targeted metabolites that will enhance microbial colonization.

Further, it will be exciting to employ CRISPR technologies for editing plant-microbe interactions to discover novel metabolic pathways and their modulation for plant immunity and fitness against abiotic as well as biotic stresses. Such metabolites possess tremendous scope in tailoring newer smart nano-based bio-formulations, besides formulating beneficial microbiomes or cocktails, which is the best alternative for climate resilient farming. The present review sheds light upon the deployment of CRISPR/Cas techniques to comprehend plant-microbe interactions, microbe-mediated abiotic and biotic stress resistance, genes edited for the development of fungal, bacterial, and viral disease resistance, nodulation process, PGP activity, CRISPR interference-based gene repression in the PGPM, metabolic pathway editing and their future implications in sustainable agriculture.

Keywords: CRISPR/Cas, Genome editing, Plant-microbe interaction, Sustainable agriculture.

INTRODUCTION

Plant-microbe interaction plays an integral role in sustaining diverse ecosystem services and sharing common ancestors, however, their survival is interdependent. The term 'Plant microbiota' or 'Plant microbiome [PM] 'has gained more significance. Microbes reside inside roots and shoots as endophytes, besides occupying the rhizosphere, phyllosphere, and spermosphere [1 - 3]. The unique endobiome of apoplastic fluid [4], guard cell [5], and nodule niche [6] in conferring plant fitness created new vistas in the route map of plant-microbe interaction studies. In general, plant-microbe trade-offs lead to unique partnerships depending on their impact on plant health and fitness, *i.e.*, mutualistic [7], neutral, commensalistic, or harmful [7, 8]. The plant-microbe interactions are bi-directional where microbes derive their nutrients from the host plants and *vice versa*. The plant system produces a nutritionally enriched environment with primary and secondary metabolites (inorganic and organic compounds), which is favorable for diverse microbial colonization. Conversely, the microbiome assists the plant to acclimatize fluctuating environmental conditions. The mechanisms include: promoting plant growth, protection against biotic and abiotic stresses, priming the immune system or induction of defense pathways, mycorrhizal symbiosis, nutrient uptake, and conversion of the unavailable nutrients into plant-accessible forms [9]. More precisely, as a direct mechanism, plant beneficial microbiome enhances plant growth through biological nitrogen fixation, phosphorous uptake, and production of phytohormones, specifically, indole-3-acetic acid (IAA), gibberellic acid (GA) and cytokinins [10 - 12]. As an indirect mechanism, plant-beneficial microbes suppress plant pathogens by producing antimicrobials and promoting induced systemic resistance in plants [13 - 15]. In contrast, many plant-pathogenic microorganisms cause devastating diseases in various crops. These PM interactions are crucial in sustainable agriculture and the environment, for food security and plant health management [16]. Consequently, profiling plant-associated microbiome [genome assemblies of all microbes] is a dawning concept in the field of molecularplant–microbe interactions. An investigation of the host plant together with the associated microbiome (holobiont) suggests the co-evolution ofplant–microbe, plant–plant, andmicrobe–microbe interactions [17]. Recent studies on plant–microbe interactions detailed Avr protein, computational strategies for protein interactions, molecular diversity, and interactions of virulence genes [18].

Next-generation sequencing [NGS], omics approaches [metagenomics, transcriptomics, proteomics, metabolomics], and other computational tools using system biology approaches shed light on the molecular aspects of plant-microbe interactions governing plant traits. Gene-level understanding of plant traits and associated microbes will be a crucial step towards unraveling microbiomes for sustainable agriculture [19, 20]. Anent to this, modern revolutionary techniques induce precise genetic modifications such as clustered regularly interspaced short palindromic repeats-based genome editing, which is an ideal platform to understand plant-microbe interactions for improving crop productivity and priming resistance [21, 22]. This review envisages the various factors shaping the formation of plant microbiota, and the applications of CRISPR-based tools in the beneficial [symbiotic] or harmful [pathogenic] plant-microbiome interactions for sustainable agricultural practices. The limitations and prospects of genome editing tools to alleviate abiotic and biotic stresses are also discussed.

FACTORS INFLUENCING PLANT-MICROBE INTERACTIONS AND THEIR COMPOSITION

In general, microbial assemblage in plants is determined by assembly rules, where plant-associated factors prefer the growth of a particular set of microbes inhibiting others [9]. Another example is the sequential decrease in microbial diversity from bulk soil to rhizosphere. The rhizosphere and phyllosphere communities vary significantly with plant species [23]. Both the biotic and abiotic factors exert profound effects on plant-microbe interaction, microbial community structure, and composition.

Biotic Factors

Plant factors include geographical location, host genotype, age, root phenomics, root and plant secretomes, and the inherent immune system, of which many reports have confirmed plant genotype as a major intriguing factor governing microbial composition [24]. Plant genotype decides the root metabolome that acts as chemical cues and ascertains the nutrient availability for microbes [19]. For instance, coumarins and lectin compounds influence host microbiota and root microbiome besides acting as semiochemicals [25]. More recently, our studies also showed that indole compounds, hydroxyl coumarin, and galactopyranose act as signaling molecules in recruiting beneficial microbes in rice [6, 26]. Plant volatiles also act as cues in tailoring the core microbiome [27]. The interaction of plant–plant, microbe–microbe, and plant-microbe determines the distinct microbiome assemblies [19]. Apart from symbiotic microbial abundance in the rhizosphere and phyllosphere, plant-pathogenic microbes alter the population of antagonistic microbes as well as plant immune responses [28, 29]. The impact of intensive agriculture on microbiome structure and composition is tremendous. Previous reports demonstrated the effect of agricultural practices, including pesticide application, indiscriminate fertilizer doses, cultivation practices, and other anthropogenic interventions in plant-microbe assemblages leading to disturbance in soil structure [30, 31]. The composition of root microbiota in petunia and *Arabidopsis* varied significantly in response to phosphorus application [32]. Likewise, changes in leaf microbiota in response to nitrogen were observed in maize and soybean [33, 34].

Abiotic Factors

Soil shapes the microbial composition in the rhizosphere and acts as a microbial seed bank. Physico-chemical properties such as soil pH, texture and structure, macro and micronutrients, soil organic matter (SOM), porosity, salinity, and moisture content drive the microbial community and its composition [35, 36]. Interestingly, different plant species/genotypes recruit distinct microbial communities in the rhizosphere, rhizoplane, and endosphere. In contrast, certain plant species/genotype recruits the appropriate microbial community irrespective of environmental and soil conditions, which are known as the **core plant microbiome** [37]. Environmental factors such as climate, light, water, ultraviolet radiation, and geographic location significantly influence the phyllosphere microbiota [38 - 41]. It can be concluded that the plant phenome is the outcome of plant genotype-microbiota-environment [GxMxE] interactions instead of GxE [42]. Crop improvement scientists are now keen on the core microbiome, and their role in breeding resilient genotypes is being studied.

In general, plant microbiome is acquired either vertically [through seed, propagation material] or horizontally [through soil, air], and resides on or inside the plant tissues. Plant microbiomes may exist either as a free-living or an endophytic state. Such endophytes represent soil microbiota that colonizes the internal tissues of the plants. In order to colonize host plants, the soil bacteria must be competent to bind to the root surface and to establish in plant tissues.

Upon invading the plant, the endophytes are covered by a cell membrane and become either intracellular or extracellular [43]. Bacterial cells that transform from free-living to endophyte require motility and secretion of extracellular enzymes like cellulases and pectinases. However, the colonization of endophytes could not induce cellular damage or cause detrimental effects to the plants. Besides, endophytic colonization in plants, it mediates biotic and abiotic stress alleviation in their respective hosts [44]. The endophytic microbiome arm diverse metabolic pathways than those simply colonizing the rhizosphere [45].

SIGNIFICANCE OF PLANT-MICROBE INTERACTION IN SUSTAINABLE AGRICULTURE

Healthy Plant-microbe Interactions

Plant microbiome consists of beneficial, neutral, or pathogenic microbial species. Among them, the beneficial plant microbiota is vital for plant growth, flowering, crop fitness and yield directly or indirectly [24, 46, 47]. Plant–microbe interactions regulate soil carbon sequestration by modulating the terrestrial carbon cycle and driving the impact of climate changes on agriculture [48]. The consequence of a specific plant-microbe interaction is beneficial under a distinct set of conditions, whereas damaging on the other side [49]. For instance, symbiotic interactions of legumes with N-fixing rhizobia and arbuscular mycorrhizal [AM] fungal taxa help host plants to access N and P, respectively, under nutrient-deficit environments [49, 50]. More recently, certain non-rhizobial endophytes [NRE], such as yeast and bacteria, in assisting nodulation by N-fixing rhizobia were reported in urd bean [6, 51]. Plant growth-promoting bacteria produces phytohormones [auxin, cytokinin, gibberellin] and plant-beneficial enzymes [1-aminocyclopropane-1-carboxylate deaminase], whereas some are beneficial under heavy-metal stress through enhanced uptake, and detoxification by either or both the partners, *i.e.*, plant or microbe [52].

Harmful Plant-microbe Interactions

Despite beneficial plant microbiota, some microbes are harmful, causing disease symptoms in plants. Plant-pathogenic microbes (*e.g.*, *Pseudomonas syringae*, *Erwinia amylovora*, *Ralstonia solanacearum*, *Xanthomonas* sp., and *Xylella fastidiosa*) infect the plant tissues through natural openings or wounds for nutrient acquisition and trigger the immune responses [53]. Factors like population size, host vulnerability, climate, and biotic factors like plant microbiota regulate the outcome of plant–pathogen interaction [54]. Few non-pathogenic microbes act as pathogens under some circumstances when there is an alteration in microbial population or change of host plants [55]. Therefore, for obtaining precise information on plant-microbe interactions at the molecular level, modern tools are considered an ideal platform.

Plant-Pathogen Interaction

Epigenetics approach is the 'bridge between genotype and phenotype' during development which constitutes chromatin remodeling, DNA methylation, and siRNA. Histone remodeling includes acetylation, methylation, phosphorylation, ubiquitination, sumoylation, carbonylation, and glycosylation, which, in turn, results in the repression or expression of specific genes [56]. RNA-dependent DNA methylation, a special event in epigenetic resistance, ensures resistance against viral diseases. DNA methylations impart disease resistance against various plant diseases. Plant Transcriptional Gene Silencing [PTGS] is the most important epigenetic approach for the management of plant diseases. However, epigenetics is an emerging field in plant pathology for disease management.

Innate immunity is the only line of defense in invertebrates and plants [57]. The general elicitors of plant pathogens that resemble pathogen-associated molecular patterns [PAMPs] are recognized as molecules involved in triggering innate immunity [58]. Microbe-associated molecular patterns [MAMPs] are molecules that bind to pattern recognition receptors [PRRs] and trigger innate immune responses in plants [59]. MAMPs in plant defense include eubacterial flagellin, elongation factors, lipopolysaccharides [LPS] from gram-negative bacteria, peptidoglycans from gram-positive bacteria, viral and bacterial nucleic acids, and fungal cell wall-derived chitins, glucans, mannans, and proteins [60, 61]. MAMPs are perceived by plants as danger cues and trigger a network of signaling systems activating defense responses. PRRs are regarded as specific receptors in the plant cell plasma membrane for the recognition of PAMPs [62]. Activated PRRs in the plant system further trigger various genes, including BAK1 [Brassinosteroid insensitive1 Associated Kinase1], BIK1 [Botrytis-Induced Kinase1], 5 BIR1 [Branching Inhibiting Receptor1] [63] followed by downstream activities: i] ROS production by the activation of NAD kinase; ii] Activation of WRKY transcription factors [64] and nitric oxide production [key enzyme in the biosynthesis of phenolics] [65] iii] chalcone synthase [CHS] involved in the synthesis of isoflavonoid phytoalexins and transcription is modulated by NO [66]; iv] Activation of SID2 gene [isochorismate synthase], responsible for salicylic acid signaling system; v] Activation of NADPH oxidase, responsible for jasmonic acid signaling system; vi] Activation of ACC synthase and oxidase responsible for ethylene signaling system; vii] Activation of Mitogen Activated Protein Kinase [MAPK], responsible for various signaling systems involved in plant defense [67].

Microbe Induced Systemic Tolerance [MIST] for Enhanced Crop Resilience

Plant-microbial interactions are the prime keys to the adaption and tolerance of plants under adverse environmental conditions. Microbe-mediated biotic or abiotic stress responses in crop plants are commonly termed induced systemic tolerance. Plant-associated microbiomes with their inherent intrinsic genetic and metabolic attributes aid in mitigating abiotic stress. Plant-associated microbes of the genera *Bacillus* [26, 68], *Pseudomonas* [69], *Azospirillum* [70], *Enterobacter,* and *Methylobacterium* [71] are known for stimulating plant growth and stress resilience in crop plants. The plausible plant-associated bacteria-induced stress resilience mechanism includes 1) root architecture modification, 2) phytohormones like gibberellic acid, cytokinin, auxins, and abscisic acid [ABA], 3) ACC deaminase to reduce the level of ethylene stress, 4) microbial volatile organic compounds (mVOCs) induced systemic tolerance, 5) exopolysaccharides (EPS) [72 - 78]. Plant-associated microbes also increase stress tolerance in plants by induction and expression of drought stress-responsive genes, synthesis of antioxidants, and osmolytes such as proline, sugar alcohols, betaine, pyrimidine derivatives, and metabolites [26, 79]. However, PGPM-mediated moisture stress resilience is highly influenced by the intensity and duration of the stress as well as the developmental stage of plants during drought exposure [80]. The mechanisms are detailed in sections forward and represented in Fig. (**1**).

Osmolytes Tussles for Stress Resilience

Abiotic stress alters the osmolality concentration in plants, which hinders their growth and survival. Under drought stress, osmotic adjustment is one of the cellular adaptations that aid in plant tolerance [81], such as the accumulation of compatible solutes such as proline, sugars, betaines, polyamines, quaternary ammonium compounds, moisture stress-related protein (dehydrins), and amino acids. Osmolytes accumulation lowers the cell water potential, maintains the cellular turgor, and protects the membrane proteins /enzymes, and cellular organelles against oxidative stress [82].

Proline is a proteinogenic amino acid accumulated during drought stress. It acts as a molecular chaperone by quenching ROS, lowering lipid peroxidation, altering the cytosolic acidity, and also guards the sub-cellular structures [83]. Proline accumulation under oxidative stress is associated with drought tolerance which was reported in many plants. Priming plants with PGPM alters the proline level, and ensures the survival of plants under drought stress. An increase in proline levels under moisture stress as a result of PGPM inoculation is previously reported [*e.g. Bacillus* sp. in maize, and *Bacillus altitudinis* in rice plants] [26, 77].

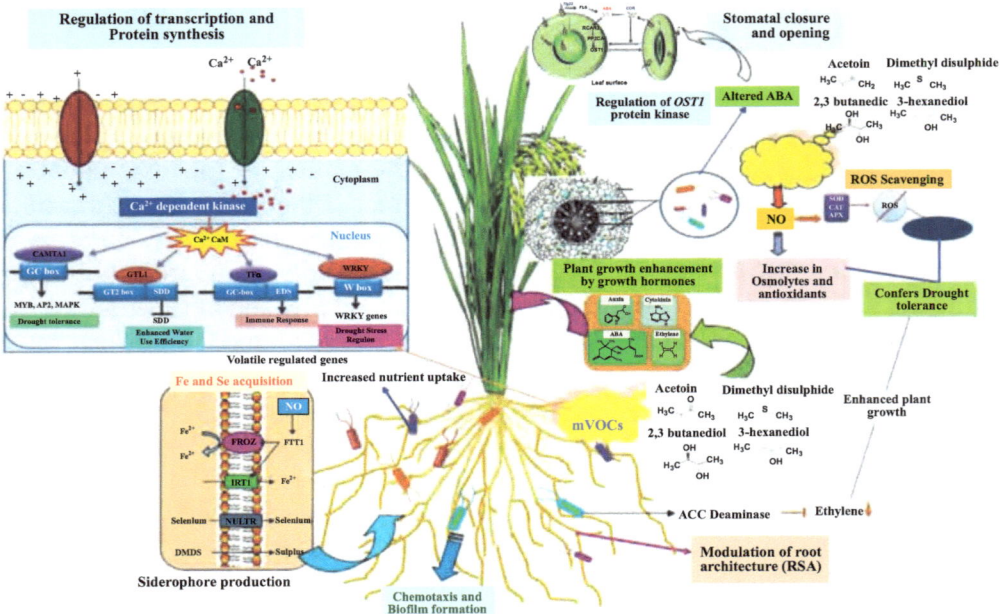

Fig. (1). Proposed mechanisms of microbe-mediated plant growth promotion and induced systemic tolerance (Source: Schematic representation by authors).

Trehalose is an osmoprotectant that plays a crucial role in the stabilization of subcellular structures and cell signaling [84]. *Phaseolus vulgaris* nodules primed with *Rhizobium*, overexpressed trehalose-6-phosphate synthase gene, and upregulated the drought stress-responsive genes [85]. Glycine betaine stabilizes membrane proteins and enzymes under induced drought stress. For example, *Bacillus subtilis* augmented *Arabidopsis* plants showed an increase in the choline and glycine betaine content in contrast to untreated control [86].

Furthermore, critical metabolites such as polyamines [cadaverine, spermidine [Spd], and putrescine] and amino acids enhance the osmotic tolerance of host plants against moisture stress. In a study, *Arabidopsis* primed with Spd-producing *Bacillus megaterium* BOFC15, increased the polyamines at cellular levels during oxidative stress. PGPM-primed plants possessed a robust root system with more lateral roots and an increase in primary root length that aid in the survival of plants under moisture stress [87]. Consequently, PGPR-induced plants exhibited an increased level of soluble sugars and free amino acids which is a complex mechanism unraveled in chickpeas [88].

Antioxidant Gadgets

Reactive Oxygen Species [ROS] is a metabolic by-product under ideal conditions. However, overproduction and detoxification of ROS in plants as a result of increased photorespiration cause skewness, which is a crucial shift during stress. The consequences of abiotic stress are the increased production of diverse ROS, that includes singlet oxygen [$^1O_2$], hydrogen peroxide [H_2O_2], superoxide radical [O_2^-], and hydroxyl radical [$HO^\cdot$] [89]. ROS affects the natural plant metabolic functions through oxidative stress to proteins, macromolecules, and other lipids which leads to it cell death [81, 90]. While the increased level of ROS triggers oxidat ive stress, the plant also requires a lower level of ROS, which is essential in signaling events to activate the defense pathways [77]. PGPM priming of plants against drought resilience altered the antioxidant levels through the MIST mechanism in *Solanum tuberosum* including maize [77, 91], wheat [92], and rice [93]. Plants primed with PGPR, trigger enhanced production of ROS scavenging enzymes, which in turn reduce the overproduced ROS, thereby conferring drought resilience. The activity of reactive oxygen species [ROS]-quenching enzymes, such as catalase, superoxide dismutase, ascorbate peroxidase, and peroxidase constitutively increased in rice treated with co-inoculants of *Bacillus altitudinis* FD48 and *Bacillus methylotrophic*us RABA6, as well as in green gram inoculated with *B. subtilis* EPB and *P. fluorescens* Pf1 [26, 94].

Root System Architecture

Plant roots are vital for nutrient-water absorption and transport, anchorage, and symbiotic relationships by recruiting beneficial microflora in the soil [95]. The root system encompasses the topology, spatial diffusion of primary and secondary roots, and changes in the number, volume, and diameter of roots. Root plasticity in plants in response to water deficit is a powerful tool to cope the drought stress [80]. PGPB alters the root growth, and elasticity, which modifies the root architecture. These bacteria-induced alterations in root structure increase the root surface area and consequently enhance the water and nutrient absorption, which has a significant influence on drought tolerance. In an experiment, *Alcaligenes feacalis* AF3 inoculated maize showed enhanced root length (10%) compared to non-inoculated plants under drought stress. Further, biotization of *B. altitudinis* FD48 modified root system architecture in rice plants and it was consequently confirmed by the expression pattern of defense-responsive genes encoding primary root formation such as *OsIAA1*, *OsIAA4*, and lateral root formation [*OsIAA11*, *OsIAA13*]. FD48 modified the endogenous IAA levels in rice by regulating the auxin-responsive genes, and thereby regulates the root system architecture [26, 96]. Several studies evidenced the modulation of rice root system traits by triggering mitogen-activated protein kinase6 and auxin signaling pathways [97, 98]. PGPRs such as *Bacillus megaterium, Serratia marcescens,* and *Pseudomonas* strains were reported to alter the root architecture in *Arabidopsis* [99].

Phytohormone Modulation and Cross-Talk: Cues in the Battle

Phytohormones pronounced by PGPM trigger the endogenous hormone level in plants, which plays a crucial role in enhancing stress tolerance.

Auxins

Auxins are accountable for lateral root development, cell division, apical dominance, and shoot and root growth orientation [100]. Approximately 80% of the PGPR produces IAA and consequently, helps in plant growth and development [101]. Likewise, *Trifolium repens* L. primed with *Pseudomonas putida* and *B. megaterium* enhanced the root and shoot biomass under drought stress. These changes are consistent with enhanced production of IAA [102]. According to Bresson *et al.*, *Arabidopsis* plants inoculated with *Phyllobacterium brassicacearum* STM196 exhibited increased root length and modulated root system architecture [103]. Besides, inoculation of auxin producing, *Azospirillum* in tomato and wheat seedlings showed enhanced root growth and lateral root formation that increased the nutrient and water uptake under moisture stress [104]. Auxin signaling is associated with the activation of mitogen-activated protein kinase 6 [MAPK6] in *Trichoderma* inoculated *Arabidopsis* plants [97].

Abscisic Acid [ABA]

Abscisic acid plays a pivotal role in the physiological processes and is essential for the survival of plants under abiotic stress such as drought and salinity [81]. ABA functions as an anti-transpirant by regulating the expression of drought stress-responsive genes under osmotic stress. It also mediates cell signaling, maintains shoot and root hydraulic conductivity, elicitation of stronger resistance responses, and stimulates stomatal closure [105]. ABA alters the root system in a stressed environment and creates deeper roots, followed by other changes in the auxiliary roots, intruding acquisition of nutrients and water for optimum plant growth. Aroca and his co-workers proposed that ABA could improve drought resilience through leaf transpiration control and hydraulic root conductivity [106]. Another theory states that ABA enhances tolerance to drought by improved aquaporin control [107]. These two hypotheses evidenced that ABA plays a vital role in plant growth and enhanced drought tolerance. ABA accumulation in *Arabidopsis* plants primed with *Phyllobacterium brassicacearum* strain STM196 altered the drought stress responses and resulted in reduced leaf transpiration [108]. *Bacillus* sp. primed in lettuce showed an elevated amount of ABA than the mock plants, thereby increasing the moisture stress tolerance [104]. Similarly, *Azospirillum brasilense* Sp245 treated *Arabidopsis* plants showed higher concentrations of ABA accumulation and enhanced tolerance to drought stress [109].

In plant-pathogen interaction, ABA acts as an effector molecule and immune regulator. ABA has a negative effect on plant pathosystems, such as the *Botrytis cinerea*—tomato [110], *Ralstonia solanacearum*—tobacco [107], *Plectosphaerella cucumerina*—*Arabidopsis* [111], and *Magnaporthe oryzae*—barley [112]. On the contrary, ABA plays an important role in a few mutualistic plant-microbe interactions with arbuscular mycorrhizal [AM] fungi and several root-associated bacteria [113, 114]. It has been demonstrated that ABA suppresses immune responses, but immune responses are also able to suppress ABA signaling. In this context, very early pathogen response signaling negatively regulates ABA responses [115].

ABA exhibits SA antibiosis against the salicylic acid hormone pathway [116]. Hence, the plants need to prioritize between many different environmental cues for an appropriate response, which could explain some of the antagonistic effects between different signaling pathways [117]. Emerging evidence on the complex interactions between multiple plant hormones and the disease resistance hormones [SA, jasmonic acid [JA], and ethylene [ET]] signaling pathways further shed light on the different plant response conflicts [118]. In general, SA is involved in resistance against biotrophic pathogens and JA is involved in resistance against necrotrophic pathogens or insects. As a part of the antagonism between resistance against biotrophs and necrotrophs, SA suppresses JA responses at multiple levels [119]. Ethylene transporters [ET], on the other hand, can enhance or influence both SA and JA responses [120]. While the SA/ABA antagonism is relatively clear in many models, the ABA influence on JA signaling is more complex. The JA/ABA regulates a different set of JA responses compared to the JA/ET branch; however, these two branches are mutually exclusive. The JA/ET branch responses rely on the ethylene-responsive transcription factors [ERF], leading to the expression of defensins and resistance against necrotrophs [121]. Similarly, the JA/ABA responses rely on the transcription factor MYC2, which regulates wounding responses, insect resistance, and suppression of JA/ET-dependent innate immunity against necrotrophs. The genes downstream of MYC2 are influenced directly by ABA through the physical interactions with one of the intracellular ABA receptors [122].

Apart from the immune suppressive effects of ABA, it can also play a positive role in pathogen resistance. One of the first indications of a positive influence from plant ABA signaling on biotic stress resistance was the reliance of ABA for a β-aminobutyric acid [BABA]-induced priming for pathogen resistance. ABA is often a positive regulator of the callose-dependent disease resistance responses. Callose is a rapidly formed β-glucan barrier, which is also in conflict with SA-dependent disease resistance responses [123]. Also, the other kinds of physical barriers interact with ABA and influence the plant disease resistance. For instance, ABA- dependent resistance responses against *R. solanacearum* were constitutively up-regulated in certain cellulose synthase mutants of *Arabidopsis* [124]. Further, ABA also plays a positive role together with JA in the resistance to *Sclerotinia sclerotiorum* in *Arabidopsis*, besides the JA/ABA pathway identified as an important one for resistance against insects. An antagonistic action of ABA on ET-dependent infection has also been suggested as a mechanism in rice resistance against *Cochliobolus miyabeanus*. However, the ABA-ET antagonism is also important for the establishment of arbuscular mycorrhiza, whereas ET suppresses successful colonization in ABA-deficient plants, most likely due to the activation of JA/ET-dependent disease resistance response against the invading fungus [114]. As an alternative mechanism, ABA can also positively influence disease resistance by regulating the stomatal closure in order to deny the pathogen entry into the plants [125].

Interestingly, the JA-dependent pathway can sometimes antagonize this ABA-induced stomatal closure, which provides another example of opposing effects from ABA and other plant disease resistance hormones. Pathogens such as *Pseudomonas syringae* and various fungi can "hijack" the inherent antagonism between different resistance pathways in order to promote the infection by using plant hormones or plant hormone-like compounds as effector molecules [126, 127].

Gibberellins

Gibberellins regulate the plant's physiological functions at different growth stages such as seed germination, flowering, fruiting, stem elongation, and senescence. Priming with GA-producing bacteria, it also ameliorates the drought stress effects in host plants. For example, GA-producing *Azospirillum lipoferum* and *P. putida* H-2-3 in maize mitigated drought stress [128]. Similarly, the exogenous application of GA enhanced the root traits such as root length, root surface area, and the number of root tips. Improved root traits in turn increased the nutrient absorption, which is responsible for altering the plant functions in stress environments [129].

Cytokinins

Cytokinins are critical hormones involved in cell division, photosynthetic activity, stomatal opening, and closure during drought. Apart from plants, the cytokinins can also be synthesized by a wide range of soil bacteria and PGPB. The plants biotized with cytokinin-producing plant-associated PGPB showed a positive effect on plant survival under water deficit conditions [108, 130].

ACC Deaminase

Ethylene acts as an essential modulator of plant fitness and development, as well as a key factor in the response of plants to a wide range of stresses. However, increased concentration of ethylene is detrimental to normal plant growth and development, as it induces leaf abscission, senescence, and chlorosis. ACC produced by plant-associated bacteria reduces the effect of ethylene in plants under stress, which helps in plant health. In ethylene biosynthesis, S-adenosylmethionine [S-Adomet] is converted into 1-aminocylopropane-1-carboxylate, which serves as a precursor of ethylene [131]. ACC deaminase [*acdS*] produced by *Rhizobacteria* cleaves ACC into ammonia and α-ketobutyrate, when the ethylene concentration is at higher levels. Consequently, the amount of ethylene is reduced, and thereby eliminating the deleterious effect on plants [132].

Microbial Volatiles [Mvocs] in Plant-microbe Interaction

The plant-associated and rhizosphere-inhabited microbes facilitate the plant to cope with biotic and abiotic stresses by various mechanisms. Studies have demonstrated that mVOCs [microbial Volatile Organic Compounds] emitted by rhizospheric microbiomes are -non-soluble metabolites. mVOCs are promising candidates for the mitigation of stress during plant growth and development. In a study, wheat seedlings grown under drought stress and exposed to volatiles emitted by *Bacillus thuringiensis* AZP2 exhibited five-fold higher survival under drought stress. Significant increases in plant biomass and increased photosynthesis rate were also observed [76]. 'Benzaldehyde, β-pinene, and geranyl compounds' emitted by *B. thuringensis* increased the drought tolerance in wheat compared to non-primed plants. Likely, mVOCs such as 2R, 3R-butanediol emitted by *Pesudomonas chlororaphis* O6 induced drought resistance in *Arabidopsis* through stomatal closure, besides inducing hormone signaling pathways. Further studies disclosed that the production of nitric oxide (NO) and hydrogen peroxide was triggered by 2,3-butanediol in plants, wherein NO signaling exerts drought tolerance [133]. In addition, mVOCs produced by *Bacillus subtilis* GB03 such as 2,3-butanediol showed higher levels of PEAMT transcripts, which is essential for choline and glycine betaine biosynthesis. Certain mVOCs like acetic acid produced by bacteria stimulated the biofilm formation where the exopolysaccharides are more prevalent [134]. In another study, *Bacillus safensis* W10 and *Ochrabacterium* IP8 inoculation in wheat revealed elevated levels of antioxidants and metabolites, suggesting their significant role in drought mitigation [135]. Proline accumulation and expression of ROS scavenging enzymes were triggered in PGPR-treated potato plants, which enhanced tolerance against various abiotic stresses including salinity, drought, and heavy metal toxicity [91]. Further, mVOCs play a crucial role in plant-pathogen interaction also. In our previous studies, the mVOC blend of *Trichoderma longibrachiatum* such as cedrene, longifolene, and longipinene, *etc.*, pronounced growth promotion and induced systemic resistance [ISR] in rice [27].

Exopolysaccharide [EPS] Production

EPS produced by rhizosphere bacteria have been well explored, and their role in moisture-holding ability was contemplated [136]. The two forms of EPS comprise capsular EPS and slime materials. EPS are absorbed in clay surfaces through cation bridges, Vander Waals force, anion absorption, and hydrogen bonding mechanisms. EPS produced by rhizosphere bacteria improved the soil impermeability, and soil aggregation and eventually sustained elevated water potential across the root region [137]. The nutrient uptake was improved which inturn influenced the plant growth and drought stress tolerance. *Pseudomonas* sp. strain GAP-P45 inoculated sunflower seedlings, when exposed to moisture stress was found to have more survival rate, due to increased plant biomass and root/shoot [RS] ratio. The structural and biochemical properties of the microbial polysaccharides (matrixome provide biofilm properties including 'adhesion, spatial and chemical heterogeneities) possess, synergistic/competitive polymicrobial interactions, antimicrobial recalcitrance, and biofilm virulence' [138].

TOOLS TO EXPLORE PLANT-MICROBE INTERACTIONS

Plant-microbe interactions are complex within the cellular system of the plants. Nevertheless, a plant-specific microbiome is a suitable approach for agricultural application, since the plant-associated microbiota influences the host's phenotype. More precisely, understanding the microbial and plant genes involved in plant-microbe interactions is vital for more insights. Molecular biology, omics tools (genomics, transcriptomics, proteomics, metabolomics), and NGS technologies have been employed to elucidate the plant microbiomes (both PGPM and plant-associated pathogens) and their regulatory metabolic networks [139]. A shift in the composition of the beneficial PGPM community was reported in the plant–pathogenic interactions [140]. Advances in high – end instrumentation, computational integration, and, data analytics, decode the specific signal molecules, proteins, genes, and gene cascades. Moreover, it is easier to understand the gene editing, RNAi- mediated gene silencing, gene mutation technology, proteomic analysis, and metabolite profiling to disclose massive molecular data that provides more insights on microbe- mediated abiotic stress mitigation in plants [141]. However, efforts are persistent in the development of durable resilience to biotic and abiotic stresses. The recent genome editing tool, Clustered Regularly Interspaced Short Palindromic Repeats [CRISPR]/Cas9 [CRISPR-associated protein 9] is highly prominent for understanding the functional genes in plants.

Genome editing (GE) creates a double-strand break [DSB] using engineered endonucleases that undergo DNA repair and generate mutations. DSB repair mechanisms occur through two major pathways: a] non-homologous end-joining (NHEJ); b] homology-directed repair (HDR). HDR is more precise than NHEJ, which is applicable in specific donor-dependent gene replacement. Targeted genetic modifications are accomplished through three mega-nucleases (site-specific nucleases or site-directed nucleases) that consist of transcription activator-like effector nucleases (TALENs), zinc finger nucleases (ZFNs) and CRISPR/Cas -associated system [142]. In general, ZNFs and TALENs are based on the ability of DNA-binding domains to recognize any specific target DNA sequence. However, the CRISPR/Cas system is advantageous over ZFNs and TALENs in terms of simple designing, versatility, cost-effectiveness, efficiency, multiplexing, and specificity. Therefore, exploring the molecular components of plant-microbe interactions using CRISPR-Cas9-based genome editing tools assists in developing durable, eco-friendly, and sustainable agricultural solutions (Fig. **2**).

CLUSTERED REGULARLY INTERSPACED SHORT PALINDROMIC REPEATS

CRISPR/Cas System and Orthologs

The PAM recognition site of Cas9 constitutes, N, R, M, W, V, and Y, where N is any nucleotide, R is A/G, M is A/C, W is A/T, V is G/C/A, R is A/G and Y is C/T [143 - 153]. Further, Cpf1 is another Cas9 ortholog, where the trans-activating RNA [tracrRNA] is not required for pre-crRNA processing. Cas9 cleavage produces blunt ends, whereas Cpf1 generates staggered ends (Table **1**).

Furthermore, CRISPR/Cas systems are categorized into: Class 1 (types I, III, and IV), and Class 2 (types II, V, and VI). Both the classes possess distinct mechanisms of guide RNA biogenesis and target interference, based on phylogenetic, structural, and functional characteristics of Cas proteins [154]. Among the two categories of CRISPR/Cas systems, Class 2 nucleases are extensively utilized for nucleic acid manipulation

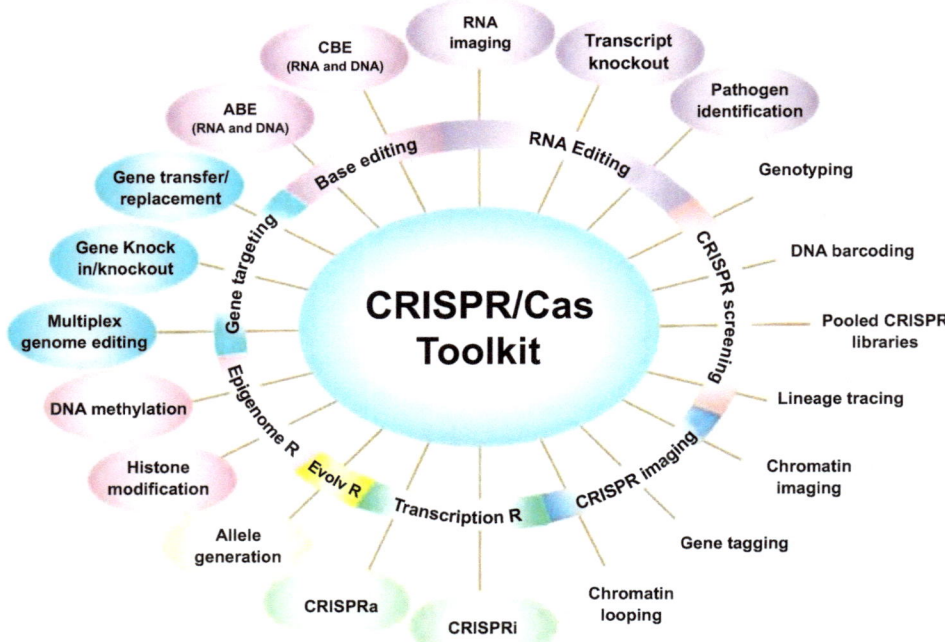

Fig. (2). Schematic representation of CRISPR toolkit designed for various applications in diverse fields. The SpCas9-mediated knock-out or knock-in strategy is used in prokaryotes and eukaryotes. Catalytically dead Cas9 [dCas9] or nickase Cas9 [nCas9] and Cas9 orthologs fused with specific modulators have been reprogrammed to perform base editing [ABE, adenine base editor; CBE, cytidine base editor], RNA editing, screening libraries, chromatin imaging, transcription regulation, allele generation using the EvolvR system, and epigenome editing. [Figure adapted from Shelake *et al.*, 2019 [52]].

Table 1. Cas9 and cpf1 orthologs with their respective PAM recognition sequences.

Cas9 Orthologs	Microbe	PAM Recognition Site	References
SpCas9	*Streptococcus pyogenes*	50-NGG-30	[143]
StCas9	*Streptococcus thermophiles*	[50-NNAGAAW-30]	[144]
NmCas9	*Neisseria meningitidis*	50-NNNGMTT-30	[145]
SaCas9	*Staphylococcus aureus*	50-NNNRRT-30	[145]
FnCas9	*Francisella novivida*	50-NGG-30	[146]
CjCas9	*Campylobacter jejuni*	50-NNNVRYAC-30	[147]
ScCas9	*Streptococcus canis*	50-NNG-30	[148]
CasX	*Streptococcus canis*	50-TTCN-30	[149]
FnCpf1	*Francisella tularensis* subsp. *novicida* U112	50-TTV/TTTV/KYTV-30	[150]
LbCpf1	*Lachnospiraceae bacterium* ND2006	50-TTTV-30	[150]

(Table 1) cont.....

Cas9 Orthologs	Microbe	PAM Recognition Site	References
AsCpf1	Acidaminococcus sp. BV3L6	50-TTTV-30	[150]
MbCpf1	Moraxella bovoculi 237	50-TTV/TTTV-30	[151]
Cas9	Bacillus halodurans	50-NNNN-30	[152]
Base editing	Sinorhizobium meliloti	30-NGG-20	[153]

CRISPR/Cas12 and Cas14

CRISPR/Cas12 is classified as type V and is the second well-documented CRISPR system. It has distinct evolutionary origins and structural architecture when compared to Cas9. RuvC-like domain-containing Cas12 (Cas12a, b, c, d, e, g, h, i, k, *etc.*) with diverse functions have been reported so far. CRISPR/Cas12a (Cpf1) and CRISPR/Cas12b (C2c1), are successful in editing plants [150]. Here, Cas12a is guided by a single mature crRNA in DNA targeting, whereas Cas12b requires both tracrRNA and crRNA. However, both Cas12a and Cas12b recognize a T-rich protospacer adjacent motif (PAM) and generate 4–5 nt long staggered ends, distal to the PAM site [155].

Cas14, another type V protein isolated from non-culturable archaea, cleaves single-stranded DNA (ssDNA) with no restrictive sequences. Hence, Cas14 is regarded as an ideal tool for engineering resistance against ssDNA plant viruses. Previous studies reported high-fidelity SNP genotyping using Cas14, due to its unrestricted cleavage and sequence-independent nature [88].

RNA-Targeting Endonucleases

RNA-targeting endonucleases belong to the type VI CRISPR/Cas13 systems, that function as a ribonuclease. CRISPR/Cas13 has been divided into four subtypes, such as type VI-A (Cas13a), VI-B (Cas13b), VI-C (Cas13c), and VI-D (Cas13d). Cas13a from *Leptotrichia shahii* or Cas13b from *Porphyromonas prevotella* are redesigned for RNA-editing from bacterial immune systems. CRISPR/Cas13 possesses two distinct HEPN RNase domains and was assembled with a single crRNA to form a crRNA-guided protein complex. Similar to PAM, in Cas9 and Cas12, the 5'- and/or 3'-protospacer-flanking site (PFS) redirects the Cas13/crRNA complex to the target site. Further, Cas13 also cleaves collateral RNAs, once activated in the presence of template targets. Hence Cas13 proteins are used to localize, detect, and track RNA molecules of different types. The Cas13-gRNA system is being applied for RNA knockdown in plant transcripts for *e.g.* rice [*Oryza sativa*] protoplasts. More recently, a specific High-Sensitivity Enzymatic Reporter Unlocking [SHERLOCK] system based on Cas13 was developed for pathogen identification, genotyping, and detection of multiple plant genes in a single reaction [156].

APPLICATIONS OF MODERN CRISPR-BASED TOOLS

CRISPR for Editing Multi-Targets

CRISPR-based programmable tools developed using modified Cas9 are useful in base editing, gene regulation, epigenetic editing, chromatin engineering, *etc*. With the advent of the next generation of CRISPR-based tools, it is possible to target precise DNA. For instance, native SpCas9 was explored to generate knock-out or knock-in mutants to study the gene functions and HR-based gene integration at specific locus [143]. Besides, the CRISPR system also allows multi-site editing. Among the targeted mutagenesis tools so far developed, EvolvR is novel and designed to incorporate semi-random mutations at a dCas9/gRNA-targeted site in a plasmid or prokaryotic genome [157]. More recently, cytidine-base editor [CBE] and adenine-base editor [ADE] that make nucleotide substitutions from C-to-T and A-to-G, were generated by fusion of impaired Cas9 (nCas9 or dCas9) using cytidine deaminase or adenine deaminase. The CBE and ABE-based editing is being attempted in plants and microbes as well [158].

CRISPR-MEDIATED PLANT –MICROBE INTERACTION AND ITS APPLICATION IN AGRICULTURE

So far, the use of CRISPR-based tools onplant–microbe interactions to develop disease resistance has been documented by several workers [189]. The interference of CRISPR technology in plant-microbe interactions is highly vital to meet the increasing food demand by unraveling the critical mechanisms of plant-microbe interactions in developing disease-resistant varieties, and plant growth promotion traits as detailed in section 3.

CRISPR in Understanding Plant-microbe Interactions

The plant microbiome encompasses beneficial, neutral, or pathogenic microbial species wherein, the beneficial plant microbiota is vital for plant health and yield [24, 47]. The well-known beneficial plant-microbe interactions are symbiotic interaction of legumes with N-fixing rhizobia; and arbuscular mycorrhizal (AM) fungi [52]. This symbiotic behavior favors plant growth- promoting microbes to dominate the community structure. The plant growth-promoting microbes influence the crop through several mechanisms such as: phytohormone production (auxin, ABA, cytokinin, and gibberellin), plant-beneficial enzymes (1-aminocyclopropane-1-carboxylate deaminase), enhanced uptake of heavy metals, and detoxification. However, successful plant-microbe interactions solely depend upon the genetics of both the microbiome and host [42]. The host immune system recognizes certain molecules secreted by its associated microbiota, that serve as cues to trigger the host protection against phytopathogens. Signaling hormones such as ethylene, jasmonate, and salicylic acid, facilitate plant-microbe interactions. Understanding the co-evolutionary cycles between the pathogens and plants '(Red Queen' dynamics) will enhance our knowledge about their vital evolutionary principles [159]. Though the recent, multiple omics tools shed light on the community-level functions of plant-associated genes and pathways, these tools cannot determine the harmful, neutral, or beneficial interaction with the host plants. Also, omics data cannot determine spatio-temporal dynamic interplays. Probing essential genes in regulating specific agronomic traits will help to improve the specific plant trait for sustainable agriculture. CRISPR-based GE tools aim to explore the gene functions at the molecular level, by a complete knockdown of the target gene, compared to the partial gene silencing by RNAi technique that produces partial phenotypes [160]. Multiple experimental evidence on model microbes, root-rhizobia, and phytopathogen *Pseudomonas syringae*, revealed the genetic factors responsible for mutual and pathogenic interactions with hosts, respectively [161].

CRISPR/Cas system along with a single-stranded DNA for different genetic modifications, including gene deletion, insertion, replacement, and transcription repression was established in the rhizospheric bacterium *Pseudomonas putida* KT2440 [162]. Therefore, genome editing of non-model microorganisms with robust CRISPR/Cas tools, enables to establish links between the genes and their functions. A recent trend of using biomaterials either in DNA, mRNA, or protein form, offers a unique solution for CRISPR/Cas delivery in organisms that cannot be possible by conventional methods [163].

CRISPR in Understanding Plant Growth Promotion [PGP] and Nutrient Uptake

In general, the beneficial plant-microbe interaction is the product of complex processes. Successful plant-microbe interaction facilitates nutrient accessibility through Fe uptake, N-fixation, and K or P solubilization. Plant root-associated AM fungi, more predominantly the phylum Glomeromycota inhabit 80–90% of the terrestrial plants and mobilize nutrients from soil to plant. Similarly, root endophytic fungi (*e.g.*, *Colletotrichum tofieldiae*) from *Arabidopsis* exhibits PGP activity and mobilize P to the host plant under phosphate-deficient conditions. However, little is known about PGP microbes involved in the beneficial interactions, whilst plant-pathogen interactions have been intensely investigated. The beneficial effects of plant signaling mechanisms are very similar to the pathogenic plant-microbe interactions. This complex phenomenon plays a puzzle in the scientist's box to draw a conclusion on the outcome of a particular plant-microbe interaction, whether beneficial, neutral, or pathogenic. The activation of plant -defense pathways against biotic stresses by these signaling cues, causes systemic acquired resistance (SAR) and induced systemic resistance (ISR) [7].

In legume symbiosis, nodule organogenesis offers insights to engineer non-legume crops and also to host N-fixing bacteria. Gene editing tools are deployed in some model legumes such as, *Lotus japonicas*, *Medicago truncatula*, *Glycine max*, and *Vigna unguiculata*, L.Walp [164]. Rhizospheric PGPMs such as *Bacillus mycoides* and *B. subtilis* were edited using CRISPR protocols. Expanding the CRISPR toolbox revolutionizes the breeding of food legumes and non-legume genotypes to acquire efficient N-fixing rhizobia and P, K, S, Fe, Si, and Zn - solubilizing microbes. Further, GE-modified beneficial microbes are the best alternative to agrochemicals. Therefore, PGPM is one of the better options to decrease the cost in an eco-friendly way. The application of such genetically edited microbial inoculants avoids the rapid decline in the introduced allochthonus microbial population and enhances the bio-augmentation which subsequently benefits the crops.

CRISPR in Priming Plant Disease Resistance

Average yield losses due to pathogenic plant-microbe andplant–insect interactions range from 11 to 30%. To overcome the time-consuming and laborious job of conventional methods, host- genetic engineering and genome editing techniques have been employed recently to develop disease resistance in plants. Genome editing targets blocking pathogen entry, alteration of the plant- defense system, modulation of recessive traits (S-genes), triggering resistance genes (R-genes), and expression of antimicrobial peptides and RNAi [165]. Eoh and his co-workers suggested that CRISPR-mediated genetic modifications of either plant or pathogens (bacteria, fungi, oocytes, and viruses) improve disease resistance, which is possible by targeting the gene either from the host plant or pathogen, based on the whole genome sequencing data of crop and its associated microbiota [165]. Plant systems normally lack an adaptive immune system, whereas, the interconnected two-tier innate immune system encompasses two different modes of actions: i] cell surface pattern-recognition receptors (PRRs) to distinguish between "non-self" and "self," *i.e.*, pathogen-associated molecular patterns (PAMPs) and plant-derived damage-associated molecular patterns (DAMPs), initiating the PAMP-triggered immunity (PTI); ii] Resistant proteins (R-proteins) tackle pathogen-derived effectors to prevent the pathogen entry into the host cell, and activate effector-triggered immunity (ETI) [166]. R-genes encode intracellular nucleotide-binding leucine-rich repeat receptor (NLRs) proteins whereas, the receptor –like kinases and receptor-like proteins act as PRRs. Genome editing by CRISPR/Cas targets either of the above candidate plant genes

(PTI or ETI) or pathogen genes to confer resistance against bacteria, fungi and oocytes, and DNA/RNA viruses [167].

GE in Bacterial Pathogens

Genome modification of plant pathogen genomes offers desirable phenotypes. In previous studies, the application of CRISPR/SpCas9 and CRISPR/FnCas12a systems in *P. putida* KT2440 achieved efficient gene deletions, gene insertions, and gene replacements [162]. Further, the pnCasPA-BEC base editing system developed by engineering cytidine deaminase APOBEC1 to the Cas9 nickase produced more efficient C > T substitutions in *P. aeruginosa, P. putida, P. fluorescens, Staphylococcus aureus* and *P. syringae*. CRISPRi (dCas9-based transcription inhibition system) using a deactivated SpCas9 was applied for target gene repression in *Pseudomonas* spp [168].

GE in Fungal Pathogens

CRISPR/SpCas9 technology was well established in fungal pathogens, such as *Magnaporthe oryzae, Alternaria alternate, Leptosphaeria maculans Fusarium oxysporum, Fusarium graminearum, Fusarium proliferatum, Sclerotinia sclerotiorum, Colletotrichum sansevieriae, Fusarium fujikuroi, B. cinerea, Sporisorium scitamineum, Ustilaginoidea virens, Ustilago maydis* and *Ustilago tricophora* [169]. Among the CRISPR/Cas systems, SpCas9/sgRNA ribonucleoprotein (RNP) complex was found to be more successful in *M. oryzae* and *F. oxysporum* [170], and for editing in oomycetes, (*P. sojae, P. capsici, P. palmivora* and *P. litchi*) through PEG-mediated protoplast transformation. In the case of *P. sojae,* disruption of the RXLR effector protein Avr4/6 prevented its recognition by the corresponding soybean R proteins (Rps4 and Rps6). The *P. capsici* knock-outs (*PcAvh1* gene which is RXLR effector gene,) confirmed that virulence in tobacco was solely due to the gene *PcAvh1*. Further, CRISPR/SpCas9 generated homozygous PpalEPIC8 mutants, showed reduced pathogenicity of *P. palmivora* in papaya fruits, by inhibiting the papain [171]. Likewise, in oomycetes, the gene PcMuORP1 for oxathiapiprolin-resistance was optimized as a high-efficiency selection marker [172].

Genome Editing for Plant Disease Resistance against Bacterial and Fungal Pathogens

Approximately, 30% of emerging plant diseases are caused by fungal pathogens, which can easily overcome the innate immunity armed by *R* genes due to high evolutionary flexibility and genetic diversity. Therefore, gene-specific editing by the CRISPR/Cas9 system will aid in the deployment of durable resistance to fungal diseases. For instance, CRISPR/Cas9 based targeted editing of the

OsERF922 gene in rice [*Oryza sativa*] reduced blast lesion symptoms in comparison to its wild variant in both seedling and tillering stages. Further, the gene *ERF922* was up-regulated during infection of *M. oryzae* but also in abiotic stresses. Besides, editing of the other three blasts inducible *ERF* [*OsBIERF1, OsBIERF3,* and *OsBIERF4*] and *Pi21* genes may provide additional insights into the role of ethylene response factors in regulating blast resistance in rice [173]. *Pi21* knockout lines demonstrated that the pathogen-associated molecular patterns as one of the key mechanisms conferring immunity. Interestingly, manipulation of exocyst subunit *OsSEC3A* and *OsMPK5* via CRISPR/Cas9 exhibited dwarf phenotypes, and enhanced resistance against *M. oryzae*. However, the knock-out mutants showed susceptibility to abiotic stresses. Therefore, the studies suggest that alternate editing strategies are required for genes playing multiple roles in plants.

Likewise, *Erysiphe necator* infection induces susceptibility, by up-regulating the *mildew locus O* [*MLO*] gene. Previously, *MLO S-genes* (*VvMLO7, VvMLO6,* and *VvMLO11*) silenced through RNAi, drastically reduced the severity of powdery mildew disease by 77 per cent in grapes [174]. *MLO-S* genes are transmembrane domain proteins that are evolutionarily conserved, and form distinct clades, which negatively regulates the penetration of *Erysiphe necator*. The edited *MLO-7* gene achieved resistance to downy mildew in grapes [175]. Besides, MLO signaling is regulated by a medley of genes, especially MYB and NAC which are involved in cell wall thickening. In cucumber, it was evident that cell wall thickening subdued the function of *MLO1* gene. CRISPR-based knock-down of downy mildew resistance gene *DMR6* conferred resistance to downy mildew disease in grapevine. Further, multiplex editing of *MLO* and *DMR6* genes provided resistance to both downy and powdery mildew in the Vitis. The *dmr6* gene increased the salicyclic acid levels, and it was concluded that targeting the *dmr6* orthologs in monocots and dicots provides information on broad-spectrum resistance development and subsequent pleiotropic effects [176]. Similarly, editing of the tomato *SlDmr6*-1 gene also conferred resistance to different pathogens such as *Phytophthora capsici, Pseudomonas syringae*, and *Xanthomonas* species. However, a slight reduction in the plant height and its associated yield penalty due to DMR6 gene editing needs to be better understood, considering its beneficial effects of plant immunity. CRISPR/SpCas9-mediated knockout of homologs TaMLOs and SlMLO1 resulted in resistance against the powdery mildew fungal pathogens *Blumeria graminis f.sp. tritici* and *Oidium neolycopersici* in wheat and tomato respectively [177].

During bacterial infection, type III effectors interrupt the host's defense pathways and/or activate the *S* genes for disease development. To improve plant resistance, good target sites for CRISPR/Cas9-mediated gene editing are *S* genes and other negative regulators of plant innate immune response. For instance, endogenous OsSWEET11gene knockouts, where the TALEs are targeted, showed an increase in resistance against *X. oryzae pv. Oryzae* (*Xoo*) in rice [178]. TALE-binding elements (EBEs) in the promoter region of both OsSWEET11 and OsSWEET14, carrying indels conferred robust resistance to most of the *Xoo* strains. The results suggested that multiplex genome editing of the promoter regions of OsSWEET11, OsSWEET13, and OsSWEET14 in rice varieties confer broad-spectrum resistance to *Xoo* strians [179].

Likewise, genome editing of the CsLOB1 (promoter region of the *S* gene) in citrus, conferred a high degree of resistance to the citrus canker disease caused by *X. citri pv. citri* (Xcc). The gene *DMR6* is responsible for resistance to downy mildew in *A. thaliana*, however the knockout of its homolog SIDMR6–1 in tomato by CRISPR/SpCas9 rendered plants resistant to different bacterial pathogens, including *P. syringae, P. capsici,* and *Xanthomonas* spp. Besides, the tomato germplasm (SlJAZ2Δjas) generated by CRISPR/SpCas9 provided resistance to *P. syringae* pv. and its defense response to the necrotrophic fungal pathogen *Botrytis cinerea* was not changed [180]. More recently, BSR-K1 gene knock-out in rice showed resistance to both *M. oryzae* and *Xoo* strains.

Furthermore, precise base editing of SNPs in the R gene was reported as an efficient and time-saving method to improve crop disease resistance. The rapid correction of the recessive pi-d2 gene in the rice variety Kitaake by a cytidine base editor rBE5 introduced a G > A substitution (M411I) [181]. However, CRISPR/Cas tools in editing the effectors of Oomycetes to improve the resistance are limited. Base editing of the target gene TcNPR3 by transient expression in *Theobroma cacao* leaves achieved enhanced resistance against *P. tropicalis*. Besides, the CRISPR/Cas system was also employed in beneficial microbes such as insect-pathogenic fungus *Beauveria bassiana, Purpureocillium lilacinum* controlling plant nematodes, and filamentous fungi like the *Trichoderma species* [182].

GE in Plant-Virus Interactions

In general, virus resistance is accomplished either by targeting host factors that are involved in the replications of the virus or targeting and destroying the viral genome itself. The eukaryotic translation initiation factor 4E (eIF4E), also known as a cap-binding protein, is a well-known susceptibility factor in plant-virus interactions. Hence, targeting the eIF4E gene results in the development of innate immunity against potyviruses. Chandrasekaran *et al.* reported that eIF4E-edited cucumber plants were resistant to Zucchini yellow mosaic virus (ZYMV), Papaya ringspot mosaic virus-W (PRSMV-W) and Cucumber vein yellowing virus

(CVYV) [183]. Similar, results were obtained while targeting geminiviral genomic DNA with CRISPR/SpCas9 [184]. A few other examples of CRISPR/Cas intervention in developing disease resistance against viral diseases are given in Table **2**.

Table 2. Disease resistance in plants against viral diseases using GE tools.

Crop	Alleles	GE tool	Disease	References
Potyviruses				
Cucumber	eIF4E	CRISPR/SpCas9	ZYMV, PRSMV-W and CVYV	[183]
Cassava and *A. thaliana*	eIF4E	CRISPR/SpCas9	Cassava brown streak disease; turnip mosaic TuMV infection	[185]
Rice	eIF4G	CRISPR/SpCas9	Rice tungro spherical virus (RTSV)	[186]
A. thaliana	eIF4E1	Cytidine base editor [C > G conversion [N176K]	Clover yellow vein virus (ClYVV)	[187]
Geminiviruses				
N. benthamiana and *A. thaliana*	Rep [replication-associated protein] gene and the intergenic region [IR]	SpCas9	Curly top virus (BSCTV); bean yellow dwarf virus (BeYDV)	[188]
N. benthamiana and Tomato	CP [coat protein]; Rep sequences	SpCas9	Tomato yellow leaf curl virus (TYLCV)	[189]
Barley and banana	MP, CP and other conserved regions of viral genomes	CRISPR/SpCas9	Wheat dwarf virus (WDV); Banana streak virus (BSV)	[190]
N. benthamiana	Viral RNA of TuMV	CRISPR/Cas13a	Cucumber mosaic virus (CMV) and tobacco mosaic virus (TMV)	[191]
N. benthamiana and *A. thaliana*	Viral genome	CRISPR/FnCas9	Tobacco mosaic virus (TMV)	[192]
Potato and rice	Viral genome	CRISPR/Cas13a	Potato virus Y (PVY); rice stripe mosaic virus (RSMV); southern rice black-streaked draft virus (SRBSDV)	[192]

Although consistent efforts of CRISPR/Cas resulted in the development of resistance against viral diseases in transgenic plants, the potential risks of viruses escaping the CRISPR/Cas9 cleavage leads to the loss of resistance, which is a major concern. This is caused due to the fast-evolving nature of the virus. So far, the edited viral genomes (33 and 48%) evolved a conserved single nucleotide mutation which conferred resistance to CRISPR/Cas9 cleavage, resulting in failed resistance to the Geminivirus [193].

GE in Unraveling Novel Metabolic Pathways and Metabolome

Rhizobiome act as microbial cell factories and are capable of secreting a large repertoire of bioactive molecules such as antimicrobials, signaling factors, small molecules, *etc*. These bioactive molecules fall under three categories: a) polyketides derived from acyl-CoAs, b) terpenes produced from acyl-CoAs, and c) small peptides from amino acids. The plant-associated microbes produce different bioactive molecules as secondary metabolites, while they are critical in plant or microbial defense mechanisms [194]. Most of the genes involved in the biosynthesis of bioactive molecules are clustered together on chromosomes and are not constitutively expressed. The actively expressed genes will become quiescent upon repeated sub-culturing under standard laboratory conditions [195]. Moreover, hunting for novel bioactive compounds is hindered due to complex genetic architecture and poor efficacy in specific gene targets. With the recent advancement of high-throughput sequencing, the elucidation of biosynthetic pathways of bioactive molecules has become more vivid. Interestingly, several plant microbiome studies documented an insight into certain microbial taxa supporting the activity of plant secondary metabolite production. For example, the seed microbiome of the medicinal plant *Salvia miltiorrhiza* showed an overlapping set of bacterial and fungal genera with that of maize, bean, rice, and rapeseed [168]. Further, plant-associated microbes share common terpenoid metabolic pathways with the host plant, signifying their potential as a repository of secondary metabolite–related genes. As a crucial note, the plant secondary metabolite varies depending on their geographical location, partly due to altered microbiota suggesting a direct relationship between the plant metabolome and their associated microbiome. Instead of recombinant technology and transgenics, CRISPR-mediated genome editing of plant-microbe-associated genes involved in the pathways may provide an innovative approach to accomplish higher production of stable bioactive molecules (Table **3**). Recent reports suggest a paramount success in CRISPR-based metabolic engineering in either microbes or plants, based on the understanding of metabolic pathways and enhanced production of targeted metabolites [196].

Table 3. The metabolic pathways so far edited using CRISPR tools.

Microbe	Tools Used	Metabolic Pathway/Targeted Metabolite	References
Plant-associated Microbiome			
Beauveria bassiana	Codon-optimized Cas9, gfp/ura5/bar, gpdA	Uridine synthesis	[197]
Trichoderma reesei	CRISPR/Cas9	Uridine synthesis	[198]
Escherichia coli	CRISPR/Cas9	Flavonoid synthesis	[199]
Myceliophthora	CRISPR/Cas9	Cellulase production	[198]
Aspergillus niger	pCas9, hyg/pyrG	Galactaric acid production	[200]
Aspergillus oryzae	Aspergillus-optimized codons Cas9, pyrG, U6	Pigment production	[201]
Shiraia bambusicola	Codon-optimized Cas9, hph, TrpC	Hypocrellin production	[202]
Aspergillus fumigatus	Functional reconstitution using CRISPR/Cas9	Trypacidin biosynthesis	[202]
Fusarium fujikuroi	CRISPR/Cas9	Giberellic acid GA4 and GA7	[2]
Glarea lozoyensis	CRISPR/Cas9 – homology-directed repair	Pneumocandin B0	[203]
A. nidulans	CRISPRa and CRISPRi	Microperfuranone biosynthesis	[204]
Plant Associated (Microbe-plant)			
Opium poppy	CRISPR-based metabolic engineering	Morphine biosynthesis	[205]
Tomato	CRISPR-based metabolic engineering	Aminobutyric acid, GABA	[206]
S. miltiorrhiza	CRISPR-based metabolic engineering	Tanshinone biosynthesis; phenolic acid	[207]
Dendrobium sp	CRISPR-based metabolic engineering	Lignocellulose biosynthesis	[208]
Camelina sativa	CRISPR-based metabolic engineering	Triacylglycerol synthesis	[209]
Tobacco	CRISPR-based metabolic engineering	Glycan biosynthesis	[210]

Hence, CRISPR technology for editing, multiplexing, and transcription activation in plant-microbe interactions is a reliable tool to discover the reporter metabolic pathways and signature metabolites as a mandate for sustainable agriculture.

Soil Health

CRISPR-Cas9 is an exciting tool to understand soil microbiome-mediated biogeochemical processes including lignocellulose valorization, nitrification, and metagenomics-assisted root microbiome. Several nutrient transporters have been explored so far. For instance, nitrate transporter gene *NRT1.1 B* regulates the root microbiome in *indica* varieties of rice [211]. Interestingly, targeted allelic replacement of *NRT1.1* significantly enhanced the nitrogen use efficiency (NUE) in japonica rice. Moreover, the edited rice plants served as biomarkers for understanding root microbiome. CRISPR-Cas9 was adopted for altering the plant cell wall components by specifically targeting the *OSH15* and *OsAt10* genes [212]. Further, it was evident that CRISPR elements are more abundant in polar soil than tropical soil, which correlates with the greater disease pressure in the tropics [213]. Accordingly, the risk of horizontal gene transfers is greatly reduced in CRISPR-Cas9 edited *Bacillus subtilis* employing an integrative plasmid [214]. Therefore, the presence of CRISPR repeats in plant-associated bacteria may provide an evolutionary advantage for the better adaptation and exploitation of beneficial genes.

Legume- *Rhizobium* Symbiosis

Biological nitrogen fixation (BNF) in legume-*Rhizobium* interactions contributes to sustainable legume productivity and soil health. More recently, rhizobial association with other non-rhizobial endophytes such as yeast [6] and bacteria [51] provided more insights on the nodule microbiome and their metabolic interventions. Various genes involved in nodulation and nitrogen fixation, including the nod factors, leghaemoglobin, *etc*., have been elucidated using CRISPR/Cas9 approach. The nodulation-specific promoter region of leghaemoglobin genes (*LjLb1, LjLb2, LjLb3*) was successfully employed for the expression of the guide RNA in *Lotus japonica* nodules. In contrast, the *symbiotic receptor-like kinase* (*VuSYMRK*) gene developed through CRISPRCas9, exhibited loss-of-function by blocking the nodule formation in cowpeas [164]. The dominant gene responsible for the restriction of nodule formation in the soybean -*Sinorhizobium fredii association* was identified through the editing of *Rfg1* and *Rj4* genes [215]. Conservation of the nodulation process in a tropical tree, *Parasponia andersonii* by the genes related to hormonal regulation during the nodulation process (*PanHK4, PanEIN2, PanNSP1*, and *PanNSP2*) was evident through gene editing [216]. Furthermore, small fragments of tRNA [tRFs] synthesized in *Rhizobia*, regulated 52 soybean genes involved in the nodulation process. CRISPR-Cas9-based editing of a few selected genes also promoted nodulation patterns in soybeans. However, in *Medicago*, CRISPR/Cas9-based editing of *gibberlin oxidase* (*MtGA2Ox10*) suppressed the infection thread formation in the initial nodulation process. CRISPR/Cas9 approach was used to identify the *MtNFS2* gene responsible for strain specificity in *Medicago* sp [178]. Therefore, CRISPR technologies not only assist in the identification of target genes in legume–rhizobia symbiosis but also facilitate with mechanistic insights on nodule formation, which can be further exploited to enhance the productivity of legume crops.

CONCLUSION AND FUTURE PERSPECTIVES

CRISPR/Cas system and its derivatives provide an opportunity to explore and engineer complex plant-microbe interactions. Recent insights on molecular plant-microbe interactions, especially microbe-mediated plant health have widened the horizon of plant-microbe engineering and its application in an appropriate way to meet the global food demand. Indeed, there remain limitations in plant-microbe engineering, such as off-target effects and therefore it requires precise editing. CRISPR/Cas approaches enable us to understand the basics of plant-microbe interaction and to develop an ideal plant microbial community (otherwise called synCom) relevant for agricultural application. In the current scenario, plant-specific microbiome is capturing the attention of researchers. The plant or microbial genes governing desirable agronomic traits such as yield, and abiotic and biotic stress tolerance, will facilitate CRISPR-based editing strategies in sustainable agricultural practices. Since there exists a direct link between the agronomic traits and the genes of plants/microbes, genome editing technologies will enable to design synthetic microbial communities for higher microbial productivity.

Furthermore, susceptible genes can be edited for imparting durable resistance in plants. With recent research advances, a well-designed gRNA sequence and specific Cas9/12/13 variants can be developed to avoid the off-target effects. Apart from CRISPR/Cas genome editing strategies, alternative methods such as CRISPRi and CRISPRa, which do not depend on DSBs, were rarely reported in plant-microbe interaction, particularly in reporter metabolites. Targeting genes involved in different pathways conferring resistance through CRISPR/Cas9 approach could assist in multiplexing resistance to multiple pathogens. Therefore, a combinatorial approach of computational biology and genomic tools have proven supportive of understanding the communication and metabolic pathways, providing an alternative to regulate these pathways, to get a beneficial effect on plants with ecological sustainability. Pathogens are recognized by plants extracellularly and/or intracellularly, as a 'spatial-immunity model'. Therefore, CRISPR-Cas9-based targeting of host genes localized in different organelles (spatial) involved in regulating colonization will provide multilayered resistance against diseases and abiotic stresses. Recent reports on increased disease resistance by triggering specific transcription factors are indeed a challenging task. Additionally, the intervention of CRISPR/Cas9 in nodulation by small tRNA, and the loss of a few genes can be exploited to fish out the key orthologs. However, the nodulation process can be better understood by editing pooled gRNA libraries specific for nodulation-related genes which is the most imperative. Besides, CRISPR-Cas-based gene editing system was established only in a few crops. The recent tools used to create high throughput mutant libraries, SNPs and SNP-typed QTLs may create powerful approaches to accelerate the gene functions. To feed the ever-increasing population, amidst the threat of climate changes, exploring microbiome-mediated climate resilience might be the only possible means of future farming and sustainable productivity. It can be said that "*Microbes Feed the World*" in the near future.

ACKNOWLEDGEMENTS

The authors would like to thank the Dean, SAS, KITS, and TNAU for providing the facilities.

REFERENCES

[1] Ran FA, Cong L, Yan WX, *et al.* In vivo genome editing using *Staphylococcus aureus* Cas9. Nature 2015; 520(7546): 186-91.
[http://dx.doi.org/10.1038/nature14299] [PMID: 25830891]

[2] Shi TQ, Gao J, Wang WJ, *et al.* CRISPR/Cas9-based genome editing in the filamentous fungus *Fusarium fujikuroi* and its application in strain engineering for gibberellic acid production. ACS Synth Biol 2019; 8(2): 445-54.
[http://dx.doi.org/10.1021/acssynbio.8b00478] [PMID: 30616338]

[3] Waghunde RR, Shelake RM, Shinde MS, Hayashi H. Endophyte microbes: a weapon for plant health management. Microorganisms for green revolution. Springer 2017; pp. 303-25.

[4] Punitha S, Kalvathi K, Uthandi S. Apoplast associated *Bacillus methylotrophicus* RABA6 induced growth and yield attributes leading to drought tolerance in rice. Madras Agric J 2019; 106.

[5] Vibitha Bala B. Effect of *Klebseilla oxytoca* and *Acinetobacter* Sp. on growth of rice genotype under moisture sress condition. International Journal of Agriculture Sciences 2019; 0975-3710.

[6] Geetha Thanuja K, Annadurai B, Thankappan S, Uthandi S. Non-rhizobial endophytic (NRE) yeasts assist nodulation of Rhizobium in root nodules of blackgram (*Vigna mungo* L.). Arch Microbiol 2020; 202(10): 2739-49.
[http://dx.doi.org/10.1007/s00203-020-01983-z] [PMID: 32737540]

[7] Rodriguez PA, Rothballer M, Chowdhury SP, Nussbaumer T, Gutjahr C, Falter-Braun P. Systems biology of plant-microbiome interactions. Mol Plant 2019; 12(6): 804-21.
[http://dx.doi.org/10.1016/j.molp.2019.05.006] [PMID: 31128275]

[8] Thrall PH, Hochberg ME, Burdon JJ, Bever JD. Coevolution of symbiotic mutualists and parasites in a community context. Trends Ecol Evol 2007; 22(3): 120-6.
[http://dx.doi.org/10.1016/j.tree.2006.11.007] [PMID: 17137675]

[9] Hassani MA, Durán P, Hacquard S. Microbial interactions within the plant holobiont. Microbiome 2018; 6(1): 58.
[http://dx.doi.org/10.1186/s40168-018-0445-0] [PMID: 29587885]

[10] Bhattacharjee RB, Singh A, Mukhopadhyay SN. Use of nitrogen-fixing bacteria as biofertiliser for non-legumes: prospects and challenges. Appl Microbiol Biotechnol 2008; 80(2): 199-209.
[http://dx.doi.org/10.1007/s00253-008-1567-2] [PMID: 18600321]

[11] Santner A, Calderon-Villalobos LIA, Estelle M. Plant hormones are versatile chemical regulators of plant growth. Nat Chem Biol 2009; 5(5): 301-7.
[http://dx.doi.org/10.1038/nchembio.165] [PMID: 19377456]

[12] Yazdani M, Bahmanyar MA, Pirdashti H, Esmaili MA. Effect of phosphate solubilization microorganisms [PSM] and plant growth promoting *Rhizobacteria* [PGPR] on yield and yield components of corn. World Acad Sci Eng Technol 2009; 49(1): 90-2. [*Zea mays* L.].

[13] Elanchezhiyan K, Keerthana U, Nagendran K, *et al.* Multifaceted benefits of *Bacillus amyloliquefaciens* strain FBZ24 in the management of wilt disease in tomato caused by *Fusarium oxysporum* f. sp. *lycopersici*. Physiol Mol Plant Pathol 2018; 103: 92-101.
[http://dx.doi.org/10.1016/j.pmpp.2018.05.008]

[14] Meyer SL, Everts KL, Gardener BM, Masler EP, Abdelnabby HM, Skantar AM. Assessment of DAPG-producing nematodes and management of Watermelon. J Nematol 2016; 48(1): 43-53.
[http://dx.doi.org/10.21307/jofnem-2017-008] [PMID: 27168652]

[15] Prabhukarthikeyan SR, Keerthana U, Raguchander T. Antibiotic-producing *Pseudomonas fluorescens* mediates rhizome rot disease resistance and promotes plant growth in turmeric plants. Microbiol Res 2018; 210: 65-73.
[http://dx.doi.org/10.1016/j.micres.2018.03.009] [PMID: 29625661]

[16] Saraf M, Pandya U, Thakkar A. Role of allelochemicals in plant growth promoting *Rhizobacteria* for biocontrol of phytopathogens. Microbiol Res 2014; 169(1): 18-29.
[http://dx.doi.org/10.1016/j.micres.2013.08.009] [PMID: 24176815]

[17] Rosenberg E, Zilber-Rosenberg I. The hologenome concept of evolution after 10 years. Microbiome 2018; 6(1): 78.
[http://dx.doi.org/10.1186/s40168-018-0457-9] [PMID: 29695294]

[18] Imam J, Mahto D, Mandal NP, Maiti D, Shukla P, Variar M. Molecular analysis of Indian rice germplasm accessions with resistance to blast pathogen. J Crop Improv 2014; 28(6): 729-39.
[http://dx.doi.org/10.1080/15427528.2014.921261]

[19] Bulgarelli D, Garrido-Oter R, Münch PC, *et al.* Structure and function of the bacterial root microbiota in wild and domesticated barley. Cell Host Microbe 2015; 17(3): 392-403.
[http://dx.doi.org/10.1016/j.chom.2015.01.011] [PMID: 25732064]

[20] Cavicchioli R, Ripple WJ, Timmis KN, *et al.* Scientists' warning to humanity: microorganisms and climate change. Nat Rev Microbiol 2019; 17(9): 569-86.
[http://dx.doi.org/10.1038/s41579-019-0222-5] [PMID: 31213707]

[21] Knott GJ, Doudna JA. CRISPR-Cas guides the future of genetic engineering. Science 2018; 361(6405): 866-9.
[http://dx.doi.org/10.1126/science.aat5011] [PMID: 30166482]

[22] Zaidi SSA, Mukhtar MS, Mansoor S. Genome editing: Targeting susceptibility genes for plant disease resistance. Trends Biotechnol 2018; 36(9): 898-906.
[http://dx.doi.org/10.1016/j.tibtech.2018.04.005] [PMID: 29752192]

[23] Müller DB, Vogel C, Bai Y, Vorholt JA. The plant microbiota: Systems-level insights and perspectives. Annu Rev Genet 2016; 50(1): 211-34.
[http://dx.doi.org/10.1146/annurev-genet-120215-034952] [PMID: 27648643]

[24] Philippot L, Raaijmakers JM, Lemanceau P, Van der Putten WH. Going back to the roots: the microbial ecology of the rhizosphere. Nat Rev Microbiol 2013; 11(11): 789-99.
[http://dx.doi.org/10.1038/nrmicro3109] [PMID: 24056930]

[25] Stringlis IA, De Jonge R, Pieterse CMJ. The age of coumarins in plant–microbe interactions. Plant Cell Physiol 2019; 60(7): 1405-19.
[http://dx.doi.org/10.1093/pcp/pcz076] [PMID: 31076771]

[26] Narayanasamy S, Thangappan S, Uthandi S. Plant growth-promoting *Bacillus* sp. cahoots moisture stress alleviation in rice genotypes by triggering antioxidant defense system. Microbiol Res 2020; 239: 126518.
[http://dx.doi.org/10.1016/j.micres.2020.126518] [PMID: 32604045]

[27] A P S, Thankappan S, G K, Uthandi S. Comprehensive profiling of the VOCs of *Trichoderma longibrachiatum* EF5 while interacting with *Sclerotium rolfsii* and *Macrophomina phaseolina*. Microbiol Res 2020; 236: 126436.
[http://dx.doi.org/10.1016/j.micres.2020.126436] [PMID: 32179388]

[28] Chapelle E, Mendes R, Bakker PAHM, Raaijmakers JM. Fungal invasion of the rhizosphere microbiome. ISME J 2016; 10(1): 265-8.
[http://dx.doi.org/10.1038/ismej.2015.82] [PMID: 26023875]

[29] Durán P, Thiergart T, Garrido-Oter R, *et al.* Microbial interkingdom interactions in roots promote m*Arabidopsis* survival. Cell 2018; 175(4): 973-83. e14.

[30] Castrillo G, Teixeira PJPL, Paredes SH, *et al.* Root microbiota drive direct integration of phosphate stress and immunity. Nature 2017; 543(7646): 513-8.
[http://dx.doi.org/10.1038/nature21417] [PMID: 28297714]

[31] Edwards J, Johnson C, Santos-Medellín C, *et al.* Structure, variation, and assembly of the root-associated microbiomes of rice. Proc Natl Acad Sci USA 2015; 112(8): E911-20.
[http://dx.doi.org/10.1073/pnas.1414592112] [PMID: 25605935]

[32] Bodenhausen N, Somerville V, Desirò A, *et al.* Petunia-and Arabidopsis-specific root microbiota responses to phosphate supplementation. Phytobiomes J 2019; 3(2): 112-24.
[http://dx.doi.org/10.1094/PBIOMES-12-18-0057-R]

[33] Ikeda S, Anda M, Inaba S, *et al.* Autoregulation of nodulation interferes with impacts of nitrogen fertilization levels on the leaf-associated bacterial community in soybeans. Appl Environ Microbiol 2011; 77(6): 1973-80.
[http://dx.doi.org/10.1128/AEM.02567-10] [PMID: 21239540]

[34] Manching HC, Balint-Kurti PJ, Stapleton AE. Southern leaf blight disease severity is correlated with decreased maize leaf epiphytic bacterial species richness and the phyllosphere bacterial diversity decline is enhanced by nitrogen fertilization. Front Plant Sci 2014; 5: 403.
[http://dx.doi.org/10.3389/fpls.2014.00403] [PMID: 25177328]

[35] Breidenbach B, Pump J, Dumont MG. Microbial community structure in the rhizosphere of rice plants. Front Microbiol 2016; 6: 1537.
[http://dx.doi.org/10.3389/fmicb.2015.01537] [PMID: 26793175]

[36] Rascovan N, Carbonetto B, Perrig D, *et al.* Integrated analysis of root microbiomes of soybean and wheat from agricultural fields. Sci Rep 2016; 6(1): 28084.
[http://dx.doi.org/10.1038/srep28084] [PMID: 27312589]

[37] Toju H, Peay KG, Yamamichi M, *et al.* Core microbiomes for sustainable agroecosystems. Nat Plants 2018; 4(5): 247-57.
[http://dx.doi.org/10.1038/s41477-018-0139-4] [PMID: 29725101]

[38] Bokulich NA, Thorngate JH, Richardson PM, Mills DA. Microbial biogeography of wine grapes is conditioned by cultivar, vintage, and climate. Proc Natl Acad Sci USA 2014; 111(1): E139-48.
[http://dx.doi.org/10.1073/pnas.1317377110] [PMID: 24277822]

[39] Peiffer JA, Spor A, Koren O, *et al.* Diversity and heritability of the maize rhizosphere microbiome under field conditions. Proc Natl Acad Sci USA 2013; 110(16): 6548-53.
[http://dx.doi.org/10.1073/pnas.1302837110] [PMID: 23576752]

[40] Schlaeppi K, Dombrowski N, Oter RG, Ver Loren van Themaat E, Schulze-Lefert P. Quantitative divergence of the bacterial root microbiota in *Arabidopsis thaliana* relatives. Proc Natl Acad Sci USA 2014; 111(2): 585-92.
[http://dx.doi.org/10.1073/pnas.1321597111] [PMID: 24379374]

[41] Ofek-Lalzar M, Sela N, Goldman-Voronov M, Green SJ, Hadar Y, Minz D. Niche and host-associated functional signatures of the root surface microbiome. Nat Commun 2014; 5(1): 4950.
[http://dx.doi.org/10.1038/ncomms5950] [PMID: 25232638]

[42] Compant S, Samad A, Faist H, Sessitsch A. A review on the plant microbiome: Ecology, functions, and emerging trends in microbial application. J Adv Res 2019; 19: 29-37.
[http://dx.doi.org/10.1016/j.jare.2019.03.004] [PMID: 31341667]

[43] Farrar K, Bryant D, Cope-Selby N. Understanding and engineering beneficial plant–microbe interactions: Plant growth promotion in energy crops. Plant Biotechnol J 2014; 12(9): 1193-206.
[http://dx.doi.org/10.1111/pbi.12279] [PMID: 25431199]

[44] Collemare J, Lebrun MH. Fungal secondary metabolites: ancient toxins and novel effectors in plant–microbe interactions. In: Martin F, Kamoun S, Wiley , Eds. Effectors in plant-microbe interactions . 2011; pp. 377-400.
[http://dx.doi.org/10.1002/9781119949138.ch15]

[45] Wang WX, Zhang F, Chen ZL, *et al.* Responses of phytohormones and gas exchange to mycorrhizal colonization in trifoliate orange subjected to drought stress. Arch Agron Soil Sci 2017; 63(1): 14-23.
[http://dx.doi.org/10.1080/03650340.2016.1175556]

[46] Berg M, Koskella B. Nutrient-and dose-dependent microbiome-mediated protection against a plant pathogen. Current Biology 2018; 28(15): 2487-92. e3.
[http://dx.doi.org/10.1016/j.cub.2018.05.085]

[47] Busby PE, Soman C, Wagner MR, *et al.* Research priorities for harnessing plant microbiomes in sustainable agriculture. PLoS Biol 2017; 15(3): e2001793.
[http://dx.doi.org/10.1371/journal.pbio.2001793] [PMID: 28350798]

[48] Shelake RM, Waghunde RR, Verma PP, Singh C, Kim JY. Carbon sequestration for soil fertility management: microbiological perspective. In: Panpatte D, Jhala y, Eds. Soil fertility management for sustainable development:. Springer 2019; pp. 25-42.
[http://dx.doi.org/10.1007/978-981-13-5904-0_3]

[49] Cao Y, Halane MK, Gassmann W, Stacey G. The role of plant innate immunity in the legume-rhizobium symbiosis. Annu Rev Plant Biol 2017; 68(1): 535-61.
[http://dx.doi.org/10.1146/annurev-arplant-042916-041030] [PMID: 28142283]

[50] Udvardi M, Poole PS. Transport and metabolism in legume-rhizobia symbioses. Annu Rev Plant Biol 2013; 64(1): 781-805.
[http://dx.doi.org/10.1146/annurev-arplant-050312-120235] [PMID: 23451778]

[51] Raja S, Thangappan S, Uthandi S. Non-rhizobial nodule associated bacteria [NAB] from blackgram [*Vigna mungo* L.] and their possible role in plant growth promotion. Madras Agric J 2019; 106: 451-9.

[52] Shelake RM, Waghunde RR, Morita EH, Hayashi H. Plant-microbe-metal interactions: basics, recent advances, and future trends. Plant microbiome: stress response. Egamberdieva D, Ahmad P. 2018; pp. 283-305. Eds
[http://dx.doi.org/10.1007/978-981-10-5514-0_13]

[53] Zoledowska S, Presta L, Fondi M, *et al.* Metabolic modeling of *Pectobacterium parmentieri* SCC3193 provides insights into metabolic pathways of plant pathogenic bacteria. Microorganisms 2019; 7(4): 101.
[http://dx.doi.org/10.3390/microorganisms7040101] [PMID: 30959803]

[54] Brader G, Compant S, Vescio K, *et al.* Ecology and genomic insights into plant-pathogenic and plant-nonpathogenic endophytes. Annu Rev Phytopathol 2017; 55(1): 61-83.

[http://dx.doi.org/10.1146/annurev-phyto-080516-035641] [PMID: 28489497]

[55] Hardoim PR, Van Overbeek LS, Berg G, *et al.* The hidden world within plants: ecological and evolutionary considerations for defining functioning of microbial endophytes. Microbiol Mol Biol Rev 2015; 79(3): 293-320.
[http://dx.doi.org/10.1128/MMBR.00050-14] [PMID: 26136581]

[56] Banerjee T, Chakravarti D. A peek into the complex realm of histone phosphorylation. Mol Cell Biol 2011; 31(24): 4858-73.
[http://dx.doi.org/10.1128/MCB.05631-11] [PMID: 22006017]

[57] Silipo A, Erbs G, Shinya T, *et al.* Glyco-conjugates as elicitors or suppressors of plant innate immunity. Glycobiology 2010; 20(4): 406-19.
[http://dx.doi.org/10.1093/glycob/cwp201] [PMID: 20018942]

[58] Nürnberger T, Brunner F. Innate immunity in plants and animals: emerging parallels between the recognition of general elicitors and pathogen-associated molecular patterns. Curr Opin Plant Biol 2002; 5(4): 318-24.
[http://dx.doi.org/10.1016/S1369-5266(02)00265-0] [PMID: 12179965]

[59] Nicaise V, Roux M, Zipfel C. Recent advances in PAMP-triggered immunity against bacteria: pattern recognition receptors watch over and raise the alarm. Plant Physiol 2009; 150(4): 1638-47.
[http://dx.doi.org/10.1104/pp.109.139709] [PMID: 19561123]

[60] Erbs G, Newman MA. The role of lipopolysaccharide and peptidoglycan, two glycosylated bacterial microbe-associated molecular patterns (MAMPs), in plant innate immunity. Mol Plant Pathol 2012; 13(1): 95-104.
[http://dx.doi.org/10.1111/j.1364-3703.2011.00730.x] [PMID: 21726397]

[61] Zipfel C, Felix G. Plants and animals: a different taste for microbes? Curr Opin Plant Biol 2005; 8(4): 353-60.
[http://dx.doi.org/10.1016/j.pbi.2005.05.004] [PMID: 15922649]

[62] Takakura Y, Ishida Y, Inoue Y, Tsutsumi F, Kuwata S. Induction of a hypersensitive response-like reaction by powdery mildew in transgenic tobacco expressing harpinpss. Physiol Mol Plant Pathol 2004; 64(2): 83-9.
[http://dx.doi.org/10.1016/j.pmpp.2004.06.002]

[63] Postel S, Küfner I, Beuter C, *et al.* The multifunctional leucine-rich repeat receptor kinase BAK1 is implicated in *Arabidopsis* development and immunity. Eur J Cell Biol 2010; 89(2-3): 169-74.
[http://dx.doi.org/10.1016/j.ejcb.2009.11.001] [PMID: 20018402]

[64] Eulgem T, Rushton PJ, Schmelzer E, Hahlbrock K, Somssich IE. Early nuclear events in plant defence signalling: rapid gene activation by WRKY transcription factors. EMBO J 1999; 18(17): 4689-99.
[http://dx.doi.org/10.1093/emboj/18.17.4689] [PMID: 10469648]

[65] Neill SJ, Desikan R, Clarke A, Hurst RD, Hancock JT. Hydrogen peroxide and nitric oxide as signalling molecules in plants. J Exp Bot 2002; 53(372): 1237-47.
[http://dx.doi.org/10.1093/jexbot/53.372.1237] [PMID: 11997372]

[66] Delledonne M, Xia Y, Dixon RA, Lamb C. Nitric oxide functions as a signal in plant disease resistance. Nature 1998; 394(6693): 585-8.
[http://dx.doi.org/10.1038/29087] [PMID: 9707120]

[67] Li C, Liu G, Xu C, *et al.* The tomato suppressor of prosystemin-mediated responses2 gene encodes a fatty acid desaturase required for the biosynthesis of jasmonic acid and the production of a systemic wound signal for defense gene expression. Plant Cell 2003; 15(7): 1646-61.
[http://dx.doi.org/10.1105/tpc.012237] [PMID: 12837953]

[68] Tiwari S, Lata C, Chauhan PS, Nautiyal CS. *Pseudomonas putida* attunes morphophysiological, biochemical and molecular responses in *Cicer arietinum* L. during drought stress and recovery. Plant Physiol Biochem 2016; 99: 108-17.

[http://dx.doi.org/10.1016/j.plaphy.2015.11.001] [PMID: 26744996]

[69] Grichko VP, Glick BR. Amelioration of flooding stress by ACC deaminase-containingplant growth-promoting bacteria. Plant Physiol Biochem 2001; 39(1): 11-7.
[http://dx.doi.org/10.1016/S0981-9428(00)01212-2]

[70] Creus CM, Sueldo RJ, Barassi CA. Water relations and yield in *Azospirillum*- inoculated wheat exposed to drought in the field. Can J Bot 2004; 82(2): 273-81.
[http://dx.doi.org/10.1139/b03-119]

[71] Madhaiyan M, Poonguzhali S, Lee HS, Hari K, Sundaram SP, Sa TM. Pink-pigmented facultative methylotrophic bacteria accelerate germination, growth and yield of sugarcane clone Co86032 (Saccharum officinarum L.). Biol Fertil Soils 2005; 41(5): 350-8. [Saccharum officinarum L.].
[http://dx.doi.org/10.1007/s00374-005-0838-7]

[72] Dimkpa C, Weinand T, Asch F. Plant-*rhizobacteria* interactions alleviate abiotic stress conditions. Plant Cell Environ 2009; 32(12): 1682-94.
[http://dx.doi.org/10.1111/j.1365-3040.2009.02028.x] [PMID: 19671096]

[73] Glick BR, Todorovic B, Czarny J, Cheng Z, Duan J, McConkey B. Promotion of plant growth by bacterial ACC deaminase. Crit Rev Plant Sci 2007; 26(5-6): 227-42.
[http://dx.doi.org/10.1080/07352680701572966]

[74] Jha Y, Subramanian RB, Patel S. Combination of endophytic and rhizospheric plant growth promoting *Rhizobacteria* in *Oryza sativa* shows higher accumulation of osmoprotectant against saline stress. Acta Physiol Plant 2011; 33(3): 797-802.
[http://dx.doi.org/10.1007/s11738-010-0604-9]

[75] Kim YC, Glick BR, Bashan Y, Ryu CM. Enhancement of plant drought tolerance by microbes. Plant responses to drought stress:. Aroca R. Springer 2012; pp. 383-413.
[http://dx.doi.org/10.1007/978-3-642-32653-0_15]

[76] Timmusk S, Abd El-Daim IA, Copolovici L, *et al.* Drought-tolerance of wheat improved by rhizosphere bacteria from harsh environments: enhanced biomass production and reduced emissions of stress volatiles. PLoS One 2014; 9(5): e96086.
[http://dx.doi.org/10.1371/journal.pone.0096086] [PMID: 24811199]

[77] Vardharajula S, Zulfikar Ali S, Grover M, Reddy G, Bandi V. Drought-tolerant plant growth promoting *Bacillus* spp.: effect on growth, osmolytes, and antioxidant status of maize under drought stress. J Plant Interact 2011; 6(1): 1-14.
[http://dx.doi.org/10.1080/17429145.2010.535178]

[78] Wang CJ, Yang W, Wang C, *et al.* Induction of drought tolerance in cucumber plants by a consortium of three plant growth-promoting rhizobacterium strains. PLoS One 2012; 7(12): e52565.
[http://dx.doi.org/10.1371/journal.pone.0052565] [PMID: 23285089]

[79] Zörb C, Geilfus CM, Dietz KJ. Salinity and crop yield. Plant Biol 2019; 21(S1) (Suppl. 1): 31-8.
[http://dx.doi.org/10.1111/plb.12884] [PMID: 30059606]

[80] Ngumbi E, Kloepper J. Bacterial-mediated drought tolerance: Current and future prospects. Appl Soil Ecol 2016; 105: 109-25.
[http://dx.doi.org/10.1016/j.apsoil.2016.04.009]

[81] Farooq M, Wahid A, Kobayashi N, Fujita D, Basra S. Plant drought stress: effects, mechanisms and management. Sustainable agriculture:. Springer: Lichtfouse, E, Navarrete, M, Debaeke, P, Véronique, S, Alberola, C. 2009; pp. 153-88.
[http://dx.doi.org/10.1007/978-90-481-2666-8_12]

[82] Huang B, DaCosta M, Jiang Y. Research advances in mechanisms of turfgrass tolerance to abiotic stresses: from physiology to molecular biology. Crit Rev Plant Sci 2014; 33(2-3): 141-89.
[http://dx.doi.org/10.1080/07352689.2014.870411]

[83] Zouari M, Hassena AB, Trabelsi L, Rouina BB, Decou R, Labrousse P. Exogenous proline-mediated

abiotic stress tolerance in plants: Possible mechanisms. Osmoprotectant-mediated abiotic stress tolerance in plants. Springer: Hossain, M, Kumar, V, Burritt, D, Fujita, M, Mäkelä, P 2019; pp. 99-121.

[84] Yang J, Kloepper JW, Ryu CM. Rhizosphere bacteria help plants tolerate abiotic stress. Trends Plant Sci 2009; 14(1): 1-4.
[http://dx.doi.org/10.1016/j.tplants.2008.10.004] [PMID: 19056309]

[85] Suárez R, Wong A, Ramírez M, et al. Improvement of drought tolerance and grain yield in common bean by overexpressing trehalose-6-phosphate synthase in rhizobia. Mol Plant Microbe Interact 2008; 21(7): 958-66.
[http://dx.doi.org/10.1094/MPMI-21-7-0958] [PMID: 18533836]

[86] Zhang H, Murzello C, Sun Y, et al. Choline and osmotic-stress tolerance induced in Arabidopsis by the soil microbe *Bacillus subtilis* (GB03). Mol Plant Microbe Interact 2010; 23(8): 1097-104.
[http://dx.doi.org/10.1094/MPMI-23-8-1097] [PMID: 20615119]

[87] Zhou C, Ma Z, Zhu L, et al. Rhizobacterial strain *Bacillus megaterium* BOFC15 induces cellular polyamine changes that improve plant growth and drought resistance. Int J Mol Sci 2016; 17(6): 976.
[http://dx.doi.org/10.3390/ijms17060976] [PMID: 27338359]

[88] Khan N, Bano A, Rahman MA, Guo J, Kang Z, Babar MA. Comparative physiological and metabolic analysis reveals a complex mechanism involved in drought tolerance in chickpea [*Cicer arietinum* L.] induced by PGPR and PGRs. Sci Rep 2019; 9(1): 2097.
[http://dx.doi.org/10.1038/s41598-019-38702-8] [PMID: 30765803]

[89] Cruz de Carvalho MH. Drought stress and reactive oxygen species. Plant Signal Behav 2008; 3(3): 156-65.
[http://dx.doi.org/10.4161/psb.3.3.5536] [PMID: 19513210]

[90] Hasanuzzaman M, Nahar K, Hossain MS, Anee TI, Parvin K, Fujita M. Nitric oxide pretreatment enhances antioxidant defense and glyoxalase systems to confer PEG-induced oxidative stress in rapeseed. J Plant Interact 2017; 12(1): 323-31.
[http://dx.doi.org/10.1080/17429145.2017.1362052]

[91] Gururani MA, Upadhyaya CP, Baskar V, Venkatesh J, Nookaraju A, Park SW. Plant growth-promoting *Rhizobacteria* enhance abiotic stress tolerance in *Solanum tuberosum* through inducing changes in the expression of ROS-scavenging enzymes and improved photosynthetic performance. J Plant Growth Regul 2013; 32(2): 245-58.
[http://dx.doi.org/10.1007/s00344-012-9292-6]

[92] Kasim WA, Osman ME, Omar MN, Abd El-Daim IA, Bejai S, Meijer J. Control of drought stress in wheat using plant-growth-promoting bacteria. J Plant Growth Regul 2013; 32(1): 122-30.
[http://dx.doi.org/10.1007/s00344-012-9283-7]

[93] Yogendra SG, U SS, A KS. Bacterial mediated amelioration of drought stress in drought tolerant and susceptible cultivars of rice (*Oryza sativa* L.). Afr J Biotechnol 2015; 14(9): 764-73.
[http://dx.doi.org/10.5897/AJB2015.14405]

[94] Saravanakumar D, Kavino M, Raguchander T, Subbian P, Samiyappan R. Plant growth promoting bacteria enhance water stress resistance in green gram plants. Acta Physiol Plant 2011; 33(1): 203-9.
[http://dx.doi.org/10.1007/s11738-010-0539-1]

[95] Hodge A, Berta G, Doussan C, Merchan F, Crespi M. Plant root growth, architecture and function. Plant Soil 2009; 321(1-2): 153-87.
[http://dx.doi.org/10.1007/s11104-009-9929-9]

[96] Ambreetha S, Chinnadurai C, Marimuthu P, Balachandar D. Plant-associated *Bacillus* modulates the expression of auxin-responsive genes of rice and modifies the root architecture. Rhizosphere 2018; 5: 57-66.
[http://dx.doi.org/10.1016/j.rhisph.2017.12.001]

[97]　Contreras-Cornejo HA, López-Bucio JS, Méndez-Bravo A, *et al.* Mitogen-activated protein kinase 6 and ethylene and auxin signaling pathways are involved in *Arabidopsis* root-system architecture alterations by *Trichoderma atroviride*. Mol Plant Microbe Interact 2015; 28(6): 701-10.
[http://dx.doi.org/10.1094/MPMI-01-15-0005-R] [PMID: 26067203]

[98]　Steele KA, Virk DS, Kumar R, Prasad SC, Witcombe JR. Field evaluation of upland rice lines selected for QTLs controlling root traits. Field Crops Res 2007; 101(2): 180-6.
[http://dx.doi.org/10.1016/j.fcr.2006.11.002]

[99]　Zamioudis C, Mastranesti P, Dhonukshe P, Blilou I, Pieterse CMJ. Unraveling root developmental programs initiated by beneficial *Pseudomonas* spp. bacteria. Plant Physiol 2013; 162(1): 304-18.
[http://dx.doi.org/10.1104/pp.112.212597] [PMID: 23542149]

[100]　Glick BR. The enhancement of plant growth by free-living bacteria. Can J Microbiol 1995; 41(2): 109-17.
[http://dx.doi.org/10.1139/m95-015]

[101]　Olanrewaju OS, Glick BR, Babalola OO. Mechanisms of action of plant growth promoting bacteria. World J Microbiol Biotechnol 2017; 33(11): 197.
[http://dx.doi.org/10.1007/s11274-017-2364-9] [PMID: 28986676]

[102]　Marulanda A, Azcón R, Chaumont F, Ruiz-Lozano JM, Aroca R. Regulation of plasma membrane aquaporins by inoculation with a *Bacillus megaterium* strain in maize (*Zea mays* L.) plants under unstressed and salt-stressed conditions. Planta 2010; 232(2): 533-43.
[http://dx.doi.org/10.1007/s00425-010-1196-8] [PMID: 20499084]

[103]　Bresson J, Varoquaux F, Bontpart T, Touraine B, Vile D. The PGPR strain *Phyllobacterium brassicacearum* STM 196 induces a reproductive delay and physiological changes that result in improved drought tolerance in *A rabidopsis*. New Phytol 2013; 200(2): 558-69.
[http://dx.doi.org/10.1111/nph.12383] [PMID: 23822616]

[104]　Vurukonda SSKP, Vardharajula S, Shrivastava M, SkZ A. Enhancement of drought stress tolerance in crops by plant growth promoting *Rhizobacteria*. Microbiol Res 2016; 184: 13-24.
[http://dx.doi.org/10.1016/j.micres.2015.12.003] [PMID: 26856449]

[105]　Egamberdieva D, Wirth SJ, Shurigin VV, Hashem A, Abd Allah EF. Abd_Allah EF. Endophytic bacteria improve plant growth, symbiotic performance of chickpea [*Cicer arietinum* L.] and induce suppression of root rot caused by Fusarium solani under salt stress. Front Microbiol 2017; 8: 1887.
[http://dx.doi.org/10.3389/fmicb.2017.01887] [PMID: 29033922]

[106]　Aroca R. Exogenous catalase and ascorbate modify the effects of abscisic acid [ABA] on root hydraulic properties in *Phaseolus vulgaris* L. plants. J Plant Growth Regul 2006; 25(1): 10-7.
[http://dx.doi.org/10.1007/s00344-005-0075-1]

[107]　Zhou N, Yao Y, Ye H, Zhu W, Chen L, Mao Y. Abscisic-acid-induced cellular apoptosis and differentiation in glioma *via* the retinoid acid signaling pathway. Int J Cancer 2016; 138(8): 1947-58.
[http://dx.doi.org/10.1002/ijc.29935] [PMID: 26594836]

[108]　Arkhipova TN, Prinsen E, Veselov SU, Martinenko EV, Melentiev AI, Kudoyarova GR. Cytokinin producing bacteria enhance plant growth in drying soil. Plant Soil 2007; 292(1-2): 305-15.
[http://dx.doi.org/10.1007/s11104-007-9233-5]

[109]　Cohen AC, Bottini R, Piccoli PN. *Azospirillum brasilense* Sp 245 produces ABA in chemically-defined culture medium and increases ABA content in arabidopsis plants. Plant Growth Regul 2008; 54(2): 97-103.
[http://dx.doi.org/10.1007/s10725-007-9232-9]

[110]　Sivakumaran A, Akinyemi A, Mandon J, *et al.* ABA *suppresses Botrytis cinerea* elicited NO production in tomato to influence H_2O_2 generation and increase host susceptibility. Front Plant Sci 2016; 7: 709.
[http://dx.doi.org/10.3389/fpls.2016.00709] [PMID: 27252724]

[111] Sánchez-Vallet A, López G, Ramos B, *et al*. Disruption of abscisic acid signaling constitutively activates *Arabidopsis* resistance to the necrotrophic fungus Plectosphaerella cucumerina. Plant Physiol 2012; 160(4): 2109-24.
[http://dx.doi.org/10.1104/pp.112.200154] [PMID: 23037505]

[112] Ulferts S, Delventhal R, Splivallo R, Karlovsky P, Schaffrath U. Abscisic acid negatively interferes with basal defence of barley against *Magnaporthe oryzae*. BMC Plant Biol 2015; 15(1): 7.
[http://dx.doi.org/10.1186/s12870-014-0409-x] [PMID: 25604965]

[113] Stec N, Banasiak J, Jasiński M. Abscisic acid - an overlooked player in plant-microbe symbioses formation? Acta Biochim Pol 2016; 63(1): 53-8.
[http://dx.doi.org/10.18388/abp.2015_1210] [PMID: 26828669]

[114] Fracetto GGM, Peres LEP, Lambais MR. Gene expression analyses in tomato near isogenic lines provide evidence for ethylene and abscisic acid biosynthesis fine-tuning during arbuscular mycorrhiza development. Arch Microbiol 2017; 199(5): 787-98.
[http://dx.doi.org/10.1007/s00203-017-1354-5] [PMID: 28283681]

[115] Desclos-Theveniau M, Arnaud D, Huang TY, *et al*. The Arabidopsis lectin receptor kinase LecRK-V.5 represses stomatal immunity induced by *Pseudomonas syringae* pv. tomato DC3000. PLoS Pathog 2012; 8(2): e1002513.
[http://dx.doi.org/10.1371/journal.ppat.1002513] [PMID: 22346749]

[116] Jiang CJ, Shimono M, Sugano S, *et al*. Abscisic acid interacts antagonistically with salicylic acid signaling pathway in rice-*Magnaporthe grisea* interaction. Mol Plant Microbe Interact 2010; 23(6): 791-8.
[http://dx.doi.org/10.1094/MPMI-23-6-0791] [PMID: 20459318]

[117] Kissoudis C, Seifi A, Yan Z, *et al*. Ethylene and abscisic acid signaling pathways differentially influence tomato resistance to combined powdery mildew and salt stress. Front Plant Sci 2017; 7: 2009.
[http://dx.doi.org/10.3389/fpls.2016.02009] [PMID: 28119708]

[118] Shigenaga AM, Argueso CT. No hormone to rule them all: Interactions of plant hormones during the responses of plants to pathogens. Semin Cell Dev Biol 2016; 56: 174-89.
[http://dx.doi.org/10.1016/j.semcdb.2016.06.005] [PMID: 27312082]

[119] Caarls L, Pieterse CMJ, Van Wees SCM. How salicylic acid takes transcriptional control over jasmonic acid signaling. Front Plant Sci 2015; 6: 170.
[http://dx.doi.org/10.3389/fpls.2015.00170] [PMID: 25859250]

[120] Broekgaarden C, Caarls L, Vos IA, Pieterse CMJ, Van Wees SCM. Ethylene: traffic controller on hormonal crossroads to defense. Plant Physiol 2015; 169(4): pp.01020.2015.
[http://dx.doi.org/10.1104/pp.15.01020] [PMID: 26482888]

[121] Müller M, Munné-Bosch S. Ethylene response factors: a key regulatory hub in hormone and stress signaling. Plant Physiol 2015; 169(1): 32-41.
[http://dx.doi.org/10.1104/pp.15.00677] [PMID: 26103991]

[122] Aleman F, Yazaki J, Lee M, *et al*. An ABA-increased interaction of the PYL6 ABA receptor with MYC2 Transcription Factor: A putative link of ABA and JA signaling. Sci Rep 2016; 6(1): 28941.
[http://dx.doi.org/10.1038/srep28941] [PMID: 27357749]

[123] Oide S, Bejai S, Staal J, Guan N, Kaliff M, Dixelius C. A novel role of PR2 in abscisic acid (ABA) mediated, pathogen-induced callose deposition in *Arabidopsis thaliana*. New Phytol 2013; 200(4): 1187-99.
[http://dx.doi.org/10.1111/nph.12436] [PMID: 23952213]

[124] Feng DX, Tasset C, Hanemian M, *et al*. Biological control of bacterial wilt in *Arabidopsis thaliana* involves abscissic acid signalling. New Phytol 2012; 194(4): 1035-45.
[http://dx.doi.org/10.1111/j.1469-8137.2012.04113.x] [PMID: 22432714]

[125] Lim C, Baek W, Jung J, Kim JH, Lee S, Jung J, Kim J-H, Lee SC. Function of ABA in stomatal defense against biotic and drought stresses. Int J Mol Sci 2015; 16(12): 15251-70.
[http://dx.doi.org/10.3390/ijms160715251] [PMID: 26154766]

[126] Goossens J, Fernández-Calvo P, Schweizer F, Goossens A. Jasmonates: signal transduction components and their roles in environmental stress responses. Plant Mol Biol 2016; 91(6): 673-89.
[http://dx.doi.org/10.1007/s11103-016-0480-9] [PMID: 27086135]

[127] Toum L, Torres PS, Gallego SM, Benavídes MP, Vojnov AA, Gudesblat GE. Coronatine inhibits stomatal closure through guard cell-specific inhibition of NADPH oxidase-dependent ROS production. Front Plant Sci 2016; 7: 1851.
[http://dx.doi.org/10.3389/fpls.2016.01851] [PMID: 28018388]

[128] Kaushal M, Wani SP. Rhizobacterial-plant interactions: Strategies ensuring plant growth promotion under drought and salinity stress. Agric Ecosyst Environ 2016; 231: 68-78.
[http://dx.doi.org/10.1016/j.agee.2016.06.031]

[129] Vacheron J, Desbrosses G, Bouffaud ML, et al. Plant growth-promoting *Rhizobacteria* and root system functioning. Front Plant Sci 2013; 4: 356.
[http://dx.doi.org/10.3389/fpls.2013.00356] [PMID: 24062756]

[130] Liu F, Xing S, Ma H, Du Z, Ma B. Cytokinin-producing, plant growth-promoting *Rhizobacteria* that confer resistance to drought stress in *Platycladus orientalis* container seedlings. Appl Microbiol Biotechnol 2013; 97(20): 9155-64.
[http://dx.doi.org/10.1007/s00253-013-5193-2] [PMID: 23982328]

[131] Saleem M, Arshad M, Hussain S, Bhatti AS. Perspective of plant growth promoting *Rhizobacteria* (PGPR) containing ACC deaminase in stress agriculture. J Ind Microbiol Biotechnol 2007; 34(10): 635-48.
[http://dx.doi.org/10.1007/s10295-007-0240-6] [PMID: 17665234]

[132] Glick BR. Bacterial ACC deaminase and the alleviation of plant stress. Adv Appl Microbiol 2004; 56: 291-312.
[http://dx.doi.org/10.1016/S0065-2164(04)56009-4] [PMID: 15566983]

[133] Cho S, Kim KD, Ahn JH, Lee J, Kim SW, Um Y. Selective production of 2,3-butanediol and acetoin by a newly isolated bacterium *Klebsiella oxytoca* M1. Appl Biochem Biotechnol 2013; 170(8): 1922-33.
[http://dx.doi.org/10.1007/s12010-013-0291-2] [PMID: 23793864]

[134] Chen Y, Gozzi K, Yan F, Chai Y. Acetic acid acts as a volatile signal to stimulate bacterial biofilm formation. MBio 2015; 6(3): e00392-15.
[http://dx.doi.org/10.1128/mBio.00392-15] [PMID: 26060272]

[135] Chakraborty U, Chakraborty BN, Chakraborty AP, Dey PL. Water stress amelioration and plant growth promotion in wheat plants by osmotic stress tolerant bacteria. World J Microbiol Biotechnol 2013; 29(5): 789-803.
[http://dx.doi.org/10.1007/s11274-012-1234-8] [PMID: 23239372]

[136] Ali SZ, Sandhya V, Venkateswar Rao L. Isolation and characterization of drought-tolerant ACC deaminase and exopolysaccharide-producing fluorescent *Pseudomonas* sp. Ann Microbiol 2014; 64(2): 493-502.
[http://dx.doi.org/10.1007/s13213-013-0680-3]

[137] Selvakumar G, Panneerselvam P, Ganeshamurthy AN. Bacterial mediated alleviation of abiotic stress in crops. Bacteria in agrobiology: stress management. Springer: Maheswari D. 2012; pp. 205-24.
[http://dx.doi.org/10.1007/978-3-642-23465-1_10]

[138] Karygianni L, Ren Z, Koo H, Thurnheer T. Biofilm matrixome: extracellular components in structured microbial communities. Trends Microbiol 2020; 28(8): 668-81.
[http://dx.doi.org/10.1016/j.tim.2020.03.016] [PMID: 32663461]

[139] Basu S, Rabara RC, Negi S, Shukla P. Engineering PGPMOs through gene editing and systems biology: a solution for phytoremediation? Trends Biotechnol 2018; 36(5): 499-510.
[http://dx.doi.org/10.1016/j.tibtech.2018.01.011] [PMID: 29455935]

[140] Niu B, Paulson JN, Zheng X, Kolter R. Simplified and representative bacterial community of maize roots. Proc Natl Acad Sci USA 2017; 114(12): E2450-9.
[http://dx.doi.org/10.1073/pnas.1616148114] [PMID: 28275097]

[141] Biswas D, Saha SC, Dey A. CRISPR-Cas genome-editing tool in plant abiotic stress-tolerance. Plant Gene 2021; 26: 100286.
[http://dx.doi.org/10.1016/j.plgene.2021.100286]

[142] Brandt K, Barrangou R. Applications of CRISPR technologies across the food supply chain. Annu Rev Food Sci Technol 2019; 10(1): 133-50.
[http://dx.doi.org/10.1146/annurev-food-032818-121204] [PMID: 30908954]

[143] Jinek M, Chylinski K, Fonfara I, Hauer M, Doudna JA, Charpentier E. A programmable dual-RNA-guided DNA endonuclease in adaptive bacterial immunity. Science 2012; 337(6096): 816-21.
[http://dx.doi.org/10.1126/science.1225829] [PMID: 22745249]

[144] Karvelis T, Gasiunas G, Miksys A, Barrangou R, Horvath P, Siksnys V. crRNA and tracrRNA guide Cas9-mediated DNA interference in *Streptococcus thermophilus*. RNA Biol 2013; 10(5): 841-51.
[http://dx.doi.org/10.4161/rna.24203] [PMID: 23535272]

[145] Hou Z, Zhang Y, Propson NE, *et al.* Efficient genome engineering in human pluripotent stem cells using Cas9 from *Neisseria meningitidis*. Proc Natl Acad Sci USA 2013; 110(39): 15644-9.
[http://dx.doi.org/10.1073/pnas.1313587110] [PMID: 23940360]

[146] Price AA, Sampson TR, Ratner HK, Grakoui A, Weiss DS. Cas9-mediated targeting of viral RNA in eukaryotic cells. Proc Natl Acad Sci USA 2015; 112(19): 6164-9.
[http://dx.doi.org/10.1073/pnas.1422340112] [PMID: 25918406]

[147] Kim E, Koo T, Park SW, *et al. In vivo* genome editing with a small Cas9 orthologue derived from *Campylobacter jejuni*. Nat Commun 2017; 8(1): 14500.
[http://dx.doi.org/10.1038/ncomms14500] [PMID: 28220790]

[148] Chatterjee P, Jakimo N, Jacobson JM. Minimal PAM specificity of a highly similar SpCas9 ortholog. Sci Adv 2018; 4(10): eaau0766.
[http://dx.doi.org/10.1126/sciadv.aau0766] [PMID: 30397647]

[149] Liu D, Huang C, Guo J, *et al.* Development and characterization of a CRISPR/Cas9n-based multiplex genome editing system for *Bacillus subtilis*. Biotechnol Biofuels 2019; 12(1): 197.
[http://dx.doi.org/10.1186/s13068-019-1537-1] [PMID: 31572493]

[150] Zetsche B, Gootenberg JS, Abudayyeh OO, *et al.* Cpf1 is a single RNA-guided endonuclease of a class 2 CRISPR-Cas system. Cell 2015; 163(3): 759-71.
[http://dx.doi.org/10.1016/j.cell.2015.09.038] [PMID: 26422227]

[151] Tóth E, Czene BC, Kulcsár PI, *et al.* Mb- and FnCpf1 nucleases are active in mammalian cells: activities and PAM preferences of four wild-type Cpf1 nucleases and of their altered PAM specificity variants. Nucleic Acids Res 2018; 46(19): 10272-85.
[http://dx.doi.org/10.1093/nar/gky815] [PMID: 30239882]

[152] Leenay RT, Maksimchuk KR, Slotkowski RA, *et al.* Identifying and visualizing functional PAM diversity across CRISPR-Cas systems. Mol Cell 2016; 62(1): 137-47.
[http://dx.doi.org/10.1016/j.molcel.2016.02.031] [PMID: 27041224]

[153] Wang L, Xiao Y, Wei X, Pan J, Duanmu D. Highly Efficient CRISPR-Mediated Base Editing in *Sinorhizobium meliloti*. Front Microbiol 2021; 12: 686008.
[http://dx.doi.org/10.3389/fmicb.2021.686008] [PMID: 34220774]

[154] Makarova KS, Wolf YI, Koonin EV. Classification and nomenclature of CRISPR-Cas systems: where

from here? CRISPR J 2018; 1(5): 325-36.
[http://dx.doi.org/10.1089/crispr.2018.0033] [PMID: 31021272]

[155] Strecker J, Jones S, Koopal B, *et al.* Engineering of CRISPR-Cas12b for human genome editing. Nat Commun 2019; 10(1): 212.
[http://dx.doi.org/10.1038/s41467-018-08224-4] [PMID: 30670702]

[156] Abudayyeh OO, Gootenberg JS, Kellner MJ, Zhang F. Nucleic acid detection of plant genes using CRISPR-Cas13. CRISPR J 2019; 2(3): 165-71.
[http://dx.doi.org/10.1089/crispr.2019.0011] [PMID: 31225754]

[157] Halperin SO, Tou CJ, Wong EB, Modavi C, Schaffer DV, Dueber JE. CRISPR-guided DNA polymerases enable diversification of all nucleotides in a tunable window. Nature 2018; 560(7717): 248-52.
[http://dx.doi.org/10.1038/s41586-018-0384-8] [PMID: 30069054]

[158] Zheng K, Wang Y, Li N, *et al.* Highly efficient base editing in bacteria using a Cas9-cytidine deaminase fusion. Commun Biol 2018; 1(1): 32.
[http://dx.doi.org/10.1038/s42003-018-0035-5] [PMID: 30271918]

[159] Han GZ. Origin and evolution of the plant immune system. New Phytol 2019; 222(1): 70-83.
[http://dx.doi.org/10.1111/nph.15596] [PMID: 30575972]

[160] Bisht DS, Bhatia V, Bhattacharya R. Improving plant-resistance to insect-pests and pathogens: The new opportunities through targeted genome editing. Semin Cell Dev Biol 2019; 96: 65-76.
[http://dx.doi.org/10.1016/j.semcdb.2019.04.008] [PMID: 31039395]

[161] Xin XF, Kvitko B, He SY. *Pseudomonas syringae*: what it takes to be a pathogen. Nat Rev Microbiol 2018; 16(5): 316-28.
[http://dx.doi.org/10.1038/nrmicro.2018.17] [PMID: 29479077]

[162] Sun Q, Lin L, Liu D, *et al.* CRISPR/Cas9-mediated multiplex genome editing of the BnWRKY11 and BnWRKY70 genes in *Brassica napus* L. Int J Mol Sci 2018; 19(9): 2716.
[http://dx.doi.org/10.3390/ijms19092716] [PMID: 30208656]

[163] Eoh J, Gu L. Biomaterials as vectors for the delivery of CRISPR–Cas9. Biomater Sci 2019; 7(4): 1240-61.
[http://dx.doi.org/10.1039/C8BM01310A] [PMID: 30734775]

[164] Ji J, Zhang C, Sun Z, Wang L, Duanmu D, Fan Q. Genome editing in cowpea (*Vigna unguiculata*) using CRISPR-Cas9. Int J Mol Sci 2019; 20(10): 2471.
[http://dx.doi.org/10.3390/ijms20102471] [PMID: 31109137]

[165] Van Esse HP, Reuber L, Van der Does D. GM approaches to improve disease resistance in crops. New Phytol 2019; 225: 70-86.
[http://dx.doi.org/10.1111/nph.15967] [PMID: 31135961]

[166] Upson JL, Zess EK, Białas A, Wu C, Kamoun S. The coming of age of EvoMPMI: evolutionary molecular plant–microbe interactions across multiple timescales. Curr Opin Plant Biol 2018; 44: 108-16.
[http://dx.doi.org/10.1016/j.pbi.2018.03.003] [PMID: 29604609]

[167] Ali Z, Mahfouz MM. CRISPR/Cas systems versus plant viruses: engineering plant immunity and beyond. Plant Physiol 2021; 186(4): 1770-85.
[http://dx.doi.org/10.1093/plphys/kiab220] [PMID: 35237805]

[168] Chen H, Wu H, Yan B, *et al.* Core microbiome of medicinal plant *Salvia miltiorrhiza* seed: a rich reservoir of beneficial microbes for secondary metabolism? Int J Mol Sci 2018; 19(3): 672.
[http://dx.doi.org/10.3390/ijms19030672] [PMID: 29495531]

[169] Huck S, Bock J, Girardello J, Gauert M, Pul Ü. Marker-free genome editing in *Ustilago trichophora* with the CRISPR-Cas9 technology. RNA Biol 2019; 16(4): 397-403.
[http://dx.doi.org/10.1080/15476286.2018.1493329] [PMID: 29996713]

[170] Schuster M, Kahmann R. CRISPR-Cas9 genome editing approaches in filamentous fungi and oomycetes. Fungal Genet Biol 2019; 130: 43-53.
[http://dx.doi.org/10.1016/j.fgb.2019.04.016] [PMID: 31048007]

[171] Gumtow R, Wu D, Uchida J, Tian M. A *Phytophthora palmivora* extracellular cystatin-like protease inhibitor targets papain to contribute to virulence on papaya. Mol Plant Microbe Interact 2018; 31(3): 363-73.
[http://dx.doi.org/10.1094/MPMI-06-17-0131-FI] [PMID: 29068239]

[172] Wang Q, Coleman JJ. CRISPR/Cas9-mediated endogenous gene tagging in *Fusarium oxysporum*. Fungal Genet Biol 2019; 126: 17-24.
[http://dx.doi.org/10.1016/j.fgb.2019.02.002] [PMID: 30738140]

[173] Li S, Shen L, Hu P, *et al.* Developing disease-resistant thermosensitive male sterile rice by multiplex gene editing. J Integr Plant Biol 2019; 61(12): 1201-5.
[http://dx.doi.org/10.1111/jipb.12774] [PMID: 30623600]

[174] Zheng Z, Nonomura T, Appiano M, *et al.* Loss of function in Mlo orthologs reduces susceptibility of pepper and tomato to powdery mildew disease caused by *Leveillula taurica*. PLoS One 2013; 8(7): e70723.
[http://dx.doi.org/10.1371/journal.pone.0070723] [PMID: 23923019]

[175] McCahill IW, Hazen SP. Regulation of cell wall thickening by a medley of mechanisms. Trends Plant Sci 2019; 24(9): 853-66.
[http://dx.doi.org/10.1016/j.tplants.2019.05.012] [PMID: 31255545]

[176] Giacomelli L, Zeilmaker T, Malnoy M, Rouppe van der Voort J, Moser C. Generation of mildew-resistant grapevine clones *via* genome editing. 2018.

[177] Nekrasov V, Wang C, Win J, Lanz C, Weigel D, Kamoun S. Rapid generation of a transgene-free powdery mildew resistant tomato by genome deletion. Sci Rep 2017; 7(1): 482.
[http://dx.doi.org/10.1038/s41598-017-00578-x] [PMID: 28352080]

[178] Kim Y-A, Moon H, Park C-J. CRISPR/Cas9-targeted mutagenesis of Os8N3 in rice to confer resistance to *Xanthomonas oryzae* pv. *oryzae*. Rice (N Y) 2019; 12(1): 1-13.
[http://dx.doi.org/10.1186/s12284-017-0196-8] [PMID: 30631971]

[179] Oliva R, Ji C, Atienza-Grande G, *et al.* Broad-spectrum resistance to bacterial blight in rice using genome editing. Nat Biotechnol 2019; 37(11): 1344-50.
[http://dx.doi.org/10.1038/s41587-019-0267-z] [PMID: 31659337]

[180] Ortigosa A, Gimenez-Ibanez S, Leonhardt N, Solano R. Design of a bacterial speck resistant tomato by CRISPR /Cas9-mediated editing of *Sl JAZ 2*. Plant Biotechnol J 2019; 17(3): 665-73.
[http://dx.doi.org/10.1111/pbi.13006] [PMID: 30183125]

[181] Ren B, Yan F, Kuang Y, *et al.* Improved base editor for efficiently inducing genetic variations in rice with CRISPR/Cas9-guided hyperactive hAID mutant. Mol Plant 2018; 11(4): 623-6.
[http://dx.doi.org/10.1016/j.molp.2018.01.005] [PMID: 29382569]

[182] Jiao Y, Li Y, Li Y, *et al.* Functional genetic analysis of the leucinostatin biosynthesis transcription regulator lcsL in *Purpureocillium lilacinum* using CRISPR-Cas9 technology. Appl Microbiol Biotechnol 2019; 103(15): 6187-94.
[http://dx.doi.org/10.1007/s00253-019-09945-2] [PMID: 31175427]

[183] Chandrasekaran J, Brumin M, Wolf D, *et al.* Development of broad virus resistance in non-transgenic cucumber using CRISPR/Cas9 technology. Mol Plant Pathol 2016; 17(7): 1140-53.
[http://dx.doi.org/10.1111/mpp.12375] [PMID: 26808139]

[184] Kalinina NO, Khromov A, Love AJ, Taliansky ME. CRISPR applications in plant virology: virus resistance and beyond. Phytopathology 2020; 110(1): 18-28.
[http://dx.doi.org/10.1094/PHYTO-07-19-0267-IA] [PMID: 31433273]

[185] Gomez MA, Lin ZD, Moll T, *et al.* Simultaneous CRISPR /Cas9-mediated editing of cassava *eIF 4E* isoforms *nCBP -1* and *nCBP -2* reduces cassava brown streak disease symptom severity and incidence. Plant Biotechnol J 2019; 17(2): 421-34.
[http://dx.doi.org/10.1111/pbi.12987] [PMID: 30019807]

[186] Macovei A, Sevilla NR, Cantos C, *et al.* Novel alleles of rice *eIF4G* generated by CRISPR/Cas9-targeted mutagenesis confer resistance to *Rice tungro spherical virus*. Plant Biotechnol J 2018; 16(11): 1918-27.
[http://dx.doi.org/10.1111/pbi.12927] [PMID: 29604159]

[187] Bastet A, Zafirov D, Giovinazzo N, *et al.* Mimicking natural polymorphism in *eIF 4E* by CRISPR - Cas9 base editing is associated with resistance to potyviruses. Plant Biotechnol J 2019; 17(9): 1736-50.
[http://dx.doi.org/10.1111/pbi.13096] [PMID: 30784179]

[188] Baltes NJ, Hummel AW, Konecna E, *et al.* Conferring resistance to geminiviruses with the CRISPR–Cas prokaryotic immune system. Nat Plants 2015; 1(10): 15145.
[http://dx.doi.org/10.1038/nplants.2015.145] [PMID: 34824864]

[189] Tashkandi M, Ali Z, Aljedaani F, Shami A, Mahfouz MM. Engineering resistance against *Tomato yellow leaf curl virusvia* the CRISPR/Cas9 system in tomato. Plant Signal Behav 2018; 13(10): e1525996.
[http://dx.doi.org/10.1080/15592324.2018.1525996] [PMID: 30289378]

[190] Tripathi JN, Ntui VO, Ron M, Muiruri SK, Britt A, Tripathi L. CRISPR/Cas9 editing of endogenous *banana streak virus* in the B genome of *Musa* spp. overcomes a major challenge in banana breeding. Commun Biol 2019; 2(1): 46.
[http://dx.doi.org/10.1038/s42003-019-0288-7] [PMID: 30729184]

[191] Aman R, Ali Z, Butt H, *et al.* RNA virus interference *via* CRISPR/Cas13a system in plants. Genome Biol 2018; 19(1): 1-9.
[http://dx.doi.org/10.1186/s13059-017-1381-1] [PMID: 29301551]

[192] Zhang T, Zhao Y, Ye J, *et al.* Establishing CRISPR /Cas13a immune system conferring RNA virus resistance in both dicot and monocot plants. Plant Biotechnol J 2019; 17(7): 1185-7.
[http://dx.doi.org/10.1111/pbi.13095] [PMID: 30785668]

[193] Mehta D, Stürchler A, Anjanappa RB, *et al.* Linking CRISPR-Cas9 interference in cassava to the evolution of editing-resistant geminiviruses. Genome Biol 2019; 20(1): 80.
[http://dx.doi.org/10.1186/s13059-019-1678-3] [PMID: 31018865]

[194] Pyne ME, Narcross L, Martin VJJ. Engineering plant secondary metabolism in microbial systems. Plant Physiol 2019; 179(3): 844-61.
[http://dx.doi.org/10.1104/pp.18.01291] [PMID: 30643013]

[195] Jiang C, Lv G, Tu Y, *et al.* Applications of CRISPR/Cas9 in the synthesis of secondary metabolites in filamentous fungi. Front Microbiol 2021; 12: 638096.
[http://dx.doi.org/10.3389/fmicb.2021.638096] [PMID: 33643273]

[196] Shanmugam K, Ramalingam S, Venkataraman G, Hariharan GN. The CRISPR/Cas9 system for targeted genome engineering in free-living fungi: advances and opportunities for lichenized fungi. Front Microbiol 2019; 10: 62.
[http://dx.doi.org/10.3389/fmicb.2019.00062] [PMID: 30792699]

[197] Chen XR, Zhang Y, Li HY, *et al.* The RXLR effector PcAvh1 is required for full virulence of *Phytophthora capsici*. Mol Plant Microbe Interact 2019; 32(8): 986-1000.
[http://dx.doi.org/10.1094/MPMI-09-18-0251-R] [PMID: 30811314]

[198] Liu Q, Gao R, Li J, *et al.* Development of a genome-editing CRISPR/Cas9 system in thermophilic fungal *Myceliophthora* species and its application to hyper-cellulase production strain engineering. Biotechnol Biofuels 2017; 10(1): 1-14.

[http://dx.doi.org/10.1186/s13068-016-0693-9] [PMID: 28053662]

[199] Wu J, Du G, Chen J, Zhou J. Enhancing flavonoid production by systematically tuning the central metabolic pathways based on a CRISPR interference system in *Escherichia coli*. Sci Rep 2015; 5(1): 13477.
[http://dx.doi.org/10.1038/srep13477] [PMID: 26323217]

[200] Kuivanen J, Wang YMJ, Richard P. Engineering *Aspergillus niger* for galactaric acid production: elimination of galactaric acid catabolism by using RNA sequencing and CRISPR/Cas9. Microb Cell Fact 2016; 15(1): 210.
[http://dx.doi.org/10.1186/s12934-016-0613-5] [PMID: 27955649]

[201] Katayama T, Tanaka Y, Okabe T, *et al*. Development of a genome editing technique using the CRISPR/Cas9 system in the industrial filamentous fungus *Aspergillus oryzae*. Biotechnol Lett 2016; 38(4): 637-42.
[http://dx.doi.org/10.1007/s10529-015-2015-x] [PMID: 26687199]

[202] Deng H, Gao R, Liao X, Cai Y. Genome editing in *Shiraia bambusicola* using CRISPR-Cas9 system. J Biotechnol 2017; 259: 228-34.
[http://dx.doi.org/10.1016/j.jbiotec.2017.06.1204] [PMID: 28690135]

[203] Wei TY, Wu YJ, Xie QP, *et al*. CRISPR/Cas9-based genome editing in the filamentous fungus *Glarea lozoyensis* and its application in manipulating gloF. ACS Synth Biol 2020; 9(8): 1968-77.
[http://dx.doi.org/10.1021/acssynbio.9b00491] [PMID: 32786921]

[204] McCarty NS, Graham AE, Studená L, Ledesma-Amaro R. Multiplexed CRISPR technologies for gene editing and transcriptional regulation. Nat Commun 2020; 11(1): 1281.
[http://dx.doi.org/10.1038/s41467-020-15053-x] [PMID: 32152313]

[205] Alagoz Y, Gurkok T, Zhang B, Unver T. Manipulating the biosynthesis of bioactive compound alkaloids for next-generation metabolic engineering in opium poppy using CRISPR-Cas9 genome editing technology. Sci Rep 2016; 6(1): 30910.
[http://dx.doi.org/10.1038/srep30910] [PMID: 27483984]

[206] Li R, Li R, Li X, *et al*. Multiplexed CRISPR/Cas9-mediated metabolic engineering of γ-aminobutyric acid levels in *Solanum lycopersicum*. Plant Biotechnol J 2018; 16(2): 415-27.
[http://dx.doi.org/10.1111/pbi.12781] [PMID: 28640983]

[207] Zhou Z, Tan H, Li Q, *et al*. CRISPR/Cas9-mediated efficient targeted mutagenesis of RAS in *Salvia miltiorrhiza*. Phytochemistry 2018; 148: 63-70.
[http://dx.doi.org/10.1016/j.phytochem.2018.01.015] [PMID: 29421512]

[208] Kui L, Chen H, Zhang W, *et al*. Building a genetic manipulation tool box for orchid biology: identification of constitutive promoters and application of CRISPR/Cas9 in the orchid, *Dendrobium officinale*. Front Plant Sci 2017; 7: 2036.
[http://dx.doi.org/10.3389/fpls.2016.02036] [PMID: 28127299]

[209] Aznar-Moreno JA, Durrett TP. Simultaneous targeting of multiple gene homeologs to alter seed oil production in *Camelina sativa*. Plant Cell Physiol 2017; 58(7): 1260-7.
[http://dx.doi.org/10.1093/pcp/pcx058] [PMID: 28444368]

[210] Mercx S, Smargiasso N, Chaumont F, De Pauw E, Boutry M, Navarre C. Inactivation of the β(1,2)-xylosyltransferase and the α(1,3)-fucosyltransferase genes in *Nicotiana tabacum* BY-2 Cells by a Multiplex CRISPR/Cas9 Strategy Results in Glycoproteins without Plant-Specific Glycans. Front Plant Sci 2017; 8: 403.
[http://dx.doi.org/10.3389/fpls.2017.00403] [PMID: 28396675]

[211] Li J, Zhang X, Sun Y, *et al*. Efficient allelic replacement in rice by gene editing: A case study of the *NRT1.1B* gene. J Integr Plant Biol 2018; 60(7): 536-40.
[http://dx.doi.org/10.1111/jipb.12650] [PMID: 29575650]

[212] Mohapatra S, Mishra SS, Bhalla P, Thatoi H. Engineering grass biomass for sustainable and enhanced

bioethanol production. Planta 2019; 250(2): 395-412.
[http://dx.doi.org/10.1007/s00425-019-03218-y] [PMID: 31236698]

[213] Kerfahi D, Tripathi BM, Dong K, *et al.* From the high Arctic to the equator: do soil metagenomes differ according to our expectations? Microb Ecol 2019; 77(1): 168-85.
[http://dx.doi.org/10.1007/s00248-018-1215-z] [PMID: 29882154]

[214] Santos KO, Costa-Filho J, Spagnol KL, Marins LF. Comparing methods of genetic manipulation in *Bacillus subtilis* for expression of recombinant enzyme: Replicative or integrative (CRISPR-Cas9) plasmid? J Microbiol Methods 2019; 164: 105667.
[http://dx.doi.org/10.1016/j.mimet.2019.105667] [PMID: 31295508]

[215] Fan Y, Liu J, Lyu S, Wang Q, Yang S, Zhu H. The soybean Rfg1 gene restricts nodulation by *Sinorhizobium fredii* USDA193. Front Plant Sci 2017; 8: 1548.
[http://dx.doi.org/10.3389/fpls.2017.01548] [PMID: 28936222]

[216] Van Zeijl A, Wardhani TAK, Seifi Kalhor M, *et al.* CRISPR/Cas9-mediated mutagenesis of four putative symbiosis genes of the tropical tree *Parasponia andersonii* reveals novel phenotypes. Front Plant Sci 2018; 9: 284.
[http://dx.doi.org/10.3389/fpls.2018.00284] [PMID: 29559988]

CHAPTER 5

Applications of Genome Editing in Bioremediation

Vibhuti Sharma[1], Rutika Sehgal[1], Vani Angra[1] and Reena Gupta[1,*]

[1] *Department of Biotechnology, Himachal Pradesh University, Summerhill, Shimla-171005, India*

Abstract: Excessive utilization of chemicals based substances such as pesticides, pharmaceuticals, fertilizers, inappropriate dumping of industrial materials and local wastes, *etc.*, into the environment is leading towards deliverance of high amounts of contaminants such as chlorinated hydrocarbons, dyes, toxins, petroleum and diesel spills into the soil. The mingling of these materials with soil and water is becoming one of the supreme complications associated with the environment, as these contaminants are a potential menace to human health. Bioremediation is a process that has the ability to destroy harmful contaminants and transform them into less toxic forms using living organisms such as bacteria, fungi, plants, *etc*. It is the most up-to-date nature-friendly approach to lower the extent of pollutants in the environment. With continuous developments in the scientific area, researchers are focussing on improving the process of bioremediation by using genome editing technologies. The gene editing techniques have the potential to significantly improve bioremediation processes such as xenobiotic removal, conversion of toxic compounds to less toxic compounds and pesticide degradation to simple components. The main gene editing techniques, CRISPR-Cas, ZFN and TALEN, have the potential to meet the aforementioned goals. This chapter focuses on the various gene editing tools and different genomic strategies such as gene editing, gene circuit, *etc.*, for the alteration or editing of the genome so that their potential value or applications can be seen in various areas.

Keywords: Bioremediation, Chlorinated, Contaminants, Dumping, Environment, Fertilizers, Hydrocarbons, Industrial, Nature-friendly, Pharmaceuticals, Pesticides, Pollutants, Potential, Spills.

INTRODUCTION

Environment plays a principal role in the well-living of all organisms as it furnishes food, water, shelter and other necessary things to living organisms. The standard of life on earth depends upon the quality of the environment present on earth [1]. In earlier times, all the essential resources were available in sufficient quantity, but nowadays, the excessive use of these resources is leading to their disappearance from nature and is also contaminating the environment. Today, the

---

[*] **Corresponding author Reena Gupta:** Department of Biotechnology, Himachal Pradesh University, Summerhill, Shimla-171005, India; E-mail: reenagupta_2001@yahoo.com

Prakash M. Halami & Aravind Sundararaman (Eds.)
All rights reserved-© 2024 Bentham Science Publishers

earth is experiencing a lot of changes because of development in almost every field, like agriculture, industry, transport, *etc* [2]. The increase in the standard of living is causing an increase in hazardous human activities which may include the disposal of hazardous waste, no proper sludge treatment, use of pesticides, *etc*. These human activities and accumulation of toxins and other pollutants are contaminating all the essential resources like water, soil, land, *etc.,* and causing a negative impact on the quality of our environment [2]. Some sources of environmental contaminants are shown in Fig. (**1**). There is a need to eliminate these contaminants from the environment to have a sustainable living world. This could be possible with the use of a biological technique called 'Bioremediation' [2].

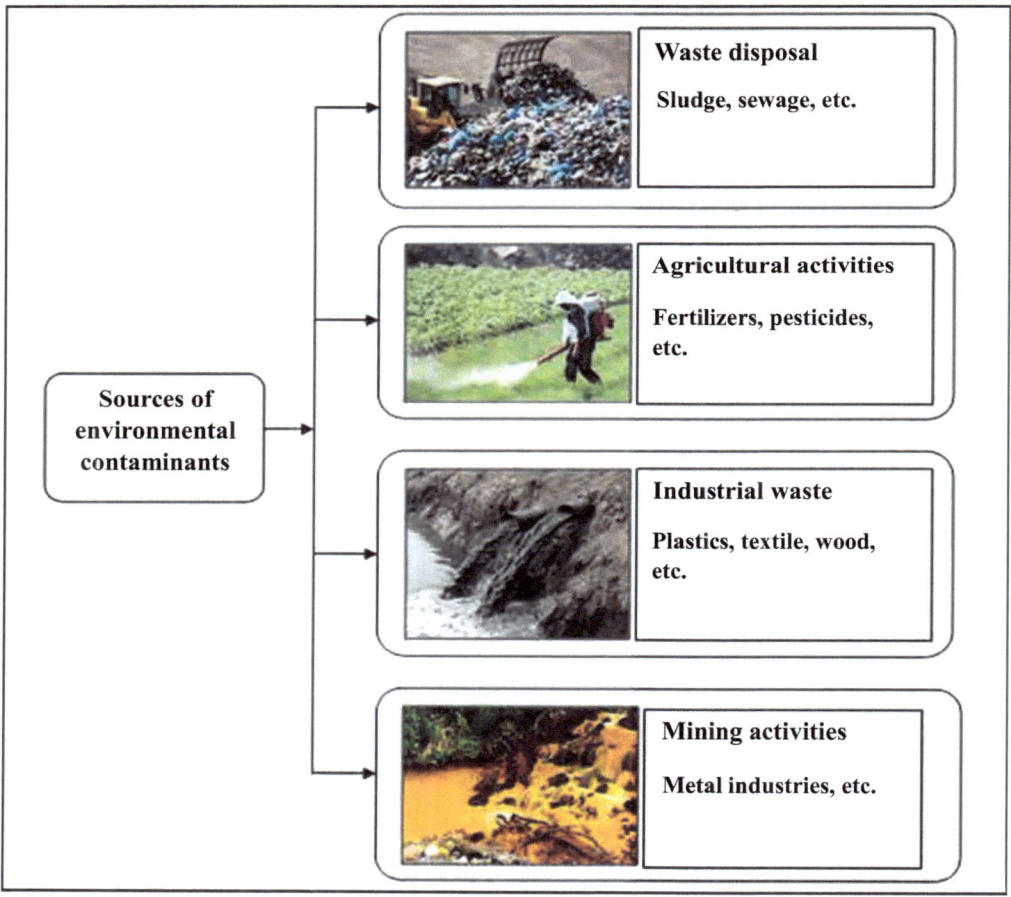

Fig. (1). Some sources of environmental contaminants [3].

Bioremediation is one of the most important practices which involve the use of microorganisms, plants, and fungi to transform hazardous compounds into less-toxic substances under specific conditions. This process is highly involved in the eradication, degradation, immobilization and detoxification of adverse chemical wastes from the environment by the action of microorganisms [4]. Though extremely specialized and diverse microbial groups are present in the environment to remove the harmful pollutants, they perform their task very slowly due to unacceptable environmental conditions, which leads to the accumulation of recalcitrant contaminants such as dichlorodiphenyltrichloroethane (DDT), hexachlorocyclohexane (HCH), *etc.* in the nature [5]. Therefore, to enhance the degradation activities of microbes, there is a need to modify them, which can be achieved by altering their genetic material.

Genomes are highly exposed to errors and changes that arise whenever a cell duplicates its DNA. These errors lead to mutations. When a mutation occurs in a particular gene, it alters its function. Genome editing is a technique of genetic engineering which generally means to change or alter the sequence of DNA or RNA of many organisms, such as microorganisms, plants, animals, *etc.*, by various methods like insertion, deletion, replacement, *etc*. This technology has the ability to treat many genetic disorders which occur due to mutations. The area of genome editing is developing at an immense rate [6]. There are various techniques used in the field of genome editing, which include recombinant DNA technology, Clustered Regularly Interspaced Short Palindromic Repeats (CRISPR)-Cas9, engineered endonucleases *etc.* These techniques can possibly make the advanced microbes with improved genes of interest that are required for the removal of recalcitrant contaminants from the environment.

BIOREMEDIATION

Bioremediation is a process that depends on some biological processes to degrade, reduce, detoxify, transform or mineralize the congregation of contaminants to a safer state [7]. The process of remediation aided by microbes present at the various polluted frameworks accounts for bioremediation [8, 9]. There are various microorganisms involved in the bioremediation of different harmful contaminants. Some microorganisms are shown in Table **1**.

Table 1. Some microorganisms involved in the bioremediation of contaminants [4].

Microorganism	Contaminants	References
Saccharomyces cerevisiae	Lead, Heavy metals, mercury, nickel, *etc.*	[10]

(Table 1) cont.....

Microorganism	Contaminants	References
Pseudomonas fluorescens and *Pseudomonas aeruginosa*	Iron, Zinc, Manganese, Lead, Copper	[11]
Enterobacter sp.	Chlorpyrifos	[12]
Staphylococcus sp., *Bacillus* sp.	Endosulfan	[13]
Bacillus subtilis	Oil based materials	[14]
Pseudomonas sp., *Ralstonia* sp. and *Microbacterium* sp.	Heterocyclic organic compounds	[15]
Bacillus safensis	Cadmium	[16]
Bacillus firmus, Bacillus macerans, Staphylococcus aureus and *Klebsiella oxytoca*	Textiles effluents	[17]

Many inorganic and organic environmental contaminants are subject to enzymatic attacks for degradation by living organisms. The production of these enzymes by microbes requires multiple metabolic pathways [18, 19]. These microbial enzymes play a very important role in the degradation pathways of xenobiotic compounds [20]. A group of enzymes called 'oxygenase' carries out the process of oxidation of chemical compounds [21]. Oxidation is liable for the splitting of aromatic rings and is also responsible for breaking down the tenacious xenobiotic compounds such as carcinogens, toxic chemicals, drugs, *etc* [22]. Biodegradation and bioconversion of recalcitrant compounds into non-toxic compounds carried out by oxygenase enzyme play a major role in the bioremediation process [18]. The objective of bioremediation is to stimulate microbes to perform their task by furnishing them with the best degree of nutrients and other chemicals which they require for their metabolism in order to detoxify substances which are perilous to the living world and environment. The microbial consortia can be maintained in many different ways, *e.g.,* by enhancing the microbial growth through nutrition supply, by managing the physical conditions like temperature and moisture, and by adding a terminal electron acceptor. Some microbes also acquire their nutrition from the contaminants [23]. The general idea of the maintenance of microbial cultures for bioremediation is shown in Fig. (**2**).

General Principle of Bioremediation

Bioremediation is an activity which participates in the deterioration of organic wastes to non-hazardous state with the help of bioremediators, under controlled conditions. Bacteria, archaea and fungi are typically essential bioremediators. Microorganisms are fit for the duty of destruction of contaminants as they contain enzymes that permit them to use environmental contaminants as food. All the metabolic reactions carried out by microbes are mediated by enzymes. These microbial enzymes belong to the class of oxidoreductases, transferases, hydrolases, lyases, isomerases and ligases. Many microbial enzymes have a capacity to degrade pollutants due to their affinity for specific and non-specific substrates. As bioremediation can be efficacious only when environmental conditions permit the growth of microorganisms and their activity, therefore, the implementation of bioremediation frequently involves the control of environmental parameters to allow microbial growth and degradation of pollutants to proceed at a faster rate [23]. The basic principles of the bioremediation process are:

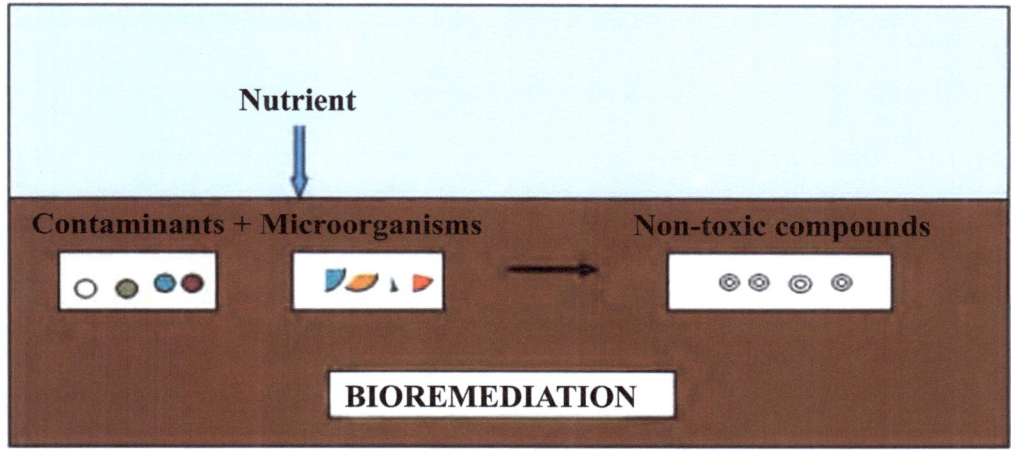

Fig. (2). The general idea of maintenance of microbial cultures for bioremediation [24].

• It utilizes various microorganisms that are capable of degrading the toxic substances into non-toxic or less toxic ones, either enzymatically or metabolically [25].

• The bioremediation can be carried out by using microflora indigenous to an area or by adding the consortia of capable microbes to the contaminated area to continue the degradation process [23].

• The microbes take sufficient nutrients from contaminants to enhance their own growth and metabolic ability to eliminate pollutants from nature.

Types of Bioremediation

There are different types of bioremediation processes. The selection of the type of bioremediation is done on the basis of many different criteria such as location, the concentration of contaminants and type of environment. On the basis of site of application, the bioremediation process is of two types: *in-situ* and *ex-situ*, which are further classified into various types [26]. Different types of bioremediation are shown in Fig. (**3**).

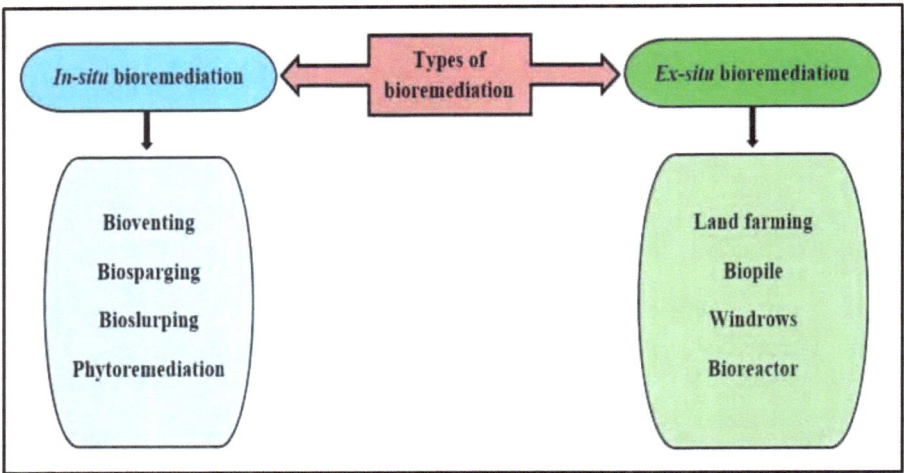

Fig. (3). Different types of bioremediation [7].

In-situ Bioremediation

In-situ bioremediation is also called natural bioremediation or intrinsic bioremediation. This type of bioremediation involves the treatment of contaminants only at the plot of contamination [27, 28]. *In-situ* bioremediation techniques are cheaper or less expensive in contrast to *ex-situ* bioremediation. These bioremediation techniques have been used profitably to treat heavy metals, dyes, chlorinated hydrocarbons and hydrocarbons polluted sites [29, 30]. Various types of *in-situ* bioremediation techniques are described below:

Bioventing

Bioventing is a process which involves the stimulation of natural *in-situ* bioremediation of contaminants in soil by providing oxygen and air to the indigenous microorganisms. Researchers modelled the process of bioventing by describing the effect of the rate of air injection on volatilization, biotransformation, and biodegradation of toluene-contaminated sites [31].

Biosparging

Biosparging is a method that is used for the remediation of Volatile Organic Compounds (VOC's). In this method, like bioventing, the air is injected into the saturated zone so that the volatile organic compounds can come upwards to the unsaturated zone due to the activities of the indigenous microbes [32]. There are some factors which affect the process of biosparging. These factors are: soil permeability, contaminant biodegradability, *etc* [33].

Bioslurping

Bioslurping is a technique which involves a combination of soil vapour extraction, vacuum-enhanced pumping and bioventing to achieve remediation of groundwater and soil by indirectly providing oxygen, thus, stimulating the contaminants to biodegradation [34].

Phytoremediation

Phytoremediation is a process in which plants are used to remove, stabilize and destroy contaminants from groundwater and soil. The plant used in the phytoremediation process is called phytoremediator. There are various factors on which the selection of phytoremediator depends. These factors include: type of root system (tap root system or fibrous root system), the toxicity of contaminants to plant, plant survival and its adaptability to prevailing environmental conditions, plant growth rate, *etc*.

Ex-situ Bioremediation

Ex-situ bioremediation involves the unearthing of contaminants from the plot of contamination and later on shifting them to any other plot for treatment [7]. There are different types of *ex-situ* bioremediation. Selection of the type of techniques to be preferred for bioremediation is done on the basis of type of pollutants, depth of pollution, cost of treatment, degree of contaminants, geographical region and geology of the polluted site. There are different types of *ex-situ* bioremediation techniques which are described below:

Landfarming

Landfarming is an *ex-situ* bioremediation technique or pollutants treatment method that is performed in the upper zone of the soil. Contaminants are transported to the landfarming site, incorporated into the surface of the soil and then turned over periodically to aerate the mixture. The basic principle behind the process of landfarming is the use of microorganisms to remove pollutants from the land by converting them into carbon dioxide and water by the process of oxidation [35, 36]. The major disadvantages of this process are: the requirement of enough space, the leaching of toxic metal ions from contaminated sites into the ground and non-biodegradability of inorganic compounds that remain adsorbed on the soil. To control these limitations, adsorbents or surfactants can be added to the soil to increase the level of bioavailability of organic pollutants for their removal [35].

Biopile

Biopiling involves the stacking of excavated pollutants above ground, which is further followed by the treatment of the stack with nutrients and air to increase the activities of microbes [36]. The technique of biopiling can minimize the vaporization of contaminants having low molecular weight. Aeration, a leachate collection system, irrigation, nutrients and a treatment bed are the major components of biopiling [37]. It can also be used successfully to remediate polluted harsh environments such as very cold regions [38, 39].

Windrows

Windrows is an *ex-situ* technique which involves the piling of contaminants similar to in the case of biopiling, but there is a periodic turning of the pile to enhance the rate of bioremediation [40]. In this process, there is uniform addition of water during the periodic turning of the pile. It increases the aeration level and makes uniform distribution of pollutants, nutrients and microbial degradative activities.

Bioreactor

The bioreactor is a container or a vessel in which the unprocessed materials are added and transformed into specific products with the help of microorganisms by following various reactions under specific suitable conditions [41 - 44]. There are many bioreactor processes which have resulted in the removal of a wide range of pollutants from the environment. Some processes are shown in Table **2**.

Table 2. Some bioreactor processes used in the removal of a wide range of pollutants from the environment [7].

Bioreactor	Sample	Contaminant	Percent Removal	References
Stir tank reactor	Lubricating oil	Saturated and aromatic hydrocarbons	65-70%	[42]
Membrane bioreactor	Coal gasification wastewater	Nitrogen	50-95%	[43]
Glass jar paddle-type impeller Reactor	Contaminated soil	2,4,6-trinitrophenylmethylnitramine	99.9%	[44]
Stir tank reactor	Crude polluted oil	Total petroleum and polyaromatic Hydrocarbons	80-95%	[45]

(Table 2) cont.....

Bioreactor	Sample	Contaminant	Percent Removal	References
Expanded granular sludge bed	Laundry Wastewater	Linear alkylbenzenesulfonate	93%	[46]
Roller slurry bioreactor	Contaminated soil	2,4-Dichlorophenoxyacetic Acid	98%	[47]

Why Bioremediation is Important?

The rapidly increasing population is causing an extended enhancement in the development of various manufacturing industries. There are many xenobiotic compounds that are released into the environment from these industries. The excessive use of resources and the release of chemicals from industries lead to the scarcity of all essential resources and the contamination of land, respectively [48]. If such problems are not treated on time, they can cause a lot of problems to the living world, and it will be the toughest task to maintain the quality of life. The bioremediation technique plays a very important role in returning the environment from its contaminated state to its original state. It is a natural crucial process which works as an alternative to old conventional techniques of pollutant removal such as incineration, use of adsorbents, *etc* [23].

Heavy metal contamination is one of the most significant environmental issues due to the following reasons:

• Metals are highly toxic to biota.

• Heavy metals decrease diversity and metabolic activity.

• They affect the quantitative and qualitative structure of microbial communities.

For treating heavy metals-contaminated areas or soils, bioremediation is a cost-effective method. Microorganisms have not evolved appropriate pathways to accumulate contaminants because of the toxicity of these compounds. A very small population of microorganisms is responsible for this bioaccumulation and is not active enough to remove these compounds completely because complex mixtures of pollutants resist their removal from the environment. The addition of exogenous microorganisms into the contaminated site is one of the efficient approaches. This process is commonly known as bio-augmentation. It can be performed either by adding natural catabolic genes containing microorganisms into the contaminated sites or by using those microbes that have been modified genetically. This plan also results in the transmission of plasmids with requisite genetic material between different populations.

Bioremediation Process: Mechanism

Bioremediation is the process of degrading environmental pollutants with the help of microbes. It is an environment-friendly procedure that makes use of microbial processes to remove the toxic contaminants completely. Microorganisms play a vital role in our life as they are very important in preserving the aerosphere [49], controlling the biogeochemical cycles [50], keeping us well and helping plants in their proper growth by suppressing plant diseases [51]. Microbes also participate in the cleaning of environmental pollutants [49]. The remediation process mediated by microbes is of great sense because it provides us with simpler, cheaper and environmentally-friendly remedial methods [52]. The minimum understanding of the mechanisms that control the growth and activity of microorganisms in contaminated environments is continuously restricting the execution of the bioremediation process [52]. Nutritional flexibility, active behaviour and ability to survive in extreme environmental conditions make microbes the most suitable life forms for toleration. These properties of microbes are much favourable and advantageous to remove toxic compounds and other pollutants from the environment [53].

Microorganisms have the ability to carry out the re-establishment of the environment through an assembly of processes which include oxidation, binding, volatilization, immobilization, and chemical transformation of the pollutants. One of the most common of these bioremediation techniques is the oxidation of toxic organic pollutants to harmless products. An extensive variety of microorganisms is able to degrade aerobic contaminants [54], especially *Pseudomonas* species. These are profoundly used due to their ability to degrade many different contaminants.

At the present time, the bioremediation process generally depends upon the sample collection from contaminated sites and further incubating them under laboratory environment, and finally, the rates of degradation and immobilization of contaminants are recorded [55]. The recorded data provides an approximation concerning the potential metabolic activities of the microbial consortia responsible for bioremediation.

GENOME EDITING IN BIOREMEDIATION

The genome editing techniques may activate new advanced development of microbial bioremediation technologies. Genome editing is the primary means of interpreting the function of genes in microorganisms. Microbial gene functions come from manipulating the genome of individual species from their natural communities. Gene editing and systemic biology tools are being applied in bioremediation with the advancement of scientific research methodologies.

Bioremediation is the degradation ability of microorganisms to dissimilate complex chemical compounds from the surrounding environment into simpler chemical compounds [8]. Genome editing is very important in making the bioremediation process more efficient. It helps to get the function of specific microorganisms with particular genes and enzymes responsible for bioremediation. Omics approaches such as metabolomics, genomics, transcriptomics and proteomics support the systems biology studies of microbes for the analysis of the regulation of the bioremediation process at the genetic level [8]. Evolution in sequencing with new techniques such as high throughput sequencing and next-generation sequencing resolves the novel genes involved in biodegradation pathways of different recalcitrant pollutants.

ROLE OF GENOME EDITING IN BIOREMEDIATION

Genome editing plays a very important role in the bioremediation process by removing pesticides, xenobiotic compounds, persistent organic pollutants (POPs), heavy metals, petroleum products, *etc* [8]. Genome editing is an astonishing approach having the ability to handle DNA with the help of molecular scissors [56 - 58]. These molecular scissors have vast applications in a wide range of research fields related to plants, animals, and microorganisms, *etc* [59, 60].

Microorganisms are broadly dispersed in the natural world as they can easily grow in a wide range of environmental conditions and their metabolic ability is very magnificent. Various microorganisms have the ability to degrade pollutants such as petroleum hydrocarbons and they can also utilize them as carbon and energy source. The specificity of the degradation process is related to the genetic ability of the specific microorganism to incorporate molecular oxygen into petroleum hydrocarbon and to give rise to the intermediates such as water, carbon dioxide, biomass, *etc.*, which further enter the general energy-yielding metabolic pathway of the cell [61]. Genome engineering by gene editing tools results in many new applications of microorganisms in various areas like feed, food, agriculture, medicine, *etc* [62]. The gene editing tools have a powerful capacity to upgrade the bioremediation processes such as the removal of xenobiotics from the environment, conversion of toxic compounds to less toxic compounds and biodegradation of pesticides to simple components [8, 63].

GENOMIC TOOLS USED FOR BIOREMEDIATION OF CONTAMINANTS

With growing developments in a branch of computational tools, it is now possible to study the genes and interaction of genes within the cellular organism. There are many different tools available at the genomic level that can be used in the bioremediation of various harmful contaminants. Different genomic tools and microorganisms involved in bioremediation of various contaminants are shown in Table 3.

Table 3. Different genomic tools and microorganisms involved in the bioremediation of various contaminants [64].

Genomic Tool	Microorganism	Contaminants	References
Second generation sequencing	*Marinomonas, Bacillus, Dietzia, Colwellia, Halomonas, Pseudoalteromonas, Acinetobacter, Alcanivorax, Salinisphaera, Cycloclasticus, Pseudomonas* and *Shewanella,* etc.	Polycyclic Aromatic Hydrocarbons (PAHs)	[65, 66]
Fluorescent *in situ* Hybridization (FISH)	*Dehalococcoides*	Chlorinated solvents	[67]
Restriction Fragment Length Polymorphism (RFLP), Fingerprinting methods	*Thermoanaerobacteraceae, Desulfobulbaceae*	Naphthalene	[68]
Cloning and sequencing of ribosomal DNA	*Stenotrophomonas maltophilia*	Pesticides, heavy metals	[69 - 71]

CRISPR-Cas, ZFN and TALEN are the main gene editing tools. The comparison between genome editing tools is shown in Table 4. The major aim of these gene editing tools is to design microorganisms with maximum quality and to create better microbes with more complex genes so that these microorganisms can contribute in various fields in different areas [8, 19].

Table 4. Comparison between different genome editing tools [72].

Features	CRISPR	TALEN	ZFN	References
Specificity	CrRNA	TALE domain	Zn finger domain	[72]
Cleavage	Cas9	FokI nuclease	Nuclease	[72, 73]
Activity	High	High	Moderate	[74]
Designing and screening	Easy	Difficult	Difficult	[19]

CRISPR-Cas9

CRISPR-Cas stands for Clustered Regularly Interspaced Short Palindromic Repeats. CRISPR-Cas systems comprise CRISPR RNA (crRNA) and Cas proteins. The crRNA is complementary to the target sequence and thus mentors the Cas proteins for sequence-specific recognition, identification and cleavage [75]. It is a technology that allows researchers to change different genes by adding, altering or deleting entire genes. Over the last few years, CRISPR-Cas technology has acquired extensive popularity because of its specificity and simplicity. The CRISPR technique is obtaining a lot of attention from biologists because of its versatility in genome editing as compared to traditional DNA engineering strategies like lambda red recombineering, *etc* [76 - 78]. Furthermore, engineering the Cas proteins to nuclease deficient Cas (dCas) further increases the power of CRISPR-Cas based systems to efficient, easy and multi-target transcriptional repression and activation. New CRISPR-Cas based technologies are continually being developed in different ways, which further increases the scope of this technology.

Various biological techniques and experiments have been re-defining the CRISPR-Cas technologies into 3 distinct types of CRISPR-Cas systems. Each system has specific Cas on the basis of model organisms [79]. Three types of CRISPR-Cas systems are as follows:

Type 1 CRISPR-Cas System

For pre-processing of crRNA, this system utilized Cas5 or Cas6; it requires Cas3 for further cleavage function, cascade and crRNA for interference.

Type 2 CRISPR-Cas System

Cas9 functions under the guidance of crRNA to target DNA, RNase III, trans activating RNA, and many other protein factors are involved in trimming at the 5' end.

Type 3 CRISPR-Cas System

This system uses Cas6 for processing crRNA 3' end trimming. The uniqueness of this system is its RNA targeting property, which is done by a specific complex called type III Csm/Cmr complex [80].

Certain species of bacteria and fungi have evolved naturally to degrade compounds found in plastics. CRISPR could be employed to increase the activity of the genetic pathways involved in the degradation process. Scientists have already begun applying CRISPR to microbes that are good candidates for producing and degrading plastics. The CRISPR-Cas9 system is mostly adopted and performed in the model organisms like *Pseudomonas, Escherichia coli, etc*. Other microorganisms like *Comomonas testosterone, Rhodococcus ruber* TH, and *Achromobacter* sp. HZO1 are suggested in the area of bioremediation [64].

There are many other types of bioremediation where CRISPR might prove handy. For instance, microbes or plants could be engineered to take up heavy metals more efficiently for cleaning up oil spills or to improve wastewater treatments.

Cas9, produced from the bacteria *Streptococcus pyogenes*, is a part of the CRISPR/Cas9 system. It is a fresh addition to the toolkit of programmable nucleases [81, 82]. Cas9 is a monomeric nuclease that forms a ribonucleoprotein (RNP) complex with a chimeric guide RNA (gRNA). The Cas9 enzyme contains two nuclease domains; each of the domains cleaves one strand of the target sequence and three nucleotides upstream of the protospacer adjacent motif (PAM) to generate blunt ends [83]. One major advantage of CRISPR/Cas9 over other genome editing tools (ZFNs and TALENs) is that there is no need to engineer Cas9 enzyme at the protein level to acknowledge different targets. Target specificity is conferred absolutely by the spacer region of the gRNA and the sequence can be altered using standard molecular biology methods [84].

TALENs

TALENs stands for Transcription Activator like-Effector Nucleases. These are one of the important tools used for genome editing. TALENs system is associated with the study of the bacteria *Xanthomonas* genus. *Xanthomonas* species cause significant damage to agricultural crops such as pepper, tomato, rice, *etc*. These bacteria are found to secrete some effector proteins called TAL proteins. These proteins make the crops more susceptible to pathogens. TALENs bind to their host DNA, act as transcription factors and activate the expression of plant genes that aid bacterial infection. This non-specific DNA-cleaving nuclease can be fused to a DNA-binding domain that can be easily engineered to target any desired sequence by a technique known as genome editing with engineered nucleases. Two protein domains, one for sequence cleavage and the second for recognizing and binding the specific site, make the TALENs a robust gene editing tool and have a great impact on research on bioremediation.

ZFNs

ZFN stands for Zinc Finger Nucleases. Zinc Finger Nucleases have Zinc Finger Proteins (ZFPs) which have the ability to act as a domain for DNA binding. The first zinc finger motif was discovered as part of the transcription factor in *Xenopus* oocytes which had a specific binding affinity to DNA. ZFPs are 30 amino acids long with alpha-helix in opposition to two antiparallel beta-sheets. ZFNs have a nucleotide cleavage domain derived from *Flavobacterium okeanokoites*. Many ZFPs surround the cleavage domain depending on the target site. ZFPs have 18bp specificity, possibly making the precise target specific gene editing. The two ZFN molecules stick to the targeted DNA in a tail-to-tail manner, separated by 5–7bp, with double-stranded DNA cleavage occurring in the spacer region [64]. The role of these tools in bioremediation is shown in Fig. (**4**).

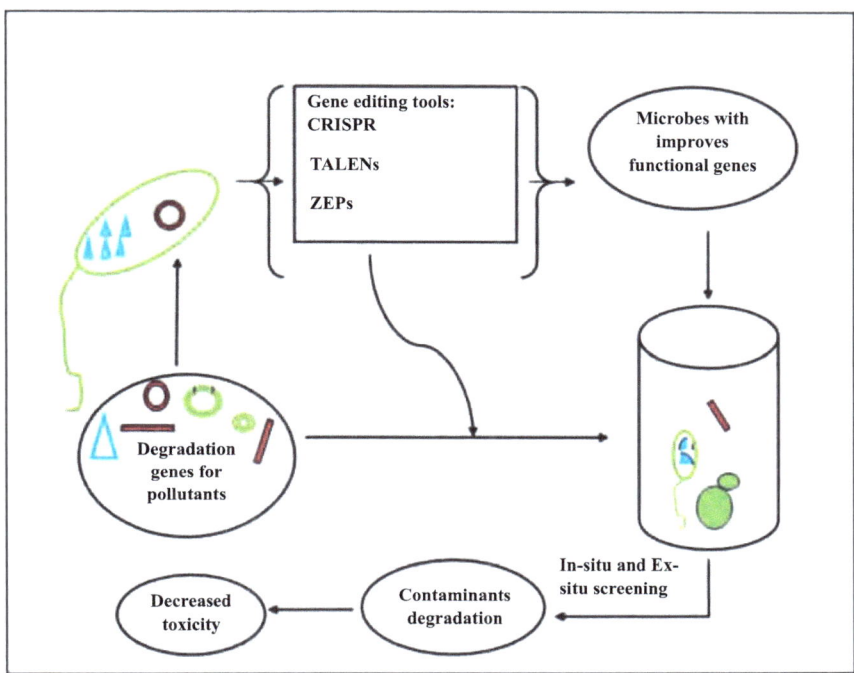

Fig. (4). Role of Gene editing tools in Bioremediation [64].

GENETIC ENGINEERING OF MICROORGANISMS

Genetically engineered microorganisms are those microorganisms whose genetic material is altered by applying genetic engineering techniques such as genome sequencing and artificial genetic exchange between microorganisms, *etc*. This kind of technical procedure is mainly known as recombinant DNA technology. Genetic engineering has been playing a vital role in the exclusion of hazardous, unwanted wastes from the environment by creating genetically modified organisms under laboratory conditions [85]. Modified living organisms can be produced by recombinant DNA techniques or by naturally switching the genetic material between organisms. There are various bacterial strains whose whole genome has been sequenced, *i.e., Rhodococcus* sp., *Pseudomonas putida, etc*.

Genetically engineered microorganisms (GEMs) have shown great potential for applications of bioremediation in soil, groundwater, and activated sludge environments, exhibiting enhanced degradative capabilities encompassing a wide range of chemical contaminants. Recently, a number of chances have been put forward for ameliorating the degradative abilities of microbes using genetic engineering plans. For example, rate-limiting steps in the metabolic pathways can be genetically modified to increase the degradation rates or totally new metabolic pathways can be incorporated into bacterial strains for the degradation of recalcitrant compounds.

Advantages of GEMs in Bioremediation

The major functions of GEMs are:

- To fasten up the recovery of waste polluted sites.

- To increase the rate of degradation of pollutants.

- To display a high catalytic capacity with a small amount of cell mass.

- To design purified and safe environmental conditions by decontamination or by counterbalancing any harmful substances.

Genetically Engineered Bacteria for Bioremediation

Bacteria have a high ability to degrade environmental contaminants. Many toxic and recalcitrant xenobiotic compounds, such as halogenated aromatic compounds, some pesticides and explosives, are usually stable and chemically inert under natural conditions and not known to be degraded efficiently by many microorganisms [86]. These limitations open the path for the development of genetically modified microorganisms. There are various strategies for the production of genetically modified bacteria:

Production of Biosurfactants

The prolonged persistence of different pollutants in the environment is due to their solubilization-limited bioavailability. A possible way to enhance the bioavailability of xenobiotics (pesticides, pharmaceuticals, petroleum compounds, polycyclic aromatic hydrocarbons, PCBs *etc.*) and their biodegradation is the application of "biosurfactants". The application of biosurfactants has shown the ability to increase the rate of degradation of hydrophobic pollutants, but the high cost of biosurfactant production restricts their application. Therefore, many important efforts are directed toward the design of recombinant biocatalysts that exhibit a desired catabolic trait and produce a suitable biosurfactant [87]. For example: A recombinant bacterium *Pseudomonas aeruginosa* has the ability to desulfurize dibenzothiophene more efficiently than the native host.

Optimizing Biocatalysts

Biocatalysts can enhance the rate of contaminant degradation processes. The proper optimization of biocatalysts requires different genetic strategies that can be combined to generate novel, improved and efficient degradation activities. One of the major strategies for designing superior biocatalysts is the rational combination of catabolic segments from different organisms within one recipient strain. Therefore, complete metabolic routes for xenobiotics can be generated and the formation of dead-end products or even toxic metabolites can be avoided. For example: the mineralization of polychlorinated biphenyls (PCBs) can be carried out by combining an oxidative pathway for chlorobiphenyl conversion into chlorobenzoate with a degradative pathway [87, 88].

Genetically Engineered Fungi for Bioremediation

Genetic engineering technology makes fungi a choice of interest for bioremediation [89]. Genes of fungus can be cloned to meet the objectives of bioremediation. Fungal mutants that over secrete specific enzymes can be produced, and various processes may be designed and scaled up in the treatment of waste and wastewater using such mutants. Noteworthy advancement related to fungi has been attained in molecular biology, especially the extraction of genetic material (RNA and DNA), gene cloning, and genetic engineering of fungi, *etc* [90]. Manipulation and genetic engineering of fungal enzymes can increase their activities for bioremediation.

Genetically Engineered Plants for Bioremediation

Generally, microorganisms show high capabilities for efficient biodegradation in laboratory conditions, but sometimes they may not perform well in actual sites of contamination [91]. Therefore, to carry out the bioremediation process at the sites of contamination, plants can be used in place of microorganisms. Plants provide protection against water and wind erosion and also prevent contaminants from spreading; they are robust in growth and are renewable resources [92, 93]. The process of phytoremediation is generally very slow, but the effectiveness of this process can be enhanced by over-expressing those genes in plants that are involved in the metabolism, uptake, or transport of specific pollutants [94 - 97].

Transgenic plants for phytoremediation were first developed for remediating heavy metal contaminated soil sites. Plants have metallothionein genes encoding cysteine rich peptides that are generally made up of 60–80 amino acids and contain 9–16 cysteine residues [98]. These metallothionein genes can protect plants from the effects of toxic heavy metal ions such as silver, cadmium, cobalt, copper, mercury, nickel, *etc.* Therefore, metallothionein genes have been cloned and introduced into several plant species. Also, the transfer of the human metallothionein gene in tobacco plants resulted in plants with enhanced cadmium tolerance [99], and the pea metallothionein gene in *Arabidopsis thaliana* enhanced copper accumulation [100].

GENOME EDITING TECHNOLOGIES USED FOR THE MODIFICATION OF MICROORGANISMS

There are various genome editing technologies that can be used to modify any microbial cell capable of degrading specific compounds and also to make it able to remain viable for many generations. The commonly used strategies to develop genetically modified microorganisms for bioremediation applications have been discussed below:

Rational Designing

The process of setting up of microorganisms having the assembly of all essential biodegradation enzymes or pathways from different organisms to perform specific reactions is known as rational designing [101]. The modification of microorganisms requires complete knowledge of the dynamic, mechanical and structural properties of proteins. For example: The construction of a strain of *Pseudomonas putida* to degrade an organophosphorus compound paraoxon which is capable of utilizing this compound as a sole carbon and nitrogen source after the mutations [102].

Genome Shuffling

The process of recombination of the chromosomes from several bacteria to produce a bacterium which has the improved activity of a desired trait is called genome shuffling. For example: The degradation of the pesticide 'pentachlorophenol' by *Sphingobium chlorophenolicum* has been enhanced by genomic shuffling up to tenfold in comparison to wild-type strain [103, 104].

Family Shuffling

Family shuffling involves the shuffling of the DNA of the related groups of genes to accelerate the bioremediation process [87]. For example: The family shuffling of the genes of biphenyl dioxygenase from *Pseudomonas pseudoalcaligenes* and *Bacillus cepacia* is one of the major applications of protein engineering for bioremediation [87].

GENOMIC STRATEGIES AND OMICS APPROACHES USED IN THE PROCESS OF BIOREMEDIATION

Different genomic strategies, such as gene editing, gene circuit, *etc.,* can maintain the natural metabolic potential of microorganisms [19]. Moreover, the controlled regulation of metabolic pathways can be achieved for regulating the bioremediation processes [104, 105]. The changes achieved by utilizing these synthetic biology applications for remediation purposes would also improve the bioremediation processes *via* the involvement of potent dissimilating contaminants [106, 107]. Different strategies used in the process of bioremediation are shown in Fig. (**5**).

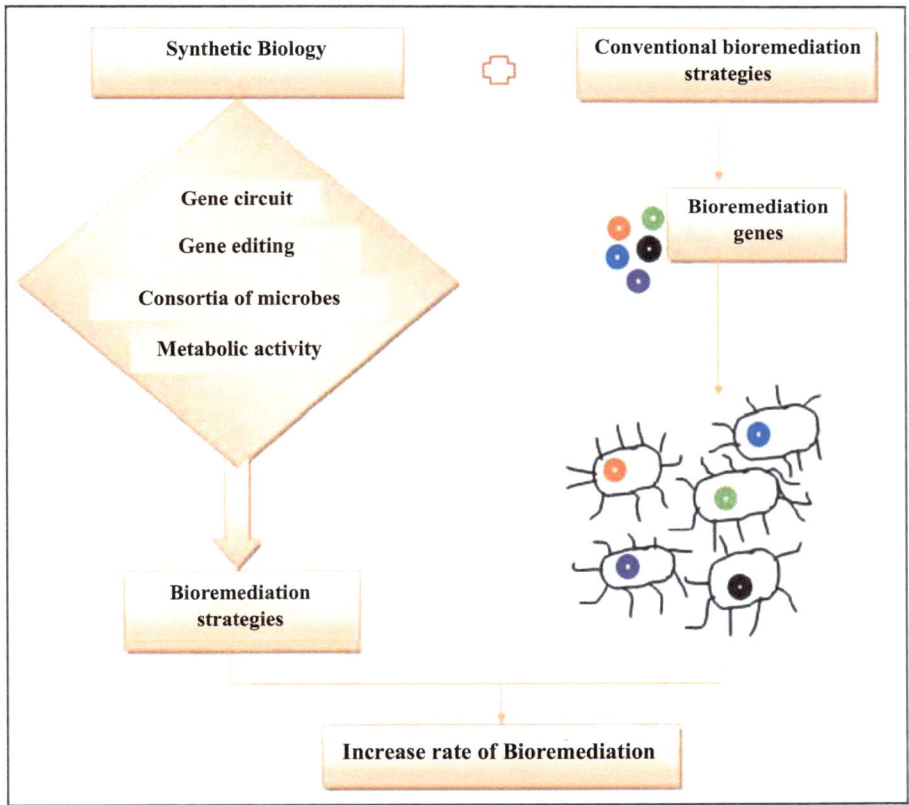

Fig. (5). Different strategies used in the process of bioremediation [72].

In a natural community, it is difficult to find out which species are actually taking part in bioremediation [108]. Thus, a synthetic microbial community is a favourable method for establishing a specialised community with function-specific species for bioremediation purposes. These communities may work as a model representative for the study of ecological, functional and structural characteristics in a controlled manner. These patterns of microbial interaction are metabolism-driven and responsible for community interaction [109].

The microbial bioremediation process makes use of native communities of microorganisms to clean up the contaminants from the environment. The rate of decontamination of contaminants depends on various factors such as the composition of the native communities of microbes, their nature, the type of the pollutant and environmental conditions [110]. Thus, the enhancement of the bioremediation process requires the combination of various complex variables to recognize and predict the fate of environmental pollutants.

Genomics, proteomics, transcriptomics and metabolomics are various metabolic approaches which aid the systems biology studies of microbes for analyzing the regulation of bioremediation at the genetic level [111]. These omics approaches make us understand the genotype and phenotype of biodegrading microbes and also allow us to study the functional aspects of the microbial communities from contaminated sites which have led to the finding of some new bacteria that were not approachable by using the conventional techniques of culturing. Recently, new highly productive omics approaches have been employed to know more about the system biology of the microbial consortia present in the environment. The applications of these bioremediation approaches give a complete idea of the factors that govern the metabolism, growth, structure, functions and dynamics of the microbial consortia of contaminated sites. The implementation of these bioremediation approaches has made it practicable and monetarily viable to explore the metagenomes of contaminated environmental samples. This has not only provided a discernment concerning the diversity of microbes but also provides acknowledged information and data about the functionality of the microbial population inhabiting the contaminated environments. The application of genomic models in the examination of pure cultures and environmental samples makes it easy to construct the systems that are required to estimate the activity of microbes under diverse bioremediation strategies. Genome-based approaches that contribute toward the development of models of how microbes function in polluted environments are shown in Fig. (**6**).

Various types of toxic compounds mostly contaminate the environment where human activities are very common [112]. Pollutants affect the most important aspects of our ecosystem, such as the soil, air and water [113]. Anthropogenic interferences, such as contamination by heavy metals [114] and hydrocarbons [115], can have a strong impact on microbial diversity and structural composition.

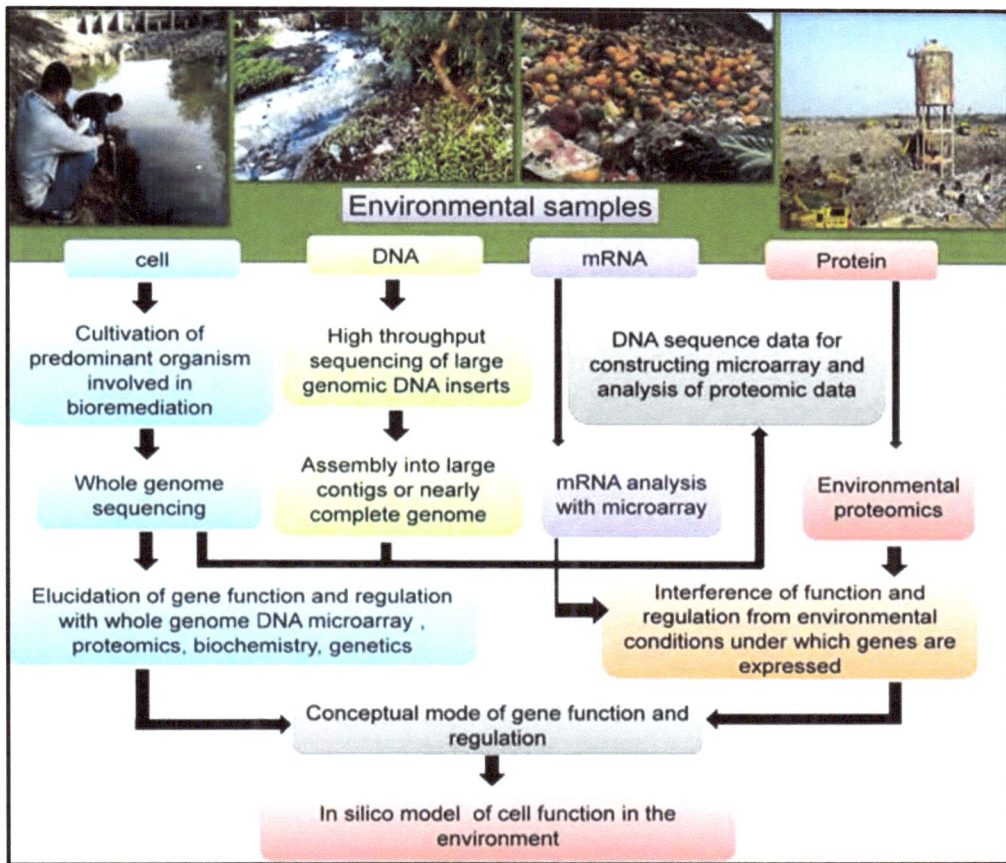

Fig. (6). Genome-based approaches that contribute toward the development of models of how microbes function in the polluted environment [58].

Metagenomics in Bioremediation

Next generation sequencing provides a chance for substantial inspection of environmental genomes. Molecular techniques such as metagenomics have modified the area of microbiology by centralising mainly on microbial diversity [116]. Metagenomics helps to shape the approach to bioremediation in a number of ways like the detection of the interacting bacterial species in a community [117, 118]. Metagenomic bioremediation provides positive results with better degradation ratios when compared to other approaches to bioremediation [119]. There are various advantages to using metagenomics approach for bioremediation.

• Metagenomics provides us a better understanding of how microorganisms develop microbial transportation chains for the degradation of xenobiotic compounds. This approach allows the differentiation of contaminated sites into areas where the native microbial community is able to remediate the environmental status by using *ex situ* or *in situ* bioaugmentation.

• Metagenomics approach can help to recognize the microbial processes and can specify how the microbial community composition could be complemented to enable the mineralization of a pollutant that is carried out by bacterial consortia rather than by individual species [120 - 122].

• Metagenomics can provide suitable metagenomic databases that can provide a rich stock of genes for the development of novel microbial strains for targeted use in bioremediation.

Therefore, microbiologists consider the metagenomics-based bioremediation approaches to be one of the most powerful tools for the removal of pollutants from the environment [123 - 125].

Metatranscriptomics and Proteomics in Bioremediation

The speedy evolution of metatranscriptomics and metaproteomics has made it possible to approximate the functional activities of microbial consortia [126, 127]. Metatranscriptomics is of great interest for research related to environmental remediation. Metatranscriptomics can be used to find out the activity of microbial genes by determining the expression of the functional genomes within environmental samples.

Proteomics approaches have often been employed to obtain more precise details of the physiological responses of microbes to changes in temperature, degree of xenobiotics and other stressors [128]. Proteomics approaches are also useful in examining the physiological changes that microorganisms undergo during bioremediation.

DISADVANTAGES OF GENOMIC APPROACHES

Despite many different advantages of genomic approaches with regard to bioremediation, there are some disadvantages also [4]. The use of genetically modified organisms is still restricted in the environment because of the instability of inserted genetic material. There are two major reasons for this limitation:

i. The efficiency of genetically modified organisms is dependent on their ability to carry the genetic material in a stable manner.

ii. The transfer of genetic material into the indigenous organism is perceived to be a negative attribute.

These factors have induced the level of competition and persistence of genetically modified organisms in the environment, as well as the potential risk involved in their use. Apart from the notable advances that have already been made concerning the development and utilization of genetically modified organisms for the bioremediation of contaminants in the environment, many more new challenges are still there. Some genetic approaches have been developed and are used to optimize enzymes, metabolic pathways and organisms that are pertinent to biodegradation.

FACTORS AFFECTING BIOREMEDIATION

There are many different factors on which the efficiency of the bioremediation process depends. These factors include: environmental factors such as type of soil, electron acceptors, temperature, pH, the presence of oxygen and nutrients and many other factors such as the concentration and chemical nature of pollutants, the availability of pollutants to microorganisms and the physico-chemical characteristics of the environment, *etc.* The major environmental factors that affect the rate of bioremediation are:

Environmental Factors

Temperature

Temperature has the ability to fasten up or slow down the rate of the bioremediation process because it highly affects the physiological properties of microbes. The rate of microbial activities generally increases with an increase in temperature and extends to its maximum level at an optimum temperature. It decreases instantaneously with a further increase or decrease in temperature and eventually stops after reaching a specific temperature [4].

Oxygen Concentration

Some organisms require oxygen and some microbes do not require oxygen for survival. Biodegradation can be carried out in aerobic as well as anaerobic conditions. The presence of oxygen, in most cases, can enhance hydrocarbon metabolism [129]. Anaerobic conditions are used to degrade highly halogenated contaminants, though some petroleum hydrocarbons may also be biodegraded anaerobically.

pH

The pH of the compound gives us an idea about the acidic, basic and alkaline nature of the compound. pH has a great impact on the metabolic activity of microbes and may have a positive or negative impact on the contaminants removal process. The analysis of pH in soil could specify the potential for microbial growth and activity [130, 131].

Nutrients Availability

The availability of nutrients and their balance for microbial growth and reproduction have a great impact on the rate of biodegradation. Nutrient balancing, especially the supply of essential nutrients such as nitrogen and phosphorus, can improve the efficiency of the biodegradation process. To continue their microbial activities and to survive, microorganisms need a number of nutrients such as carbon, nitrogen, phosphorous, *etc* [4].

Toxic Compounds

High amounts of toxic compounds in some contaminants can create toxic effects to microorganisms and slow down the rate of decontamination. The degree and mechanisms of toxicity differ with specific toxicants and their concentration. Some inorganic compounds, such as lead, mercury, nitrites, chromium, fluorides, *etc.*, and organic compounds, such as phenols, azo dyes, polyaromatic hydrocarbons, polychlorinated biphenyls, *etc.*, are toxic to targeted life forms [132].

Biological Factors

Biological factors can affect the degradation of organic compounds through competition between microorganisms for limited carbon sources, antagonistic interactions between microorganisms or the predation of microorganisms by protozoans. Major biological factors that affect bioremediation include: the types of microbes present in the environment, microbial interaction, enzymes secreted by microbes, enzyme activity, *etc* [132].

FUTURE PERSPECTIVE

The increased use of harmful chemicals and contaminants is causing a threat to the environment as well as to the living world. Thus, the removal of harmful contaminants from the environment is required. 'Bioremediation' can be considered one of the best remedies to treat such threats. Different omics approaches, such as metagenomics, transcriptomics, metabolomics, *etc.*, are currently playing a very important role in decontamination processes using one to two bioremediation pathways. There is a need to increase the efficiency of these pathways to remove pollutants. The implementation of omics approaches in bioremediation will provide an opportunity to develop new efficient strains of microbes so that the metabolism of different contaminants can be improved. The efficacy of the bioremediation process will surely be increased if accurate molecular approaches are correctly used and scientifically followed.

CONCLUSION

Our environment is continuously being harmed due to the excessive release of harmful substances like xenobiotic compounds, hydrocarbons, pesticides, *etc.*, into the environment. There is a need to protect our environment from such compounds. There are various bioremediation strategies which involve the use of genome editing technology to carry out the process of pollutant removal from the environment. CRISPR-Cas9, ZFNs, and TALEN are special gene editing tools which have the ability to degrade harmful contaminants into less harmful compounds by altering the genetic material of microorganisms which participate in the bioremediation process. Various genome editing technologies, such as rational designing, genome shuffling and family shuffling, allow the development of a wide range of advanced microorganisms. Combining all these new developments and techniques can provide a basic ground level for successful interventions in environmental processes and thus may result in the growth of improved strategies for bioremediation.

ACKNOWLEDGEMENTS

The financial support from the Department of Biotechnology, Ministry of Science and Technology, Govt. of India, to the Department of Biotechnology, Himachal Pradesh University, Shimla (India), is thankfully acknowledged. The financial assistance to Ms. Rutika Sehgal from CSIR (Council of Scientific and Industrial Research), in the form of a Senior Research Fellow (SRF), is thankfully acknowledged.

REFERENCES

[1] Streimikiene D. Environmental indicators for the assessment of quality of life. Intellectual Economics 2015; 9(1): 67-79.
[http://dx.doi.org/10.1016/j.intele.2015.10.001]

[2] Gavrilescu M, Demnerova K, Aamand J, Agathos S, Fava F. Emerging pollutants in the environment: Present and future challenges in biomonitoring, ecological risks and bioremediation. N Biotechnol 2014.
[http://dx.doi.org/10.1016/j.nbt.2014.01.001] [PMID: 24462777]

[3] Abdelhafez AA, Li J. Geochemical and statistical evaluation of heavy metal status in the region around Jinxiriver. Soil Sediment Contam 2014; 23(8): 850-68.
[http://dx.doi.org/10.1080/15320383.2014.887651]

[4]   Abatenh E, Gizaw B, Tsegaye Z, Wassie M. The role of microorganisms in bioremediation : Review. Open J Environ Biol 2017; 2(1): 038-46.
[http://dx.doi.org/10.17352/ojeb.000007]

[5]   Verma JP, Jaiswal DK. Book review: Advances in biodegradation and bioremediation of industrial waste. Front Microbiol 2016; 6: 1555.
[http://dx.doi.org/10.3389/fmicb.2015.01555]

[6]   Doudna JA, Gersbach CA. Genome editing: The end of the beginning. Genome Biol 2015; 16(1): 292.
[http://dx.doi.org/10.1186/s13059-015-0860-5] [PMID: 26700220]

[7]   Azubuike CC, Chikere CB, Okpokwasili GC. Bioremediation techniques–classification based on site of application: Principles, advantages, limitations and prospects. World J Microbiol Biotechnol 2016; 32(11): 180.
[http://dx.doi.org/10.1007/s11274-016-2137-x] [PMID: 27638318]

[8]   Basu S, Rabara RC, Negi S, Shukla P. Engineering PGPMOs through gene editing and system biology: A solution for phytoremediation. Trends Biotechnol 2018; 36(5): 499-510.
[http://dx.doi.org/10.1016/j.tibtech.2018.01.011] [PMID: 29455935]

[9]   Kumar M, Jaiswal S, Sodhi JK, *et al.* Antibiotics bioremediation: Perspectives on its exotoxicity and resistance. Environ Int 2019; 124: 448-61.
[http://dx.doi.org/10.1016/j.envint.2018.12.065]

[10] Infante J C, De Arco RD, Angulo ME. Removal of lead, mercury and nickel using the yeast *Saccharomyces cerevisiae.* Rev Mvz Cordoba 2014; 19(2): 4141-9.
[http://dx.doi.org/10.21897/rmvz.107]

[11] Hassan FM, AL-Baidhani ANAR, Al-Khalidi SHH. Bioadsorption of heavy metals from industrial wastewater using some species of bacteria. Baghdad Sci J 2016; 13(3): 0435.
[http://dx.doi.org/10.21123/bsj.2016.13.3.0435]

[12] Raffa CM, Chiampo F. Bioremediation of agricultural soils polluted with pesticides: A review. Bioengineering 2021; 8(7): 92.
[http://dx.doi.org/10.3390/bioengineering8070092] [PMID: 34356199]

[13] Mohamed AT, El Hussein AA, El Siddig MA, Osman AG. Degradation of oxyflorfen herbicide by soil microorganisms: Biodegradation of herbicides. Biotechnology 2011; 10(3): 274-9.
[http://dx.doi.org/10.3923/biotech.2011.274.279]

[14] Phulpoto AH, Qazi MA, Mangi S, Ahmed S, Kanhar NA. Biodegradation of oil-based paint by *Bacillus* species monocultures isolated from the paint warehouses. Int J Environ Sci Technol 2016; 13(1): 125-34.
[http://dx.doi.org/10.1007/s13762-015-0851-9]

[15] Simarro R, González N, Bautista LF, Molina MC. Assessment of the efficiency of *in situ* bioremediation techniques in a creosote polluted soil: Change in bacterial community. J Hazard Mater 2013; 262: 158-67.
[http://dx.doi.org/10.1016/j.jhazmat.2013.08.025] [PMID: 24025312]

[16] Alotaibi BS, Khan M, Shamim S. Unraveling the underlying heavy metal detoxification mechanisms of *Bacillus* species. Microorganisms 2021; 9(8): 1628.
[http://dx.doi.org/10.3390/microorganisms9081628] [PMID: 34442707]

[17] Adebajo S, Balogun S, Akintokun A. Decolourization of vat dyes by bacterial isolates recovered from local textile mills in southwest. Microbiol Res J Int 2017; 18(1): 1-8.
[http://dx.doi.org/10.9734/MRJI/2017/29656]

[18] Sharma B, Dangi AK, Shukla P. Contemporary enzyme based technologies for bioremediation: A review. J Environ Manage 2018; 210: 10-22.
[http://dx.doi.org/10.1016/j.jenvman.2017.12.075] [PMID: 29329004]

[19] Dangi AK, Sharma B, Hill RT, Shukla P. Bioremediation through microbes: Systems biology and metabolic engineering approach. Crit Rev Biotechnol 2019; 39(1): 79-98.
[http://dx.doi.org/10.1080/07388551.2018.1500997] [PMID: 30198342]

[20] Junghare M, Spiteller D, Schink B. Anaerobic degradation of xenobiotic isophthalate by the fermenting bacterium *Syntrophorhabdus aromaticivorans*. ISME J 2019; 13(5): 1252-68.
[http://dx.doi.org/10.1038/s41396-019-0348-5] [PMID: 30647456]

[21] Guengerich FP, Yoshimoto FK. Formation and cleavage of C–C bonds by enzymatic oxidation–reduction reactions. Chem Rev 2018; 118(14): 6573-655.
[http://dx.doi.org/10.1021/acs.chemrev.8b00031] [PMID: 29932643]

[22] Tudi M, Daniel Ruan H, Wang L, *et al.* Agriculture development, pesticide application and its impact on the environment. Int J Environ Res Public Health 2021; 18(3): 1112.
[http://dx.doi.org/10.3390/ijerph18031112] [PMID: 33513796]

[23] Bala S, Garg D, Thirumalesh BV, *et al.* Recent strategies for bioremediation of emerging pollutants: A review for a green and sustainable environment. Toxics 2022; 10(8): 484.
[http://dx.doi.org/10.3390/toxics10080484] [PMID: 36006163]

[24] Sales da Silva IG, Gomes de Almeida FC, Padilha da Rocha e Silva NM, Casazza AA, Converti A, Asfora Sarubbo L. Soil bioremediation: Overview of technologies and trends. Energies 2020; 13(18): 4664.
[http://dx.doi.org/10.3390/en13184664]

[25] Vidali M. Bioremediation : An overview. Pure Appl Chem 2001; 73(7): 1163-72.
[http://dx.doi.org/10.1351/pac200173071163]

[26] Frutos FJG, Pérez R, Escolano O, *et al.* Remediation trials for hydrocarbon-contaminated sludge from a soil washing process: Evaluation of bioremediation technologies. J Hazard Mater 2012; 199-200: 262-71.
[http://dx.doi.org/10.1016/j.jhazmat.2011.11.017] [PMID: 22118850]

[27] Folch A, Vilaplana M, Amado L, Vicent T, Caminal G. Fungal permeable reactive barrier to remediate groundwater in an artificial aquifer. J Hazard Mater 2013; 262: 554-60.
[http://dx.doi.org/10.1016/j.jhazmat.2013.09.004] [PMID: 24095995]

[28] Kim S, Krajmalnik-Brown R, Kim JO, Chung J. Remediation of petroleum hydrocarbon-contaminated sites by DNA diagnosis-based bioslurping technology. Sci Total Environ 2014; 497-498: 250-9.
[http://dx.doi.org/10.1016/j.scitotenv.2014.08.002] [PMID: 25129160]

[29] Frascari D, Zanaroli G, Danko AS. *In situ* aerobic cometabolism of chlorinated solvents: A review. J Hazard Mater 2015; 283: 382-99.
[http://dx.doi.org/10.1016/j.jhazmat.2014.09.041] [PMID: 25306537]

[30] Roy M, Giri AK, Dutta S, Mukherjee P. Integrated phytobial remediation for sustainable management of arsenic in soil and water. Environ Int 2015; 75: 180-98.
[http://dx.doi.org/10.1016/j.envint.2014.11.010] [PMID: 25481297]

[31] Sui H, Li X. Modeling for volatilization and bioremediation of toluene-contaminated soil by bioventing. Chin J Chem Eng 2011; 19(2): 340-8.
[http://dx.doi.org/10.1016/S1004-9541(11)60174-2]

[32] Birhanu GT, Tesfaye AT. Mode of action, mechanism and role of microbes in bioremediation sevice for environmental pollution management. J Biotechnol Bioinforma Res 2020; 2: 1-5.
[http://dx.doi.org/10.47363/JBBR/2020(2)118]

[33] Pongratz R, Heumann KG. Production of methylated mercury, lead, and cadmium by marine bacteria as a significant natural source for atmospheric heavy metals in polar regions. Chemosphere 1999; 39(1): 89-102.
[http://dx.doi.org/10.1016/S0045-6535(98)00591-8]

[34] Gidarakos E, Aivalioti M. Large scale and long term application of bioslurping: The case of a Greek petroleum refinery site. J Hazard Mater 2007; 149(3): 574-81.
[http://dx.doi.org/10.1016/j.jhazmat.2007.06.110] [PMID: 17709182]

[35] Maila MP, Cloete TE. Bioremediation of petroleum hydrocarbons through landfarming: Are simplicity and cost-effectiveness the only advantages? Rev Environ Sci Biotechnol 2004; 3(4): 349-60.
[http://dx.doi.org/10.1007/s11157-004-6653-z]

[36] Straube WL, Nestler CC, Hansen LD, Ringleberg D, Pritchard PH, Jones-Meehan J. Remediation of polyaromatic hydrocarbons (PAHs) through land farming with biostimulation and bioaugmentation. Acta Biotechnol 2003; 23(23): 179-96.
[http://dx.doi.org/10.1002/abio.200390025]

[37] Jessica López RI, Lopez J. Bioremediation for a soil contaminated with hydrocarbons. J Pet Environ Biotechnol 2015; 6(2): 2.
[http://dx.doi.org/10.4172/2157-7463.1000208]

[38] Dias RL, Ruberto L, Calabró A, Balbo AL, Del Panno MT, Mac Cormack WP. Hydrocarbon removal and bacterial community structure in on-site biostimulated biopile systems designed for bioremediation of diesel-contaminated Antarctic soil. Polar Biol 2015; 38(5): 677-87.
[http://dx.doi.org/10.1007/s00300-014-1630-7]

[39] Gomez F, Sartaj M. Optimization of field scale biopiles for bioremediation of petroleum hydrocarbon contaminated soil at low temperature conditions by response surface methodology (RSM). Int Biodeterior Biodegradation 2014; 89: 103-9.
[http://dx.doi.org/10.1016/j.ibiod.2014.01.010]

[40] Whelan MJ, Coulon F, Hince G, et al. Fate and transport of petroleum hydrocarbons in engineered biopiles in polar regions. Chemosphere 2015; 131: 232-40.
[http://dx.doi.org/10.1016/j.chemosphere.2014.10.088] [PMID: 25563162]

[41] Dadrasnia A, Shahsavari N, Emenike UC. Remediation of Contaminated Sites.Hydrocarbon. InTech 2013.
[http://dx.doi.org/10.5772/51591]

[42] Bhattacharya M, Guchhait S, Biswas D, Datta S. Waste lubricating oil removal in a batch reactor by mixed bacterial consortium: a kinetic study. Bioprocess Biosyst Eng 2015; 38(11): 2095-106.
[http://dx.doi.org/10.1007/s00449-015-1449-9] [PMID: 26271337]

[43] Xu P, Ma W, Han H, Jia S, Hou B. Isolation of a naphthalene-degrading strain from activated sludge and bioaugmentation with it in a MBR treating coal gasification wastewater. Bull Environ Contam Toxicol 2015; 94(3): 358-64.
[http://dx.doi.org/10.1007/s00128-014-1366-7] [PMID: 25178430]

[44] Fuller ME, Kruczek J, Schuster RL, Sheehan PL, Arienti PM. Bioslurry treatment for soils contaminated with very high concentrations of 2,4,6-trinitrophenylmethylnitramine (tetryl). J Hazard Mater 2003; 100(1-3): 245-57.
[http://dx.doi.org/10.1016/S0304-3894(03)00115-8] [PMID: 12835026]

[45] Chikere CB, Okoye AU, Okpokwasili GC. Microbial community profiling of active oleophilic bacteria involved in bioreactor based crude-oil polluted sediment treatment. J Appl Environ Microbiol 2016; 4: 1-20.
[http://dx.doi.org/10.12691/jaem-4-1-1]

[46] Delforno TP, Moura AGL, Okada DY, Sakamoto IK, Varesche MBA. Microbial diversity and the implications of sulfide levels in an anaerobic reactor used to remove an anionic surfactant from laundry wastewater. Bioresour Technol 2015; 192: 37-45.
[http://dx.doi.org/10.1016/j.biortech.2015.05.050] [PMID: 26005927]

[47] Mustafa YA, Abdul-Hameed HM, Razak ZA. Biodegradation of 2,4-dichlorophenoxyacetic acid contaminated soil in a roller slurry bioreactor. Clean 2015; 43(8): 1241-7.

[http://dx.doi.org/10.1002/clen.201400623]

[48] Tasleem M, El-Sayed AAAA, Hussein WM, Alrehaily A. Bioremediation of chromium-contaminated groundwater using chromate reductase from *Pseudomonas putida*: An *In Silico* Approach. Water 2022; 15(1): 150.
[http://dx.doi.org/10.3390/w15010150]

[49] Morris CE, Sands DC, Bardin M, *et al.* Microbiology and atmospheric processes: research challenges concerning the impact of airborne micro-organisms on the atmosphere and climate. Biogeosciences 2011; 8(1): 17-25.
[http://dx.doi.org/10.5194/bg-8-17-2011]

[50] Griggs D, Stafford-Smith M, Gaffney O, *et al.* Sustainable development goals for people and planet. Nature 2013; 495(7441): 305-7.
[http://dx.doi.org/10.1038/495305a] [PMID: 23518546]

[51] Pineda A, Kaplan I, Bezemer TM. Steering soil microbiomes to suppress above ground insect pests. Trends Plant Sci 2017; 22(9): 770-8.
[http://dx.doi.org/10.1016/j.tplants.2017.07.002] [PMID: 28757147]

[52] Lovley DR. Cleaning up with genomics: Applying molecular biology to bioremediation. Nat Rev Microbiol 2003; 1(1): 35-44.
[http://dx.doi.org/10.1038/nrmicro731] [PMID: 15040178]

[53] Lovley DR, Phillips EJP, Gorby YA, Landa ER. Microbial reduction of uranium. Nature 1991; 350(6317): 413-6.
[http://dx.doi.org/10.1038/350413a0]

[54] Parales RE, Ditty JL. Laboratory evolution of catabolic enzymes and pathways. Curr Opin Biotechnol 2005; 16(3): 315-25.
[http://dx.doi.org/10.1016/j.copbio.2005.03.008] [PMID: 15961033]

[55] Bayat Z, Hassanshahian M, Cappello S. Immobilization of microbes for bioremediation of crude oil polluted environments: A mini review. Open Microbiol J 2015; 9: 48-54.
[http://dx.doi.org/10.2174/1874285801509010048] [PMID: 26668662]

[56] Dai Z, Zhang S, Yang Q, *et al.* Genetic tool development and systemic regulation in biosynthetic technology. Biotechnol Biofuels 2018; 11(1): 152.
[http://dx.doi.org/10.1186/s13068-018-1153-5] [PMID: 29881457]

[57] Gaur N, Narasimhulu K, Pydisetty Y. Recent advances in the bio-remediation of persistent organic pollutants and its effect on environment. J Clean Prod 2018; 198: 1602-31.
[http://dx.doi.org/10.1016/j.jclepro.2018.07.076]

[58] Malla MA, Dubey A, Yadav S, Kumar A, Hashem A, Abd Allah EF. Understanding and designing the strategies for the microbe-mediated remediation of environmental contaminants using omics approaches. Front Microbiol 2018; 9: 1132.
[http://dx.doi.org/10.3389/fmicb.2018.01132] [PMID: 29915565]

[59] Singh AK, Raj A. Emerging and eco-friendly approaches for waste management: A book review. Environ Sci Eur 2020; 32(1): 107.
[http://dx.doi.org/10.1186/s12302-020-00383-w]

[60] Butt H, Jamil M, Wang JY, Al-Babili S, Mahfouz M. Engineering plant architecture *via* CRISPR/Cas9-mediated alteration of strigolactone biosynthesis. BMC Plant Biol 2018; 18(1): 174.
[http://dx.doi.org/10.1186/s12870-018-1387-1] [PMID: 30157762]

[61] Millioli VS, Servulo ELC, Sobral LGS, De Carvalho DD. Bioremediation of crude oil-bearing soil: Evaluation of rhamnolipid addition as for the toxicity and crude oil biodegradation efficiency. Glob NEST J 2013; 11(2): 181-8.
[http://dx.doi.org/10.30955/gnj.000592]

[62] Yadav R, Kumar V, Baweja M, Shukla P. Gene editing and genetic engineering approaches for

advanced probiotics: A review. Crit Rev Food Sci Nutr 2018; 58(10): 1735-46.
[http://dx.doi.org/10.1080/10408398.2016.1274877] [PMID: 28071925]

[63] Hussain I, Aleti G, Naidu R, *et al.* Microbe and plant assisted-remediation of organic xenobiotics and its enhancement by genetically modified organisms and recombinant technology: A review. Sci Total Environ 2018; 628-629: 1582-99.
[http://dx.doi.org/10.1016/j.scitotenv.2018.02.037] [PMID: 30045575]

[64] Jaiswal S, Singh DK, Shukla P. Gene editing and system biology tools for pesticide bioremediation: A review. Front Microbiol 2019; 10: 87.
[http://dx.doi.org/10.3389/fmicb.2019.00087]

[65] Dong C, Bai X, Sheng H, Jiao L, Zhou H, Shao Z. Distribution of PAHs and the PAH-degrading bacteria in the deep-sea sediments of the high-latitude Arctic Ocean. Biogeosciences 2015; 12(7): 2163-77.
[http://dx.doi.org/10.5194/bg-12-2163-2015]

[66] Lozada M, Dionisi HM. Molecular biological tools for the assessment of hydrocarbon-degrading potential in coastal environments.Biology and biotechnology of Patagonian microorganisms. Cham: Springer 2016; pp. 15-29.
[http://dx.doi.org/10.1007/978-3-319-42801-7_2]

[67] Matturro B, Aulenta F, Majone M, Papini MP, Tandoi V, Rossetti S. Field distribution and activity of chlorinated solvents degrading bacteria by combining CARD-FISH and real time PCR. N Biotechnol 2012; 30(1): 23-32.
[http://dx.doi.org/10.1016/j.nbt.2012.07.006] [PMID: 22835732]

[68] Marozava S, Mouttaki H, Müller H, Laban NA, Probst AJ, Meckenstock RU. Anaerobic degradation of 1-methylnaphthalene by a member of the Thermoanaerobacteraceae contained in an iron-reducing enrichment culture. Biodegradation 2018; 29(1): 23-39.
[http://dx.doi.org/10.1007/s10532-017-9811-z] [PMID: 29177812]

[69] Raman NM, Asokan S, Shobana Sundari N, Ramasamy S. Bioremediation of chromium(VI) by *Stenotrophomonas maltophilia* isolated from tannery effluent. Int J Environ Sci Technol 2018; 15(1): 207-16.
[http://dx.doi.org/10.1007/s13762-017-1378-z]

[70] Kwaslema D, Amuri N, Tindwa H. Minjingu phosphate rock solubilization and potential for use of *Klebsiella variicola*-MdE4 and *Klebsiella variicola*-MdG1 as biofertilizer for maize production. J Cent Eur Agric 2022; 23(4): 817-31.
[http://dx.doi.org/10.5513/JCEA01/23.4.3636]

[71] Shukla A, Srivastava S, D'Souza SF. An integrative approach toward biosensing and bioremediation of metals and metalloids. Int J Environ Sci Technol 2018; 15(12): 2701-12.
[http://dx.doi.org/10.1007/s13762-018-1766-z]

[72] Jaiswal S, Shukla P. Alternative strategies for microbial remediation of pollutants *via* synthetic biology. Front Microbiol 2020; 11: 808.
[http://dx.doi.org/10.3389/fmicb.2020.00808] [PMID: 32508759]

[73] Kumar V, Dangi AK, Shukla P. Engineering thermostable microbial xylanases toward its industrial applications. Mol Biotechnol 2018; 60(3): 226-35.
[http://dx.doi.org/10.1007/s12033-018-0059-6] [PMID: 29380253]

[74] Shanmugam K, Ramalingam S, Venkataraman G, Hariharan GN. The CRISPR/Cas9 system for targeted genome engineering in free-living fungi: Advances and opportunities for lichenized fungi. Front Microbiol 2019; 10: 62.
[http://dx.doi.org/10.3389/fmicb.2019.00062] [PMID: 30792699]

[75] Hille F, Richter H, Wong SP, Bratovič M, Ressel S, Charpentier E. The biology of CRISPR-Cas: Backward and forward. Cell 2018; 172(6): 1239-59.
[http://dx.doi.org/10.1016/j.cell.2017.11.032] [PMID: 29522745]

[76] Greene AC. CRISPR-based antibacterials: Transforming bacterial defense into offense. Trends Biotechnol 2018; 36: 127-30.

[77] David F, Siewers V. Advances in yeast genome engineering. FEMS Yeast Res 2014; 15(1): n/a.
[http://dx.doi.org/10.1111/1567-1364.12200] [PMID: 25154295]

[78] Esvelt KM, Wang HH. Genome-scale engineering for systems and synthetic biology. Mol Syst Biol 2013; 9(1): 641.
[http://dx.doi.org/10.1038/msb.2012.66] [PMID: 23340847]

[79] Cooper LA, Stringer AM, Wade JT. Determining the specificity of cascade binding, interference, and primed adaptation *in vivo* in the *Escherichia coli* type IE CRISPR-cas system. MBio 2018; 9(2): e02100-17.
[http://dx.doi.org/10.1128/mBio.02100-17] [PMID: 29666291]

[80] Rath D, Amlinger L, Rath A, Lundgren M. The CRISPR-Cas immune system: Biology, mechanisms and applications. Biochimie 2015; 117: 119-28.
[http://dx.doi.org/10.1016/j.biochi.2015.03.025] [PMID: 25868999]

[81] Bortesi L, Fischer R. The CRISPR/Cas9 system for plant genome editing and beyond. Biotechnol Adv 2015; 33(1): 41-52.
[http://dx.doi.org/10.1016/j.biotechadv.2014.12.006] [PMID: 25536441]

[82] Zhang D, Zhang Z, Unver T, Zhang B. CRISPR/Cas: A powerful tool for gene function study and crop improvement. J Adv Res 2021; 29: 207-21.
[http://dx.doi.org/10.1016/j.jare.2020.10.003] [PMID: 33842017]

[83] Gasiunas G, Barrangou R, Horvath P, Siksnys V. Cas9–crRNA ribonucleoprotein complex mediates specific DNA cleavage for adaptive immunity in bacteria. Proc Natl Acad Sci 2012; 109(39): E2579-86.
[http://dx.doi.org/10.1073/pnas.1208507109] [PMID: 22949671]

[84] Mali P, Yang L, Esvelt KM, *et al.* RNA-guided human genome engineering *via* Cas9. Science 2013; 339(6121): 823-6.
[http://dx.doi.org/10.1126/science.1232033] [PMID: 23287722]

[85] Ojuederie O, Babalola O. Microbial and plant-assisted bioremediation of heavy metal polluted environments: A review. Int J Environ Res Public Health 2017; 14(12): 1504.
[http://dx.doi.org/10.3390/ijerph14121504] [PMID: 29207531]

[86] Parrilli E, Papa R, Tutino ML, Sannia G. Engineering of a psychrophilic bacterium for the bioremediation of aromatic compounds. Bioeng Bugs 2010; 1(3): 213-6.
[http://dx.doi.org/10.4161/bbug.1.3.11439] [PMID: 21326928]

[87] Pieper DH, Reineke W. Engineering bacteria for bioremediation. Curr Opin Biotechnol 2000; 11(3): 262-70.
[http://dx.doi.org/10.1016/S0958-1669(00)00094-X] [PMID: 10851148]

[88] Bosma T, Kruizinga E, De Bruin EJ, Poelarends GJ, Janssen DB. Utilization of trihalogenated propanes by *Agrobacterium radiobacter* AD1 through heterologous expression of the haloalkane dehalogenase from *Rhodococcus* sp. strain M15-3. Appl Environ Microbiol 1999; 65(10): 4575-81.
[http://dx.doi.org/10.1128/AEM.65.10.4575-4581.1999] [PMID: 10508091]

[89] Obire OE, Anyanwu C, Okigbo RN. Saprophytic and crude oil-degrading fungi from cow dung and poultry droppings as bioremediating agents. Agric Technol Thail 2008; 4: 81-9. Available from: http://ijat-aatsea.com/pdf/Nov_v4_n2_08/9%20IJAT2008-12-R.pdf

[90] Gao D, Du L, Yang J, Wu WM, Liang H. A critical review of the application of white rot fungus to environmental pollution control. Crit Rev Biotechnol 2010; 30(1): 70-7.
[http://dx.doi.org/10.3109/07388550903427272] [PMID: 20099998]

[91] Macek T, Kotrba P, Svatos A, Novakova M, Demnerova K, Mackova M. Novel roles for genetically

modified plants in environmental protection. Trends Biotechnol 2008; 26(3): 146-52.
[http://dx.doi.org/10.1016/j.tibtech.2007.11.009] [PMID: 18243383]

[92] Suresh B, Ravishankar GA. Phytoremediation : A novel and promising approach for environmental clean-up. Crit Rev Biotechnol 2004; 24(2-3): 97-124.
[http://dx.doi.org/10.1080/07388550490493627] [PMID: 15493528]

[93] Abhilash PC, Jamil S, Singh N. Transgenic plants for enhanced biodegradation and phytoremediation of organic xenobiotics. Biotechnol Adv 2009; 27(4): 474-88.
[http://dx.doi.org/10.1016/j.biotechadv.2009.04.002] [PMID: 19371778]

[94] Cherian S, Oliveira MM. Transgenic plants in phytoremediation: Recent advances and new possibilities. Environ Sci Technol 2005; 39(24): 9377-90.
[http://dx.doi.org/10.1021/es051134l] [PMID: 16475312]

[95] Doty SL. Enhancing phytoremediation through the use of transgenics and endophytes. New Phytol 2008; 179(2): 318-33.
[http://dx.doi.org/10.1111/j.1469-8137.2008.02446.x] [PMID: 19086174]

[96] Van Aken B. Transgenic plants for phytoremediation: Helping nature to clean up environmental pollution. Trends Biotechnol 2008; 26(5): 225-7.
[http://dx.doi.org/10.1016/j.tibtech.2008.02.001] [PMID: 18353473]

[97] Chatthai M, Kaukinen KH, Tranbarger TJ, Gupta PK, Misra S. The isolation of a novel metallothionein-related cDNA expressed in somatic and zygotic embryos of Douglas-fir: Regulation by ABA, osmoticum, and metal ions. Plant Mol Biol 1997; 34(2): 243-54.
[http://dx.doi.org/10.1023/A:1005839832096] [PMID: 9207840]

[98] Misra S, Gedamu L. Heavy metal tolerant transgenic *Brassica napus* L. and *Nicotiana tabacum* L. plants. Theor Appl Genet 1989; 78(2): 161-8.
[http://dx.doi.org/10.1007/BF00288793] [PMID: 24227139]

[99] Evans KM, Gatehouse JA, Lindsay WP, Shi J, Tommey AM, Robinson NJ. Expression of the pea metallothionein-like gene PsMT A in *Escherichia coli* and *Arabidopsis thaliana* and analysis of trace metal ion accumulation: Implications for PsMT A function. Plant Mol Biol 1992; 20(6): 1019-28.
[http://dx.doi.org/10.1007/BF00028889] [PMID: 1463837]

[100] Ang EL, Zhao H, Obbard JP. Recent advances in the bioremediation of persistent organic pollutants *via* biomolecular engineering. Enzyme Microb Technol 2005; 37(5): 487-96.
[http://dx.doi.org/10.1016/j.enzmictec.2004.07.024]

[101] Kumar S, Dagar VK, Khasa YP, Kuhad RC. Genetically modified organisms (GMOs) for bioremediation.Biotechnology for Environmental Management and Resource Recovery. Springer India 2013; pp. 191-218.
[http://dx.doi.org/10.1007/978-81-322-0876-1_11]

[102] Zhang YX, Perry K, Vinci VA, Powell K, Stemmer WPC, Del Cardayré SB. Genome shuffling leads to rapid phenotypic improvement in bacteria. Nature 2002; 415(6872): 644-6.
[http://dx.doi.org/10.1038/415644a] [PMID: 11832946]

[103] Dai M, Copley SD. Genome shuffling improves degradation of the anthropogenic pesticide pentachlorophenol by *Sphingobium chlorophenolicum* ATCC 39723. Appl Environ Microbiol 2004; 70(4): 2391-7.
[http://dx.doi.org/10.1128/AEM.70.4.2391-2397.2004] [PMID: 15066836]

[104] Alves LF, Westmann CA, Lovate GL, De Siqueira GMV, Borelli TC, Guazzaroni ME. Metagenomic approaches for understanding new concepts in microbial science. Int J Genomics 2018; 2018: 1-15.
[http://dx.doi.org/10.1155/2018/2312987] [PMID: 30211213]

[105] Rochfort S. Metabolomics reviewed: A new "omics" platform technology for systems biology and implications for natural products research. J Nat Prod 2005; 68(12): 1813-20.
[http://dx.doi.org/10.1021/np050255w] [PMID: 16378385]

[106] Zhu YG, Xue XM, Kappler A, Rosen BP, Meharg AA. Linking genes to microbial biogeochemical cycling: Lessons from arsenic. Environ Sci Technol 2017; 51(13): 7326-39.
[http://dx.doi.org/10.1021/acs.est.7b00689] [PMID: 28602082]

[107] Trigo A, Valencia A, Cases I. Systemic approaches to biodegradation. FEMS Microbiol Rev 2009; 33(1): 98-108.
[http://dx.doi.org/10.1111/j.1574-6976.2008.00143.x] [PMID: 19054119]

[108] Großkopf T, Soyer OS. Synthetic microbial communities. Curr Opin Microbiol 2014; 18(100): 72-7.
[http://dx.doi.org/10.1016/j.mib.2014.02.002] [PMID: 24632350]

[109] Wintermute EH, Silver PA. Dynamics in the mixed microbial concourse. Genes Dev 2010; 24(23): 2603-14.
[http://dx.doi.org/10.1101/gad.1985210] [PMID: 21123647]

[110] Chakraborty R, Wu CH, Hazen TC. Systems biology approach to bioremediation. Curr Opin Biotechnol 2012; 23(3): 483-90.
[http://dx.doi.org/10.1016/j.copbio.2012.01.015] [PMID: 22342400]

[111] Pacwa-Płociniczak M, Płaza GA, Piotrowska-Seget Z, Cameotra SS. Environmental applications of biosurfactants: Recent advances. Int J Mol Sci 2011; 12(1): 633-54.
[http://dx.doi.org/10.3390/ijms12010633] [PMID: 21340005]

[112] Saharan BS, Sahu RK, Sharma D. A review on biosurfactants: Fermentation, current developments and perspectives. Genet Eng Biotechnol J 2012; 2011: 1-14.

[113] Pessoa-Filho M, Barreto CC, Dos Reis Junior FB, *et al.* Microbiological functioning, diversity, and structure of bacterial communities in ultramafic soils from a tropical savanna. Antonie van Leeuwenhoek 2015; 107(4): 935-49.
[http://dx.doi.org/10.1007/s10482-015-0386-6] [PMID: 25616909]

[114] Van Dorst J, Siciliano SD, Winsley T, Snape I, Ferrari BC. Bacterial targets as potential indicators of diesel fuel toxicity in subantarctic soils. Appl Environ Microbiol 2014; 80(13): 4021-33.
[http://dx.doi.org/10.1128/AEM.03939-13] [PMID: 24771028]

[115] De Sousa CS, Hassan SS, Pinto AC. Microbial omics: Applications in biotechnology.Omics Technologies and Bio-Engineering. Cambridge, MA: Academic Press 2018; pp. 3-20.
[http://dx.doi.org/10.1016/B978-0-12-815870-8.00001-2]

[116] Riesenfeld CS, Schloss PD, Handelsman J. Metagenomics: Genomic analysis of microbial communities. Annu Rev Genet 2004; 38(1): 525-52.
[http://dx.doi.org/10.1146/annurev.genet.38.072902.091216] [PMID: 15568985]

[117] Das N, Adholeya A. Role of microorganisms in remediation of contaminated soil.Microorganisms in environmental management: microbes and environment. Springer Netherlands 2012; pp. 81-111.
[http://dx.doi.org/10.1007/978-94-007-2229-3_4]

[118] Tripathi M, Singh D, Vikram S, Singh V, Kumar S. Metagenomic approach towards bioprospection of novel biomolecule(s) and environmental bioremediation. Annu Res Rev Biol 2018; 22(2): 1-12.
[http://dx.doi.org/10.9734/ARRB/2018/38385]

[119] Karlapudi AP, Venkateswarulu TC, Tammineedi J, *et al.* Role of biosurfactants in bioremediation of oil pollution : A review. Petroleum 2018; 4(3): 241-9.
[http://dx.doi.org/10.1016/j.petlm.2018.03.007]

[120] Bedard DL, Bailey JJ, Reiss BL, Jerzak GV. Development and characterization of stable sediment-free anaerobic bacterial enrichment cultures that dechlorinate aroclor 1260. Appl Environ Microbiol 2006; 72: 2460-70.
[http://dx.doi.org/10.1128/AEM.72.4.2460-2470.2006]

[121] Supaphol S, Panichsakpatana S, Trakulnaleamsai S, Tungkananuruk N, Roughjanajirapa P, O'Donnell AG. The selection of mixed microbial inocula in environmental biotechnology: Example using

petroleum contaminated tropical soils. J Microbiol Methods 2006; 65(3): 432-41.
[http://dx.doi.org/10.1016/j.mimet.2005.09.001] [PMID: 16226327]

[122] Thomas T, Gilbert J, Meyer F. Metagenomics : A guide from sampling to data analysis. Microb Inform Exp 2012; 2(1): 3.
[http://dx.doi.org/10.1186/2042-5783-2-3] [PMID: 22587947]

[123] Assadi MM, Tabatabaee MS. Biosurfactants and their use in upgrading petroleum vacuum distillation residue: A review. Int J Environ Res 2010; 4: 549-72.
[http://dx.doi.org/10.22059/ijer.2010.242]

[124] Satpute SK, Banpurkar AG, Dhakephalkar PK, Banat IM, Chopade BA. Methods for investigating biosurfactants and bioemulsifiers: A review. Crit Rev Biotechnol 2010; 30(2): 127-44.
[http://dx.doi.org/10.3109/07388550903427280] [PMID: 20210700]

[125] Chandran P, Das N. Role of sophorolipid biosurfactant in degradation of diesel oil by Candida tropicalis. Bioremediat J 2012; 16(1): 19-30.
[http://dx.doi.org/10.1080/10889868.2011.628351]

[126] Poretsky RS, Hewson I, Sun S, Allen AE, Zehr JP, Moran MA. Comparative day/night metatranscriptomic analysis of microbial communities in the North Pacific subtropical gyre. Environ Microbiol 2009; 11(6): 1358-75.
[http://dx.doi.org/10.1111/j.1462-2920.2008.01863.x] [PMID: 19207571]

[127] Verberkmoes NC, Russell AL, Shah M. Shotgun metaproteomics of the human distal gut microbiota. ISME J 2009; 3: 179-89.
[http://dx.doi.org/10.1038/ismej.2008.108]

[128] Lacerda CMR, Reardon KF. Environmental proteomics: Applications of proteome profiling in environmental microbiology and biotechnology. Brief Funct Genomics Proteomics 2008; 8(1): 75-87.
[http://dx.doi.org/10.1093/bfgp/elp005] [PMID: 19279070]

[129] Macaulay BM. Understanding the behaviour of oil-degrading microorganisms to enhance the microbial remediation of spilled petroleum. Appl Ecol Environ Res 2015; 13(1): 247-62.
[http://dx.doi.org/10.15666/aeer/1301_247262]

[130] Asira EE. Factors that determine bioremediation of organic compounds in the soil. Acad J Interdiscipl Stud 2013; 2: 125-8.
[http://dx.doi.org/10.5901/ajis.2013.v2n13p125]

[131] Wang Q, Zhang S, Li Y, Klassen W. Potential approaches to improving biodegradation of hydrocarbons for bioremediation of crude oil pollution. J Environ Prot 2011; 2(1): 47-55.
[http://dx.doi.org/10.4236/jep.2011.21005]

[132] Alori ET, Gabasawa AI, Elenwo CE, Agbeyegbe OO. Bioremediation techniques as affected by limiting factors in soil environment. Front Soil Sci 2022; 2: 937186.
[http://dx.doi.org/10.3389/fsoil.2022.937186]

CHAPTER 6

Genome Editing and Genetically Engineered Bacteria for Bioremediation of Heavy Metals

Nirmala Akoijam[1,*] **and S.R. Joshi**[1,*]

[1] *Department of Biotechnology & Bioinformatics, North-Eastern Hill University, Shillong, India*

Abstract: Genetic engineering involves the manipulation of DNA to either improve, enhance or repair a function by using recombinant DNA technology, which has contributed greatly to the fields of medicine and agriculture. In recent times, the CRISPR-Cas system of gene editing has come to the forefront of genome engineering, transforming disease treatment strategies and the production of modified crops. Industrial activities cause environmental pollution by releasing heavy metal-containing xenobiotic compounds into the environment and affect animal health by causing organ dysfunction and even cancer. Although plants utilize heavy metals from soil in small quantities for their growth, excessive exposure leads to disruption of plant cell machinery and reduces productivity. Similarly, heavy metals degrade soil health by interfering with microbial processes that contribute to soil fertility. Apart from existing methods available for the remediation of contaminated sites, bioremediation is emerging as a potent technique due to its high efficacy, cost-effectiveness and eco-friendly nature. Microbes possess a number of physiological and biochemical properties that have been exploited for the removal and detoxification of metal pollutants. This chapter elaborates on the approaches of gene editing and the development of genetically engineered bacteria to modify the expression of specific genes coding for enzymes that take part in the degradative or detoxification pathway of metals and xenobiotic compounds. It is crucial to address the scope as well as limitations involved in the use of genetically engineered microbes to ensure a safe and cost-effective method for the bioremediation of heavy metal contaminants.

Keywords: Bacteria, Bioaccumulation, Bioremediation, Biosorption, CRISPR-Cas, Genome editing.

INTRODUCTION

Amidst great apprehension in the 1970s, genetic engineering went on to revolutionize the fields of medical science and agriculture with the production of

---

[*] **Corresponding authors Nirmala Akoijam and S.R. Joshi:** Department of Biotechnology & Bioinformatics North-Eastern Hill University, Shillong, India; Tel: +94361 02171; E-mails: naocha16@gmail.com, srjoshi2006@gmail.com

Prakash M. Halami & Aravind Sundararaman (Eds.)
All rights reserved-© 2024 Bentham Science Publishers

recombinant human insulin and genetically modified crops with enhanced traits like drought-resistant maize and many others [1]. Genetic engineering has also been used to enhance livestock by expression of monoclonal antibodies against specific disease-causing pathogens, thereby inducing disease-resistance in the livestock [2]. However, genetically modified foods (GMF) remain opposed by the public on moral and ethical grounds fuelled by the general suspicion of GMF's ill effects on human health [3]. Moreover, genetic engineering of animals gave rise to limitations that ranged from technical to ethical issues [4]. Even though the genes being transferred are naturally occurring in other species, however, the recipient organism may face risks of altered metabolism and growth rates due to the foreign gene expression. These risks may transfer beyond the genetically modified organism and into the natural environment, creating a risk of exposure to other organisms [5].

Horizontal gene transfer (HGT) is a process of gene transfer that occurs naturally in prokaryotes, eukaryotes and even between prokaryotic symbionts and their eukaryotic hosts. HGT plays a significant role in the variation of gene content and contributes to the adaptation potential of the organisms that take part in HGT [6]. Unfortunately, HGT aids in the transfer of antibiotic-resistant genes in bacteria, which is alarming because multidrug resistance (MDR) threatens human and animal health, environment and food safety [7]. According to the World Health Organisation (WHO), the number of human deaths due to MDR across the world is set to escalate to about 10 million by 2050, which is far more than the estimated number of deaths due to cancer [8].

The restriction-modification (R-M) and CRISPR-Cas systems are prokaryotic defence systems against invading phages and plasmids. These two systems are compatible and act together to defend against attacks on the prokaryotic cell [9]. CRISPR (clustered regularly interspaced palindromic repeats) was first discovered in 1987 as short direct repeats interspaced with short sequences in the genome of *Escherichia coli* [10]. Similarly, it was observed that cas(CRISPR-associated) proteins possess putative nuclease and helicase domains which explain the degradation process of foreign DNA [11]. This mechanism of CRISPR-Cas has revolutionized the field of genome editing. Kang and team in 2015 [12] targeted the chemokine (C-C motif) receptor 5 (CCR5), which is an HIV-1 co-receptor, and CRISPR-Cas9 was used to disrupt expression of *CCR5*, thus protecting the cell from HIV infection. In a similar study, individuals homozygous or heterozygous for the C-C chemokine receptor type 5 gene with 32-bp deletions (CCR5Δ32) seemingly resist or show a slower progression of HIV infection, respectively. This was demonstrated by generating the CCR5Δ32 mutation using CRISPR-Cas9 and transcription activator-like effector nucleases (TALENs) in induced pluripotent stem cells (iPSCs) [13]. In cervical carcinoma cells, CRISPR-

Cas9 was used to inactivate the E6 or E7 oncogene in the human papilloma virus (HPV). E6 and E7 function as disruptors or degraders of tumour suppressor genes like p53 and the retinoblastoma (Rb) protein. Hence, inactivation of these oncogenes leads to cancer cell death [14].

The applications of CRISPR-Cas9 extend beyond therapeutic uses and into the food and agricultural industry. Viral infections in plants reduce their yield and create economic constraints [15]. Geminiviruses are associated with a number of plant infections like the yellow mosaic disease, curly top, leaf curling, stunting, and streaks, which ultimately lead to reduced yields [16]. In a study conducted by Baltes and team, the genome of bean yellow dwarf virus (BeYDV) was targeted at six different regions using the CRISPR-Cas9 mechanism to inhibit the replication process of the virus, thereby conferring resistance to the virus in a plant model [17]. In addition to diseases, plants undergo physical and chemical stresses, which potentially reduce crop yield. Herbicides, although useful in removing unwanted weeds and invasive plants, can sometimes damage crops with low resistance to these chemicals [18]. Acetolactate synthase (ALS) is an enzyme involved in the biosynthesis of branched amino acids like valine, leucine, and isoleucine and is present in many species of higher plants as well as bacteria, fungi, yeasts, and algae [19]. In fact, many commercial herbicide families like sulfonylureas, imidazolinones, triazolopyrimidines, pyrimidinylthio (or oxy)-benzoates and sulfonylamino-carbonyltriazolinones inhibit ALS [20]. In a study conducted on *Brassica napus*, point mutations were created using CRISPR/Cas9-mediated cytosine base-editing technology to produce edited *BnALS* genes which rendered the plant herbicide-resistant [21]. Nutritional improvement of crops is another significant application of CRISPR/Cas9. To promote levels of health-promoting nutrients, proanthocyanidins and anthocyanins in rice, Zhu *et al.* [22] converted three white pericarp varieties into red ones (controlled by the complementary genes, *Rc* and *Rd*). They used a CRISPR/Cas9-mediated method where the recessive *rc* allele was functionally restored to the *Rc* allele through site-specific mutagenesis. Rice rich in amylose and resistant starch (RS) is another desirable nutritional improvement as consumption of RS could lead to a reduced glycemic index that is beneficial in preventing of progression of insulin resistance [23]. High amylose content in rice was achieved by down-regulating starch branching enzymes (SBE), *SBEI* and *SBEIIb* through a CRISPR/Cas9-mediated targeted mutagenesis [24].

ENVIRONMENTAL POLLUTION: CAUSES AND IMPACTS

Environmental pollution is a grave and extensive issue that endangers the lives of every being on this planet and also threatens to destabilise human societies. Anthropological activities have aggravated this problem even further by disrupting the natural environment. The Anthropocene is a proposed epoch where human activities have impacted and altered the earth's geological and other natural processes to the extent of species extinction and even climate change [25]. Pollution is of three major types: air, water and soil pollution. Fossil fuel combustion, manufacturing facilities, mining operations and deforestation are major sources of air pollution. Ozone and air particulate matter adversely affect human and animal health and even agricultural crops [26, 27]. Rapidly developing countries like China and India nest numerous cities with air quality index (AQI) exceeding the WHO limit. AQI is measured based on the measurement of particulate matter [(PM2.5 and emissions of gases like Carbon Monoxide (CO), Ozone (O_3), Sulfur Dioxide (SO_2) and Nitrogen Dioxide (NO_2) [28]. Moreover, rural households in developing countries rely on unprocessed biomass fuels like wood, cow dung and agricultural wastes, which contributes to indoor air pollution and presents a great health risk of pulmonary diseases, especially among women and children [29, 30]. Water and soil pollution is an intertwined problem that has a large impact on food safety and the health of all living organisms [31]. Water and soil pollutants usually have anthropogenic sources, *i.e.*, oil spills, overuse of pesticides, industrial effluents and wastes which lead to the accumulation of heavy metals like lead (Pb), cadmium (Cd), mercury (Hg), arsenic (As) and nickel (Ni) and even radioactive materials in soil and water bodies [32, 33].

Agricultural pesticides are undeniably important for pest control and to prevent yield reduction caused by pest infestation [34]. However, the risk that it entails to human health and the natural ecosystem is another harrowing aspect of its usage. In a study, the micronucleus (MN) assay was done to evaluate the genotoxicity in human peripheral lymphocytes exposed *in vitro* to alloxydim sodium herbicide. This assay revealed a reduction in the nuclear division index (NDI, a marker of cell proliferation in cultures) and an increase in the MN frequency, indicating the cytotoxicity and genotoxicity potential of alloxydim sodium [35]. The intensive and widespread use of organochlorine pesticides (OCPs) from the 1940s through the 1960s in agriculture has had a tremendous detrimental effect on certain populations of wildlife, leading to their decline [36]. Major OCPs like aldrin, chlordane, DDT, diazion, methoxychlor and others have numerous health effects on humans and animals, ranging from neurotoxic effects, convulsions, skin diseases in humans to reduced fertility and liver damage in mice and wing spasms in birds [37, 38].

Heavy metals are present in soil in various forms as soluble metal complexes, free ions and insoluble compounds. The bioavailability of these metals (governed by factors like pH, solubility, organic matter and others) determines their toxicity towards other living organisms. High concentrations of heavy metals prove to be toxic for plants, animals and soil microbes [39, 40]. In plants, heavy metals have been established by numerous studies to have effects on growth, cell division, cell membrane and structure, photosynthesis system and nutrient absorption [41]. Likewise, microbial diversity and soil chemistry, which are contributing factors to soil health, are drastically affected in heavy metal-contaminated areas [42, 43].

Heavy Metal Tolerance in Plants and Microbes

Despite the adverse effects of heavy metal pollution, some of these autochthonous or indigenous organisms in polluted habitats have adapted and evolved tolerance mechanisms as a response to increasing concentrations of heavy metals. Upon exposure to metal stress, several biochemical and physiological changes are induced, and plants and microbes respond to these changes in several ways [44]. In the case of plants, metal stress triggers mechanisms that can regulate physiological, biochemical and molecular processes. The primary target of metal toxicity is mostly roots due to the proximity of interaction with soil, which acts as the source. Several reports suggest that morphological structures like mycorrhizas, root exudates and even biomolecules like phytochelatins, metallothioneins and organic acids can function as ligands to help chelate excess metal ions while heat shock proteins (HSPs) help to protect and repair proteins that have been under heavy metal stress [45].

However, there are plants known as metallophytes or hyperaccumulators, which have the ability to take up heavy metals. The term hyperaccumulation was termed by Jaffre *et al.* [46] in regard to *Sebertia acuminata*, a tree species that has the ability to hyperaccumulate nickel from the soil. Heavy metal hyperaccumulators usually grow on metal-rich soils having slow growth, low biomass and metal selectivity [47]. They are found in tropical and temperate climates and are distributed throughout southern and central Europe, South East Asia, South Africa, New Caledonia, Latin America and North America [48]. In contrast to non-hyperacculumator plants, which store the absorbed heavy metals in the roots, hyperaccumulators transport the same into shoots by means of the xylem, which is the transport system in plants. A number of proteins are involved in the transport and sequestration of heavy metals belonging to various transporter families like metal ion transporters natural resistance-associated macrophage (Nramp), cation diffusion family proteins (CDF), zinc iron permease (ZIP), membrane-bound heavy metal-transporting ATPases (CPx-type ATPases) and multidrug and toxin efflux (MATE) as reviewed by Singh *et al.* [49]. The uptake is dependent on various factors like pH, redox potential, water content and other rhizospheric components [50].

Despite the evident toxic effects of heavy metals, several microbes are able to thrive in metal-contaminated areas and develop resistance mechanisms against the pollutants [51, 52]. Microbial activities that are able to change metal speciation through redox reactions and the production of metal-binding proteins like metallothioneins affect the bioavailability of metals and their respective minerals in soil [53]. Some microbial processes can influence the speciation of metals and affect their bioavailability, thereby possibly having significant bioremediation potential. Microorganisms are able to mobilize metals by leaching *via* autotrophic or chemoorganotrophic processes, which produce acids able to solubilize metal complexes, thereby making available free metal ions in the environment [54]. A study by Gholami *et al.* [55] on spent catalysts from oil refineries determined the bioleaching potential of two bacterial species, *Acidithiobacillus ferrooxidans* and *Acidithiobacillus thiooxidans*. At optimum conditions in batch cultures, *A. ferrooxidans* and *A. thiooxidans* were able to extract significant amounts of aluminium, cobalt, molybdenum and nickel, emphasizing the significance of bioleaching in waste treatment. *A. ferrooxidans* showed a similar result in a separate study where it could leach up to 100%, 88%, and 20% of lithium, cobalt and manganese, respectively, from spent coin cells [56]. Microorganisms also produce low molecular weight compounds called siderophores mainly to act as ligands for iron chelation; however, their ability to bind to other metals has garnered attention for their potential to remove other heavy metals from contaminated areas [57]. Siderophores produced by *Pseudomonas azotoformans* efficiently removed arsenic from arsenic-contaminated soil in laboratory conditions in comparison to EDTA and citric acid, which are also metal chelators. In addition, it was observed that siderophores removed both bound and unbound/leachable fractions of arsenic [58].

In view of the widespread effects of heavy metal contamination, this chapter describes the importance of genetically engineered microorganisms (GEMs) and the advances in genomics and genome editing technology by highlighting their prospects in bioremediation of contaminated areas.

Existing Tools to Combat Heavy Metal Pollution

Different types of techniques belonging to physical, chemical and biological categories are in function so far to remove, reduce or neutralize pollutants from contaminated areas. Generally, these techniques are employed in *in situ* or *ex situ* modes depending on the type of contaminant, site characteristics, costs, regulatory laws and time constraints [59, 60]. In this section, some of the remediation techniques are described from the perspective of *in situ* and *ex situ* strategies or a combination of both in certain cases.

In situ Remediation

In situ remediation methods, as the name suggests, comprise utilising remediation methods in the contaminated sites without having to excavate or extract contaminated soil or groundwater, thereby considerably reducing the carbon footprint arising from transportation and the risk of contamination of other sites as well [61].

Capping

Capping is the most commonly used method for *in situ* remediation of sediments, involving the installation of a physical structure or "cap" over the sample to be treated. Passive or "inactive" capping makes use of neutral or unreactive material (sand, silt, clay, rock debris) to cover or isolate the contaminated sediment from its environment. However, a downside of passive capping is the risk of leakage of contaminants. Active capping, on the other hand, uses materials like apatite, activated carbon, organoclay, zeolite, *etc.*, that react and sequestrate the contaminants to limit their bioavailability and toxicity [62]. Cho *et al.* [63] reported effective sequestration of polychlorinated biphenyls (PCBs) by activated carbon (AC) in contaminated sediments. Contrastingly, in another study, the use of AC for capping contaminated sediments resulted in reduced nutrient fluxes and denitrification. This was mainly due to the sequestration of nutrients and organic material in addition to the contaminants on the surface of AC particles [64].

Heating

This technique involves the use of steam/hot air, electrical resistance, microwave and thermal conduction, as well as an extensive electrical heating process known as vitrification to volatilize compounds by decreasing adsorption, viscosity and increasing solubility [65 - 68]. Despite the ease of the process, thermal remediation of contaminated soils is perceived as an infeasible method because of its soil-altering properties. Soil organic matter (SOM) after heat treatment degrades rapidly while the remaining SOM condenses to form hydrophobic compounds due to the high temperatures (above 220°C) [69]. Subsequent changes in pH were also observed after SOM degradation. Heat-treated soils were found to have a higher pH due to the destruction of organic acids and displacement of H^+ ions due to dehydration [70].

Soil Washing

Soil washing or soil scrubbing can be done either *in situ* or *ex situ* and involves both physical and chemical processes to separate soil from its contaminants [71]. In this technique, the pollutants are not detoxified but simply removed by utilising any one or a combination of the available processes. Physical processes comprise of mechanical screening, gravity concentration, froth flotation, magnetic separation, electrostatic separation and attrition scrubbing [72 - 74], whereas chemical processes involve extraction of metals into aqueous solutions using surfactants, acids/bases, redox agents, chelating agent and salt [75 - 77].

Ion Exchange

Ion exchange is an efficient and reversible process of heavy metal removal which can be carried out both *in situ* as well as *ex situ* wherein metal ions are exchanged for a similarly charged ion attached to an immobile solid structure (which could be either natural inorganic zeolites or synthetic organic resins) [78]. Ion exchange is influenced by various factors like the pH of the exchange medium, the amount of the exchange material, reaction time, reaction temperature, initial metal ion concentration and the presence of other similarly charged ions, which could affect the rate of the ion exchange process [79]. Numerous studies have validated the efficiency of ion exchange for heavy metal remediation while using different materials for the ion exchanger, like nylon 6,6 Zr (IV) phosphate, chalcogenide and metal sulphides in addition to the usual zeolite and resin [80, 81].

Ex situ Remediation

Ex situ remediation methods involve the excavation of contaminated soil from its original site, following which treatment of the soil is carried out either on-site or off-site. In the case of contaminated water, it is extracted and further treated at the surface [82].

Landfilling

Also known as the "dig and haul" or "dig and dump" method, landfilling is a convenient method and is widely used, where contaminated soil is excavated and taken to an assigned location for its disposal. Landfills are secure engineered structures that help prevent leakage and contamination of groundwater [74]. Nonetheless, there are several reports confirming the release of heavy metals into the atmosphere and groundwater. In an investigation done by Pinel-Raffaitin *et al.* [82], almost all arsenic species (*viz.* AsIII, AsV, MMA, DMA, TMAO and arsenobetaine, AsB) were found in landfill sites. In leachate groundwater, heavy metals were found to be associated with small-size colloidal matter and organic molecules [83].

Oxidation

Chemical oxidation of heavy metals by oxidizing agents like permanganate, ozone, peroxides, chlorine, chlorine-di-oxide, hypochlorites not only affects the metal's mobility but also causes destruction of the metal, thus detoxifying the soil [84]. At present, advanced oxidation processes (AOPs) are being widely used for the treatment of groundwater or wastewater. In a study conducted by Andreottola *et al.* [85], it was determined that chemical oxidation with modified Fenton's reagent, AOPs with hydrogen peroxide and ozone (perozone), and AOPs with hydrogen peroxide and UV radiation showed effective removal of organolead compounds from contaminated groundwater. Ozone delivery using nanobubbles also enhanced the oxidation of heavy metals in contaminated sediments [86].

Adsorption

Removal of heavy metals from contaminated water by adsorption is one of the most widely used techniques due to its cost effectiveness, high efficiency and ease of use [87]. Commercially available adsorbents are graphene, activated carbon, hydrogels, carbon nanotubes and more [88]. Heavy metals, Cu and Zn, in contaminated soils were extracted using organic acids resulting in the formation of complexes which were then electrochemically adsorbed [89]. Magnetic carboxymethylchitosan nanoparticles were observed to efficiently remove Cu^{2+}, Pb^{2+} and Zn^{2+} from aqueous solutions, which present an efficient, environmentally safe and cost-effective method for heavy metal remediation [90].

Pyrolysis

This is a technology used for soil and sludge remediation in which the sample to be treated is thermally decomposed at temperatures ranging from 400-1000°C under pressure and in the absence of oxygen, resulting in the formation of non-condensable gases (H_2, CH_4, CO and CO_2), solid (biochar and charcoal) and liquid residues [91]. Magnetic biochars made from co-precipitation of iron oxide with the char (obtained from fast pyrolysis of oak wood and oak bark) could remediate lead and cadmium from water optimally at pH 5 using batch sorption studies [92]. The surface of biochar holds a number of functional groups that are able to form metal complexes which is an important property of an adsorbent. In addition, magnesium-loaded biochars also presented high efficiency in heavy metal removal from wastewater [93].

BIOREMEDIATION AS AN ALTERNATIVE AND ENVIRONMENT-FRIENDLY TECHNIQUE

George M. Robinson first discovered bioremediation in the 1960s where he used dried bacterial cultures to treat sewage sludge and oil spills [94]. The past few decades have seen tremendous development in bioremediation strategies in order to implement environmentally friendly approaches that can safely restore the target sites. Bioremediation makes use of biological processes in organisms like bacteria, fungi, algae and plants to remediate sites contaminated by heavy metals, hydrocarbons and other toxic wastes [95 - 97]. Some commonly used bioremediation techniques are summarised in Table **1** and described in detail below:

Table 1. Bioremediation methods for sites contaminated by heavy metals and other pollutants.

Techniques	Mode	Agents	Applications	References
Biopile	*Ex situ*	Indigenous heterotrophic bacteria, soil amendments and nutrients.	Heavy metal and petroleum-hydrocarbon and contaminated soils.	[98, 211]
Windrows	*Ex situ*	Indigenous heterotrophic aerobic bacteria, soil amendments and equipment to turn soil.	Wastewater solids, mine spoils, agricultural lands.	[212]
Bioreactors	*Ex situ*	Reaction vessel containing heterotrophic aerobic bacteria, nutrients and oxygen.	Soil, sludge and wastewater.	[104]
Bioventing	*In situ*	Indigenous heterotrophic bacteria, soil amendments, nutrients and oxygen supply to the unsaturated soil zone.	Heavy metal and spilled petroleum products in the soil.	[112]
Biosparging	*In situ*	Indigenous heterotrophic bacteria, soil amendments, nutrients and oxygen supply to the saturated soil zone.	Hydrocarbon-contaminated soils and groundwater.	[140]
Bioslurping	*In situ*	Indigenous heterotrophic bacteria, soil amendments, nutrients and vacuum-assisted pumping.	Soils contaminated with petroleum hydrocarbons.	[101]
Phytoremediation	*In situ*	Hyperaccumulator plants with extensive root system like *Eichhornia crassipes*, *Pteris vittata* and others.	Soils contaminated with heavy metals and petroleum hydrocarbons.	[141]

Biopile

Also known as biocell, bioheap, biomound and compost pile. This *ex-situ* method involves piling excavated contaminated soil above ground and mixed with soil amendments like organic or inorganic matter to enhance degradation by metabolic processes of native microbes. A biopile setup allows the contaminated soil to be aerated and irrigated with nutrients in addition to a leachate collection system [71]. García-Carmona *et al.* [98] evaluated different bioremediation techniques (composting, landfarming and biopiles) in heavy metal contaminated soil and found that the highest reduction metals were observed in biopiles treatment. Toxicity bioassays for biopiles also showed a better outcome in comparison to the other techniques. This method is also feasible for different soil types and extremely cold environments [99, 100].

Windrows

This method, like the biopile method, utilizes indigenous microbes to aerobically degrade the pollutants in contaminated soil. It requires the mechanical or manual turning of the piles of contaminated soil to enhance the aeration and uniform distribution of the contaminants [101]. However, the effectiveness of windrowing depends on the soil type; in particular, heavy textured soil is more effectively remediated by this method [102]. A major disadvantage to this method is the emission of greenhouse gases and the generation of leachate [103].

Bioreactors

In this *ex-situ* method, samples (soil, sludge and wastewater) to be treated are made to react in a vessel in the presence of microorganisms, nutrients, oxygen and other requirements to stimulate the degradation processes and form non-toxic products. There are three types of bioreactors: slurry bioreactors, which consist of microbial colonies or flocs in liquid suspension, fixed bed bioreactors, where biofilm is grown and immobilized on a porous carrier, and fluidized bed carriers, where carrier particles have a biofilm layer consisting of bacteria and other microbes along with extracellular polymers [104]. Ibrahim *et al.* [105] used *Pseudomonas aeruginosa* immobilised on fixed bed reactors to efficiently remediate polluted water samples by accumulating heavy metal ions (Cu^{2+}, Zn^{2+} and Cd^{2+}) using a process called biosorption. Similarly, microbial consortiums isolated from soil contaminated with industrial wastes were biostimulated (enhanced aeration and stirring) to effectively remove heavy metals from the effluent sample [106]. In another interesting study, a novel biomaterial was developed for the immobilization of *Pseudomonas* sp. H117, for the denitrification and removal of Cd^{2+} and Mn^{2+} [107].

Bioventing

This remediation technology takes place *in situ* and stimulates indigenous microbes in the unsaturated soil zone to biodegrade contaminants by inoculating oxygen and, sometimes, nutrients in a controlled manner [108]. Soil structure and stratification decide the effectiveness of the technique; coarse-textured soil has a higher permeability, thus allowing more soil vapour (volatile compounds) flow, whereas fine-textured soil has less permeability *(e.g.*, clays) with the reduced flow of soil vapour [71].

Biosparging

Similar to bioventing, biosparging also involves stimulating the growth of native microbes, however, the difference lies in the injection of oxygen and nutrients in the saturated soil zone to remediate contaminated soil and groundwater [71]. This technique is similar in all aspects to air sparging except that while air sparging focuses on the volatilization of the contaminants, biosparging aims to degrade them [109]. Soils and groundwater contaminated by hydrocarbons are best suited for this remediation technique, as validated by several reports [110 - 112].

Bioslurping

Bioslurping is also known as multiphase extraction because it is a combination of three extraction processes: enhanced vacuum pumping, soil vapour extraction and bioventing. Both soil and contaminated groundwater can be remediated by the method in which oxygen is indirectly supplied to stimulate the biodegradation of the contaminants. This technique is employed in a systematic way to ensure effective biodegradation of the contaminants [113]. First, the vacuum system pumps out any free-floating products like the light non-aqueous phase liquids (LNAPLs) from the water table. When the LNAPLs have been extracted, soil vapour extraction is then done to remove volatile or semi-volatile contaminants. The process of bioventing then microbially degrades the remaining contaminants [101].

Phytoremediation

Phytoremediation is a green remediation technology that uses plants to improve soil and aquatic environments contaminated by heavy metals, radionuclides and organic pollutants. As discussed in the previous section, hyperaccumulator plants are ideal for heavy metal removal from contaminated soil due to their inherent metal-accumulation traits without much change to the soil, as may be the case in the other techniques [114]. The selection of plants for phytoremediation is critical, and so plants with high biomass, metal tolerance, rapid growth rate, extensive root system and a high translocation factor or shoot-root quotient (ability of plants to translocate metals from roots to shoots) are ideal for this technique [115, 116]. *Eichhornia crassipes* (water hyacinth) is an aquatic plant that has been extensively used in the treatment of wastewater [117, 118]. Depending on the environment and type of contaminant, phytoremediation has been classified into several types:

Phytostabilization

Metal-tolerant species stabilize the contaminants in the root or rhizosphere, thus reducing their bioavailability and preventing offsite contamination [119]. This technique is enhanced by the application of soil amendments that boost the growth of metal-tolerant species. Root exudates like organic and amino acids affect the dynamics of metals in the soil by forming stable complexes *via* processes like metal complexation, chelation and precipitation [120]. Soil microbes adsorb and immobilize metals on their cell wall by secreting compounds that assist in metal chelation. *Funneliformis mosseae*, an arbuscular mycorrhizal fungus (AMF), retains Cu and Zn in the rhizosphere of sunflower plants and reduces transportation of these metals to the shoots, contrary to another AMF species that transports and accumulates these metals in the shoots [121]. Li *et al.* [122] performed a similar study where two AMF species *Glomus mosseae* and *Glomus intraradices* were inoculated in the soil growing *Astragalus sinicus* L. and accumulated higher levels of Cd in the roots without transporting the same to the shoots.

Phytovolatilization

In phytovolatilization, plants absorb heavy metals from the soil and translocate to the shoots to form compounds that subsequently volatilize from the leaves into the atmosphere [123]. *Pteris vittata*, also known as the Chinese brake, is a fern species that has the ability to hyperaccumulate high levels of metals like arsenic and cadmium. In a greenhouse experiment, *P. vittata* removed almost 90% of the total As from soils contaminated with As [124]. The efficient root-shoot transport enables As to reach the fronds from which the secretory glands remove the As as a volatile compound. *Brassica juncea* (Indian mustard) is also an ideal plant used for phytoremediation due to its hyperaccumulation potential. Moreno *et al.* [125] studied the phytoextraction and phytovolatilization of mercury by *B. juncea* from mine tailings. However, volatilization of high levels of heavy metals may cause secondary contamination of the environment [126].

Phytoextraction

This method uses hyperaccumulator plants to extract heavy metals and other contaminants from soil to gradually accumulate and store in their biomass aboveground, which is then harvested and disposed of appropriately [127]. However, hyperaccumulator plants take up metals selectively with some metals being taken up more than others. Under the same conditions, hyperaccumulators can amass metals upto 100 times more than non-hyperaccumulator plant species [128]. *Corrigiola telephiifolia*, an herbaceous plant is an efficient accumulator of As due to its high translocation factor [129]. *Rumex crispus* is also another hyperaccumulator plant for Cd and Zn [130]. *Medicago sativa, Brassica juncea, Brassica nigra* and *Helianthus annuus* are commonly used plants for the hyperaccumulation of Pb (Koptsik, 2014). *Thlaspi goesingense Senecio coronatus Alyssum bertolonii* and *Alyssum lesbiacum* are Ni-hyperaccumulator plants that compartmentalize Ni in the epidermal cells [131]. Two tree species, *Salix viminalis* and *Salix dasyclados* have exhibited potential to hyperaccumulate Cd and Zn from contaminated sediments implicating bioremediation prospects using tree species [132].

Phytofiltration

Phytofiltration of heavy metals from contaminated groundwater, surface water and wastewater can take place in various forms depending on the part of the plant: roots (rhizofiltration), shoots (caulofiltration) or seedlings (blastofiltration) [133]. In rhizofiltration, pollutants like heavy metals and other organic compounds are either precipitated onto or absorbed by the roots. Certain species of both aquatic and terrestrial plants are capable of accumulating metals on their roots [134]. The plants to be used for phytofiltration are first grown hydroponically in optimal conditions to develop an extensive root system, following which the clean water is substituted by polluted water to adapt the plants to the pollutants. Thereafter, the acclimatized plants are grown in the contaminated sites to concentrate the pollutants on the roots, harvested and disposed of in suitable conditions [120]. Abhilash *et al* [135] tested the rhizofiltration efficiency of *Limnocharis flava* (L.) Buchenau in hydroponic conditions and found that it could remove Cd from contaminated water even in low concentrations of Cd. Certain aquatic mosses and liverworts have also shown great efficiency in accumulating heavy metals. *Monosolenium tenerum,* a liverwort species, exhibited high potential in removing toxic metals from a multi-contaminated (Zn, Cu, Ni, Mn and Fe) water with different biosorption selectivity and dose-dependency [136]. Similarly, numerous studies have emphasised the high adsorption capability (pH-dependent) of the shoots of alfalfa (*Medicago sativa*) from multi-contaminated aqueous solutions [137, 138]. Hydroponically grown *Micranthemum umbrosum* is a good phytofilter for the metals As and Cd as it showed approximately 70-80% removal of both metals in an aqueous solution containing low levels of both contaminants [139].

MICROBES-ASSISTED BIOREMEDIATION

Certain metals (Ni, Cu, Co, Mo, Zn and Fe) serve as micronutrients essential for cellular processes in microbes. However, at high concentrations, both essential and non-essential metals create a detrimental effect on microbial cells by disrupting essential enzymatic functions, influencing the formation of reactive oxygen species (ROS), irregular ion regulation and affecting the synthesis of both DNA and proteins [142]. Microbes found in metal-contaminated sites face selective pressure and have gradually evolved various mechanisms of heavy metal resistance consisting of extracellular barrier, intra- and extra-cellular sequestration, efflux (active transport) of metal ions, detoxification by cellular enzymes and reduction of metal ions [143]. These mechanisms are mediated mostly by plasmids and are often associated with antibiotic resistance [144]. In view of the various heavy metal mechanisms of microbes, the following methods are described with a purview for bioremediation of heavy metal contamination of terrestrial and aquatic habitats. A list of wildtype and genetically engineered microbes for heavy metal bioremediation is given in Table **2**.

Table 2. Wildtype and genetically engineered microbes for the bioremediation of heavy metals.

Microbial Process	Wild Type (WT) Microbes	Genetically Engineered (GE) Microbes and Genes Manipulated/Enhanced	References
Biosorption	*Bacillus thuringiensis, Escherichia coli, Pseudomonas fluorescens.*	GE *Escherichia coli* expressing mice metallothionein I gene (*pMt-Thio*) for biosorption of Pb^{2+} and Cd^{2+}.	[181, 210, 215]
Intracellular sequestration	*Pseudomonas marginalis, Bacillus megaterium.*	GE *Escherichia coli* co-expressing synthetic phytochelatins (ECs) and Hg^{2+} transport system intracellularly.	[182, 211, 216]
Extracellular sequestration	*Pseudomonas marginalis*, of *Synechocystis* PCC 6803, *Alcaligenes eutrophus* CH34.	GE *Saccharomyces cerevisiae* displaying metal binding motifs of bacterial P1-Type ATPases for adsorption of Pb^{2+}.	[164, 212, 217].
Permeability Barrier	*Rhodopseudomonas acidophila, Enterobacter cloaceae, Gloeocapsa calcarea, Aureobasidium pullulans.*	GE *E. coli* JM109 expressing elastin-like polymer (ELP) and bacterial metalloregulatory protein, MerR for the binding of Hg^{2+}.	[169]
Metal Methylation	*Trichoderma asperellum* SM-12F1, *Penicillium janthinellum* SM-12F4, *Fusarium oxysporum* CZ-8F1	GE *Pseudomonas putida* KT2440 expressing *arsM* gene from *Chlamydomonas reinhardtii* for methylation of inorganic arsenic into organoarsenicals which are less toxic.	[107, 213, 218]

(Table 2) cont.....

Microbial Process	Wild Type (WT) Microbes	Genetically Engineered (GE) Microbes and Genes Manipulated/Enhanced	References
Metal Reduction	Sulphate-reducing bacteria (SRB) like *Desulfuromonas alkenivorans* S-7 and iron-reducing bacterium *Desulfuromonas palmitatis*	Overexpression of the oxygen-insensitive nitroreductase *nfsA* gene in *E. coli* for reduction of Cr^{5+} to Cr^{3+}.	[178, 214, 219]

Microbial Biosorption of Metals

Biosorption in microbes occurs by sequestration of heavy metals on the cell wall by processes like chemisorption, precipitation, complexation, adsorption–complexation on surface and pores, ion exchange or by absorption of metals into the cell [145]. The negatively charged chemical functional groups (carboxyl, hydroxyl, amine and phosphonate) present on the surface of the microbial cell wall influence the mechanism of biosorption by binding to the metal cations. Gram-positive and Gram-negative bacteria exhibit different sorptive behaviour even for the same metal cation because of the difference in the cell wall structure. *Bacillus thuringiensis* (Gram-positive) was observed to sorb a greater number of copper and cadmium ions than *Escherichia coli* (Gram-negative). *B. thuringiensis* cells consist of carboxyl groups on which had better binding to Cu^{2+} and Cd^{2+} than *E. coli*, which has phosphate groups for the binding of Cu^{2+} and Cd^{2+}. Gram-positive cell walls have a higher composition of teichoic acids, which increases the reactivity of the cell by providing acidic functional groups in greater numbers like carboxyl groups [208]. Amine groups are also effective in sequestering metal ions, as they not only bind to cations but also the anions through hydrogen bonding and electrostatic interactions. In an insightful study, the surface characterization of *Pseudomonas aeruginosa* was carried out by FTIR spectroscopy and functional groups belonging to carboxyl, phosphoryl, hydroxyl, and amino groups were found. Chromium-loaded biomass of *P. aeruginosa* was also studied to determine any differences in functional groups involved in the binding of Cr (III) and Cr (VI) ions. In chromium-loaded bacteria, carboxyl groups were responsible for binding to Cr (III), whereas amino groups were involved in Cr (VI) binding [146]. The solution pH is also an important factor that influences biosorption. With increasing pH, the overall charge on the cell surface becomes negative and the majority of the functional groups get deprotonated, allowing efficient electrostatic interaction and adsorption with cations [147].

Biosorbents can be chemically modified to enhance or modify binding sites by treating the biomass with acid, alkali, ethanol and acetone [148, 149]. In addition, inactive or inefficient functional groups can be converted into active binding groups by chemical treatment methods. To increase the uptake capacity of chitosan beads for the removal of mercury, they were aminated by the reaction of ethylenediamine with carbodiimide at pH 7 [150]. In another instance, modification of carboxylic and amine groups reduced the biosorption capacity of *Saccharomyces cerevisiae*, which suggested that carboxylic and amine groups are essential functional groups that influence biosorption of Cu^{2+} ion from aqueous solution [151].

Intracellular Sequestration by Protein Binding

In intracellular sequestration, metals accumulate inside the cell by binding to proteins to prevent exposure of metal-sensitive cellular components to the metals. The mechanism of sequestration involves the working of metal-binding proteins and peptides like metallothioneins (MTs) and glutathione (GSH), respectively [152]. Margoshes and Vallee first discovered MT in the horse kidney in the year 1957 [210], following which other organisms like plants, fungi and bacteria were also found to possess MTs. These are low molecular weight, cysteine-rich proteins that preferably bind to divalent metals like Cu and Zn through the thiol group of cysteine [153]. SmtA isolated from the Gram-negative *Cyanobacteria Synechococcus* PCC 7942 is the only MT-like protein that protects the bacterial cell from Cu toxicity. It has three distinct zinc-binding sites *via* eight Cys residues [154]. Transcription of the *smtA* gene is induced in the presence of metals like Zn, Cu, Cd, Co, Ni, Pb, Cr and Hg [155]. Recently, Gold *et al*. [156] identified the presence of an MT in *Mycobacterium tuberculosis* (mycobacterial metallothionein, MymT) that could bind up to six Cu(I) ions. *Cladosporium cladosporioides*, a fungus isolated from deep-sea sediment was observed to sequestrate Mn and P intracellularly [157]. GSH, a tripeptide of L-γ-glutamy--l-cysteine-glycine, also has a high metal affinity similar to MT. It maintains homeostasis in *E. coli* by providing resistance to cells that lack a metal efflux system [158]. In a study conducted by Lima *et al*. [159], they reported that cadmium sequestration mediated by GSH occurs intracellularly.

Extracellular Sequestration

Extracellular sequestration is characterised by chelation and binding, leading to the accumulation of metal ions on the periplasm [160]. *E. coli* contains the CueO multi-copper oxidase in its extracellular periplasmic space and the CusCFBA multicomponent efflux transport system that protects the cell from the toxic effects of Cu^+ ions that are pumped out by CopA from the cytoplasm. Hence, CueO oxidises Cu^+ to the less toxic form, *i.e.*, Cu^{2+} [161]. A similar observation was made by Telwell *et al* [162], where excess Zn^{2+} is pumped out by the ZiaA-mediated efflux into the periplasmic space of *Synechocystis* PCC 6803, where Zn^{2+} is sequestered by the metallothionein, SmtA. *Alcaligenes eutrophus* CH34 contains the *czc* operon, which mediates the efflux of Cd^{2+} ions from the cytoplasm into the extracellular space where the exopolysaccharides precipitate the ions on the negatively charged cell surface [163]. *Pseudomonas marginalis* shows a promising Pb-resistance wherein it produced the extracellular polymer that sequestrated the toxic Pb^{2+} [164]. In a similar instance, the fungus *Aureobasidium pullulans* secreted EPS that trapped the Pb^{2+} ions on the surface of the cells protecting the cell from lead toxicity [165].

Permeability Barrier

The permeability barrier in microbes consists of the cytoplasmic membrane and the envelope. Production of extracellular polymeric substances (EPS) by the microbial cell helps to protect them against unfavourable environmental conditions (desiccation, presence of toxic compounds, low temperatures or high osmotic pressures) [166]. EPS (polymers of exopolysaccharide glycocalyx, enzymes, structural proteins, lipids, nucleic acids and humic acids) are essential in the formation of biofilms (sessile surface-associated microbial communities) as they assist in the attachment of bacteria to the substrata [167]. High concentrations of toxic metal ions induce the production of EPS in bacterial cells. *Rhodopseudomonas acidophila* in the presence of Cu(II), Cr(VI), Cd(II) and 2,4-dichlorophenol (2,4-DCP) produced EPS consisting mainly of proteins and carbohydrates with negative ionic functional groups that bind with the metals [168]. The marine bacterium *Enterobacter cloaceae* in the presence of Cr(VI) not only showed tolerance up to high levels of the metal but also produced exopolysaccharide and exhibited an enhanced growth [169]. Exopolysaccharides produced by two cyanobacterial strains at high concentrations of Cr(VI) solution, *Gloeocapsa calcarea* and *Nostoc punctiforme* showed a promising biosorption potential. Pb^{2+} accumulation was also reported on the EPS produced by *Aureobasidium pullulans* [170].

Microbial Methylation of Metals

Methylation of metals is a microbial mediated process, which occurs either in aerobic or anaerobic conditions. In this process, a methyl group is transferred *via* transferases produced by the microbe, and the resulting methylated metal compounds vary in their solubility, toxicity and volatility depending on the type of metal. *Humicola* sp., a soil fungus, showed tremendous potential in biomethylating arsenic into volatile forms (biovolatilization). This study manifested the remediation prospects of utilising rhizospheric bacteria to relieve plants of heavy metal stresses [171]. This process is catalyzed by the enzyme As(III) S-adenosylmethionine methyltransferase (ArsM) which is involved in the transfer of methyl group from S-adenosyl-L-methionine (SAM) to arsenite

(As(III)), which is sequentially transformed into monomethylarsonous acid (MMAs(III)), dimethylarsinous acid (DMAs(III)), and finally to the volatile trimethylarsonous (TMAs) [172]. In addition, similar properties were exhibited by fungal species *Trichoderma asperellum* SM-12F1, *Penicillium janthinellum* SM-12F4, and *Fusarium oxysporum* CZ-8F1 where arsenate (As(V)) was reduced and methylated to As (III), MMA and DMA [173]. In flooded paddy fields, sulphate-reducing bacteria and methanogens were observed for their arsenic methylation and demethylation properties [174]. Numerous studies have reported methylation of inorganic selenium compounds, selenite (SeO_3^{2-}) and selenate (SeO_4^{2-}) into the volatile dimethyl selenide (DMSe) and dimethyl diselenide (DMDSe) by microbial action [175].

Microbial Reduction of Metals

Microbes are able to carry out redox transformations of metals, converting toxic metal species into non-toxic ones and *vice-versa*. Bacteria possess electron transport systems in the plasma membrane and are able to transform heavy metals on the cell surface layer. In yeasts and fungi, electron transport systems present in the mitochondria supply electron donors to the intracellular heavy metal oxido-reduction systems [176]. A well-known detoxification system in bacteria is the *mer* operon composed of mercuric reductase (MerA) protein which catalyzes the reduction of Hg(II) to volatile Hg(0) [177]. Sulphate-reducing bacteria (SRB) have been widely studied for the bioremediation of acid mine tailings. SRB catalyses the reduction of sulphate to sulphide to form sulphuric acid and hydrogen sulphide, which in the presence of metals, react to form non-toxic precipitates of metal sulphides while simultaneously raising the pH [178]. *Desulfuromonas alkenivorans* S-7, an SRB was able to remove over 90% of Mn^{2+} and Pb^{2+} in the presence of sulphate during a seven-day treatment [179]. Likewise, the iron-reducing bacterium *Desulfuromonas palmitatis* in combination with the chelating agent EDTA, exhibited a high extraction rate for As and Pb [180]. Fig. (**1**) illustrates the various processes involved in the bioremediation of heavy metals in soil by plants and bacteria.

GENETICALLY ENGINEERED BACTERIA FOR BIOREMEDIATION OF HEAVY METALS

It is evident that microbial species in the environment are capable of degrading or detoxifying heavy metal contaminants. However, most of these microbes carry out slow degradation due to a lack of catabolic pathways required for the degradation/detoxification of the contaminants [183]. Through genetic engineering techniques known as recombinant DNA technology, it is possible to enhance the existing mechanisms of microbial species gene exchange between microbes involved in the degradation process. The resultant microbes are known as genetically modified microorganisms (GMM) or genetically engineered microorganisms (GEM). Genetically engineered microorganisms (GEMs) have been widely used for soil, activated sludge and groundwater bioremediation [184]. Following are a list of strategies that can manipulate bacterial genetic and biochemical potential to enhance their role in heavy metal bioremediation:

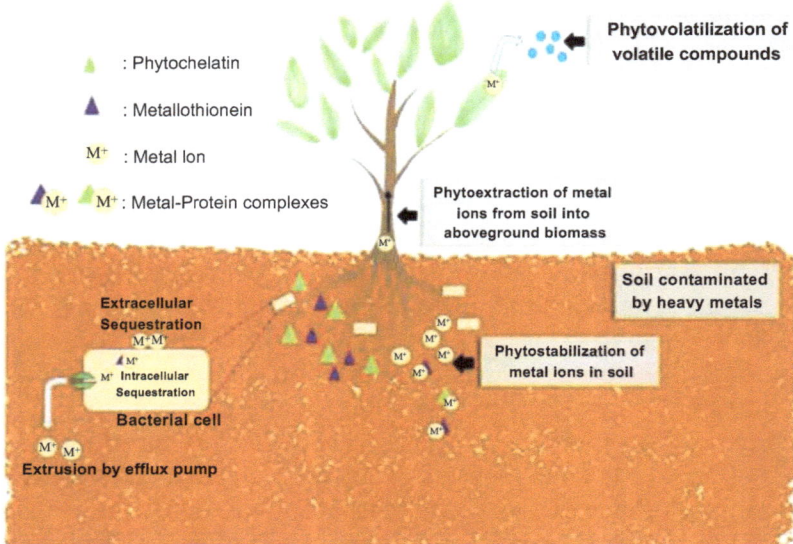

Fig. (1). Illustrates the various processes involved in the bioremediation of heavy metals in soil by plants and bacteria.

Overexpression of Gene or Operon Involved in Metal Detoxification Pathways

In this method of recombinant DNA technology, plasmids (expression vector) containing the gene of interest are introduced into and expressed in selected bacterial strains. Liu *et al.* [185], in their study, used strains *Bacillus idriensis* and *Sphingomonas desiccabilis* with high resistance to arsenic but do not have the *arsM* gene (responsible for methylation of arsenic to form organic arsenicals). A plasmid containing the *arsM* gene isolated from *Rhodopseudomonas palustris* was overexpressed *B. idriensis* and *S. desiccabilis* using a T7 promoter. It was observed that these recombinant cells could remove total arsenic from solution and form about 49% of methylated organic arsenicals. When incubated in As-contaminated soil for 30 days, about 2.2% – 4.5% of arsenic was removed by biovolatilization, with *B. idriensis* exhibiting a better production of volatile arsenic. In a similar scenario, the thiosulphate reductase gene (*phsABC*) from

Salmonella enterica serovar Typhimurium was overexpressed in *Escherichia coli*. The recombinant *E. coli* grown in a medium containing $CdCl_2$ produced hydrogen sulphide from thiosulphate and formed bright yellow cadmium sulphide precipitates [186]. Interestingly, yeast (CUP1) and mammalian (HMT-1A) metallothioneins (MTs) were expressed in *E. coli* and fused to the outer membrane protein, LamB. The resultant hybrid protein enabled the recombinant *E. coli* to bind Cd^{2+} 15- to 20-fold [187]. Kiyono and Pan-Hou [188] engineered a plasmid to transform *E. coli* to express the *mer* operon from *Pseudomonas* K62 that consist of genes (involved in binding and transport of mercury, mercuric reductase and organomercurial lyase) and an overexpression of polyphosphate which is an efficient chelator of divalent cations.

Expression of Transport Proteins and Efflux Pumps

Another feature of bacteria that confers heavy metal resistance is the presence of efflux pumps that removes metal ions from the cell to the exterior. Sometimes, a particular efflux pump may function in both heavy metal and antibiotic resistance [189]. The cadA pump in *Helicobacter pylori* confers resistance to Cd(II) as well as Zn(II) and Co(II) when expressed in *E. coli zntA* mutant [190]. The mercury transport genes (*merP* and *merT*) and a glutathione-S-transferase pea metallothionein gene (GST-PMT) were expressed on *E. coli* JM109 harboring plasmids containing the above two genes. The bacteria were seen to accumulate as well as adsorb high amounts mercury due to the presence of the membrane-associated transport protein [191].

Genome Editing by CRISPR-Cas Technology

The term 'genome editing' is interchangeable with 'genome engineering' or 'gene editing', and CRISPR-Cas is one of the recent approaches used for genome editing. There are three major types of CRISPR-Cas systems, *i.e.*, Type I, II and III containing the signature Cas proteins: Cas3, Cas9 and Cas 10, respectively [192, 193]. The CRISPR/Cas systems consist of 30–40 bp direct repeat sequences flanked by "spacer sequences", which, after processing and transcription, are converted to precursor CRISPR RNA (crRNA). Further, a complex of crRNA and tracrRNA (transactivating crRNA) is formed in the presence of Cas nuclease guided by spacer sequence, which then binds protospacer-adjacent motif (PAM) [194]. This binding creates a specific double-strand break (DSB) in the target DNA sequence, where the gene of interest can then be either deleted or inserted from the system with the help of CRISPR/Cas. The DSBs are usually repaired by error-prone pathways, which likely produce mutations like single nucleotide polymorphisms (SNPs) and small insertions or deletions (INDELs) at the site of the break. A few repair pathways include non-homologous end joining (NHEJ) or homology-directed repair (HDR), alternative end-joining (A-EJ), or micro-homology-mediated end joining (MMEJ) [194 - 196]. The implementation of CRISPR/Cas systems in bacteria for heavy metal bioremediation is still at its infancy; however, there exist quite a few studies where CRISPR-Cas technology has been executed for phytoremediation of heavy metals. Tang *et al.* [197] used the CRISPR/Cas9 system to knock out the metal transporter gene *OsNramp5* in rice plant, which resulted in a decreased accumulation of Cd in the indica rice lines without negatively influencing the yield. The advantages of this technique are immeasurable, however, there is an underlying risk of potential off-target mutations or lethal mutations which might alter gene function and lead to genome instability [198].

CRISPR-Cas systems associated with transposons are crucial tools in the present era of gene editing. Transposons or transposable elements or 'jumping genes' are mobile genetic elements that were first discovered by Barbara McClintock in the 1950s for which she was conferred the Nobel Prize in 1983 [199]. Transposons are found in a variety of organisms like bacteria, yeasts and higher eukaryotes. Based on the transposition process, transposable elements can be classified as Class I elements or retrotransposons that use an RNA intermediate before getting reverse transcribed to cDNA and integrating into a new locus; Class II elements, on the contrary, do not require an intermediate and are simply excised from the genome before integrating into another locus [200]. Depending on the target organism, several transposon systems in combination with CRISPR-Cas have been developed over the years and extensively used in gene editing in various research areas. For instance, the *piggyBac* (PB) transposon utilizes a "cut-an--paste" mechanism where the PB transposase recognizes specific sequences located on the vector and moves them from their original sites and integrates them into the targeted chromosomal sites [201]. Li *et al.* [202] developed a *piggyBac* (PB) transposon-based CRISPR activation (CRISPRa) all-in-one system which functions to activate and accelerate the differentiation of human iPSCs (induced pluripotent stem cells) into astrocytes and neurons. This development shows great promise not only in the field of neuroscience but also in agriculture, where temporary expression of CRISPR-Cas9 has been successfully carried out in rice plants [203].

A recent study describes the development of a new technique called Environmental Transformation Sequencing (ET-Seq) which makes it possible for individual bacterial species to integrate external DNA present in a community. Next, the DNA of these genetically tractable bacterial species was edited using a DNA-editing All-in-one RNA-guided CRISPR-Cas Transposase (DART) system, which signifies the possibility of gene editing in individual species of a microbial community [204]. These new findings suggest the multiple prospects that this technique entails in the context of bioremediation, where certain genes encoding for metal detoxification enzymes or metal efflux pumps can be introduced into targeted organisms or specific loci within a particular microbial consortium or microbiome existing in a heavy metal-contaminated environment.

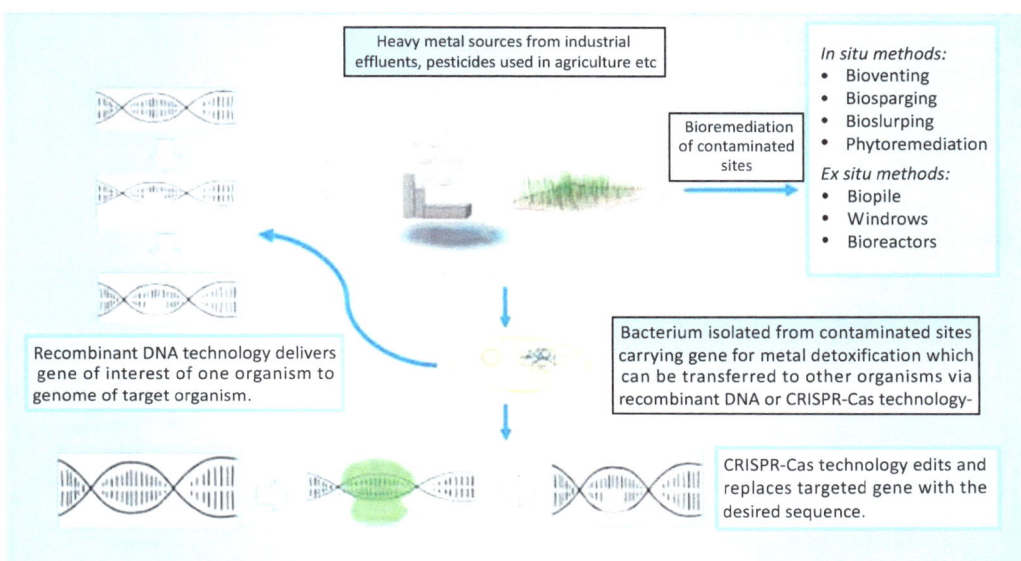

Fig. (2). Summary of bioremediation of heavy metals from contaminated sites.

CONCLUSION AND FUTURE PROSPECTS

Urbanisation and modernisation have undoubtedly made the lives of human beings convenient and less arduous. However, the dependence on synthetic chemicals like pesticides for agriculture or the uncurbed manner of exploiting the Earth's resources has put tremendous pressure on the fragile balance of the ecosystem by releasing contaminants like polyaromatic hydrocarbons, polychlorinated biphenyls, pesticides, heavy metals and organic wastes [205, 206]. As discussed in the previous sections of this chapter, there are numerous techniques available and in use for the remediation of environmental contaminants. Bioremediation techniques are ideal for detoxifying or degrading contaminants from the environment due to the natural breakdown of the contaminants using inherent processes present in plants and microbes. Moreover, bioremediation is a sustainable approach to managing environmental contaminants because of the formation of non-toxic products, economic feasibility and low-technology requirements [95]. Newer approaches like genetic engineering and gene editing are currently used to transform plant and microbial species by transferring desirable genes from hyperaccumulator species of either plant or microbes (Fig. 2). Furthermore, omics approaches coupled with systems biology hold great potential in carrying out effective bioremediation of heavy metals [207].

ACKNOWLEDGMENTS

The authors acknowledge the financial support provided by Council of Scientific & Industrial Research [09/0347(12424)/2021-EMR-I] to carry out the present work. We also acknowledge DST-FIST[SR/FST/LSI-666/2016(C)] and UGC-SAP [F.4-7/2016/DRS-1(SAP-II)] for the financial support provided to the parent department.

REFERENCES

[1] Rosenberg E. Genetic Engineering In: It's in Your DNA: from discovery to structure, function and role in Evolution, Cancer and Aging. 1st. Academic Press 2017; pp. 81-93.

[2] Laible G. Enhancing livestock through genetic engineering—Recent advances and future prospects. Comp Immunol Microbiol Infect Dis 2009; 32(2): 123-37.
[http://dx.doi.org/10.1016/j.cimid.2007.11.012] [PMID: 18243310]

[3] Royzman E, Cusimano C, Leeman RF. What lies beneath? Fear vs. disgust as affective predictors of absolutist opposition to genetically modified food and other new technologies. Judgm Decis Mak 2017; 12(5): 466-80.
[http://dx.doi.org/10.1017/S1930297500006495]

[4] Ormandy EH, Dale J, Griffin G. Genetic engineering of animals: Ethical issues, including welfare concerns. Can Vet J 2011; 52(5): 544-50.
[PMID: 22043080]

[5] Devos Y, Maeseele P, Reheul D, Van Speybroeck L, De Waele D. Ethics in the societal debate on genetically modified organisms: A (re)quest for sense and sensibility. J Agric Environ Ethics 2008; 21(1): 29-61.
[http://dx.doi.org/10.1007/s10806-007-9057-6]

[6] Soucy SM, Huang J, Gogarten JP. Horizontal gene transfer: Building the web of life. Nat Rev Genet 2015; 16(8): 472-82.
[http://dx.doi.org/10.1038/nrg3962] [PMID: 26184597]

[7] Sun D, Jeannot K, Xiao Y, Knapp CW. Editorial: Horizontal gene transfer mediated bacterial antibiotic resistance. Front Microbiol 2019; 10: 1933.
[http://dx.doi.org/10.3389/fmicb.2019.01933] [PMID: 31507555]

[8] WHO. Antimicrobial Resistance Global Report on Surveillance. World Health Organization 2016.

[9] Dupuis MÈ, Villion M, Magadán AH, Moineau S. CRISPR-Cas and restriction–modification systems are compatible and increase phage resistance. Nat Commun 2013; 4(1): 2087.
[http://dx.doi.org/10.1038/ncomms3087] [PMID: 23820428]

[10] Ishino Y, Shinagawa H, Makino K, Amemura M, Nakata A. Nucleotide sequence of the iap gene, responsible for alkaline phosphatase isozyme conversion in *Escherichia coli*, and identification of the gene product. J Bacteriol 1987; 169(12): 5429-33.
[http://dx.doi.org/10.1128/jb.169.12.5429-5433.1987] [PMID: 3316184]

[11] Makarova KS, Grishin NV, Shabalina SA, Wolf YI, Koonin EV. A putative RNA-interference-based immune system in prokaryotes: computational analysis of the predicted enzymatic machinery, functional analogies with eukaryotic RNAi, and hypothetical mechanisms of action. Biol Direct 2006; 1(1): 7.

[http://dx.doi.org/10.1186/1745-6150-1-7] [PMID: 16545108]

[12] Kang H, Minder P, Park MA, Mesquitta WT, Torbett BE, Slukvin II. CCR5 disruption in induced pluripotent stem cells using CRISPR/Cas9 provides selective resistance of immune cells to CCR5-tropic HIV-1 virus. Mol Ther Nucleic Acids 2015; 4: e268.
[http://dx.doi.org/10.1038/mtna.2015.42] [PMID: 26670276]

[13] Ye L, Wang J, Beyer AI, *et al.* Seamless modification of wild-type induced pluripotent stem cells to the natural CCR5Δ32 mutation confers resistance to HIV infection. Proc Natl Acad Sci 2014; 111(26): 9591-6.
[http://dx.doi.org/10.1073/pnas.1407473111] [PMID: 24927590]

[14] Kennedy EM, Kornepati AVR, Goldstein M, *et al.* Inactivation of the human papillomavirus E6 or E7 gene in cervical carcinoma cells by using a bacterial CRISPR/Cas RNA-guided endonuclease. J Virol 2014; 88(20): 11965-72.
[http://dx.doi.org/10.1128/JVI.01879-14] [PMID: 25100830]

[15] Alexander HM, Bruns E, Schebor H, Malmstrom CM. Crop-associated virus infection in a native perennial grass: Reduction in plant fitness and dynamic patterns of virus detection. J Ecol 2017; 105(4): 1021-31.
[http://dx.doi.org/10.1111/1365-2745.12723]

[16] Inoue-Nagata AK, Lima MF, Gilbertson RL. A review of geminivirus diseases in vegetables and other crops in Brazil: Current status and approaches for management. Hortic Bras 2016; 34(1): 8-18.
[http://dx.doi.org/10.1590/S0102-053620160000100002]

[17] Baltes NJ, Hummel AW, Konecna E, *et al.* Conferring resistance to geminiviruses with the CRISPR–Cas prokaryotic immune system. Nat Plants 2015; 1(10): 15145.
[http://dx.doi.org/10.1038/nplants.2015.145] [PMID: 34824864]

[18] Eş I, Gavahian M, Marti-Quijal FJ, *et al.* The application of the CRISPR-Cas9 genome editing machinery in food and agricultural science: Current status, future perspectives, and associated challenges. Biotechnol Adv 2019; 37(3): 410-21.
[http://dx.doi.org/10.1016/j.biotechadv.2019.02.006] [PMID: 30779952]

[19] Whitcomb CE. An introduction to ALS-inhibiting herbicides. Toxicol Ind Health 1999; 15(1): 231-9.
[http://dx.doi.org/10.1191/074823399678846592] [PMID: 10188205]

[20] Garcia MD, Nouwens A, Lonhienne TG, Guddat LW. Comprehensive understanding of acetohydroxyacid synthase inhibition by different herbicide families. Proc Natl Acad Sci 2017; 114(7): E1091-100.
[http://dx.doi.org/10.1073/pnas.1616142114] [PMID: 28137884]

[21] Wu J, Chen C, Xian G, *et al.* Engineering herbicide resistant oilseed rape by CRISPR/Cas9 mediated cytosine base editing. Plant Biotechnol J 2020; 18(9): 1857-9.
[http://dx.doi.org/10.1111/pbi.13368] [PMID: 32096325]

[22] Zhu Y, Lin Y, Chen S, *et al.* CRISPR /Cas9 mediated functional recovery of the recessive *rc* allele to develop red rice. Plant Biotechnol J 2019; 17(11): 2096-105.
[http://dx.doi.org/10.1111/pbi.13125] [PMID: 31002444]

[23] Giacco R, Clemente G, Brighenti F, *et al.* Metabolic effects of resistant starch in patients with type 2 diabetes. Diabetes Nutr Metab 1998; 11(6): 330-5.

[24] Sun Y, Jiao G, Liu Z, *et al.* Generation of high-amylose rice through CRISPR/Cas9-Mediated targeted mutagenesis of starch branching enzymes. Front Plant Sci 2017; 8: 298.
[http://dx.doi.org/10.3389/fpls.2017.00298] [PMID: 28326091]

[25] Steffen W, Grinevald J, Crutzen P, McNeill J. The Anthropocene: conceptual and historical perspectives. Philos Trans- Royal Soc, Math Phys Eng Sci 2011; 369(1938): 842-67.
[http://dx.doi.org/10.1098/rsta.2010.0327] [PMID: 21282150]

[26] West JJ, Cohen A, Dentener F, *et al.* What we breathe impacts our health: Improving understanding of

the link between air pollution and health. Environ Sci Technol 2016; 50(10): 4895-904.
[http://dx.doi.org/10.1021/acs.est.5b03827] [PMID: 27010639]

[27] Fishman J, Creilson JK, Parker PA, *et al.* An investigation of widespread ozone damage to the soybean crop in the upper Midwest determined from ground-based and satellite measurements. Atmos Environ 2010; 44(18): 2248-56.
[http://dx.doi.org/10.1016/j.atmosenv.2010.01.015]

[28] WHO Air quality guidelines for particulate matter, ozone, nitrogen dioxide and sulfur dioxide. World Health Organization 2006.

[29] Begum BA, Paul SK, Dildar Hossain M, Biswas SK, Hopke PK. Indoor air pollution from particulate matter emissions in different households in rural areas of Bangladesh. Build Environ 2009; 44(5): 898-903.
[http://dx.doi.org/10.1016/j.buildenv.2008.06.005]

[30] Mestl H, Aunan K, Seip H, Wang S, Zhao Y, Zhang D. Urban and rural exposure to indoor air pollution from domestic biomass and coal burning across China. Sci Total Environ 2007; 377(1): 12-26.
[http://dx.doi.org/10.1016/j.scitotenv.2007.01.087] [PMID: 17343898]

[31] Lu Y, Song S, Wang R, *et al.* Impacts of soil and water pollution on food safety and health risks in China. Environ Int 2015; 77: 5-15.
[http://dx.doi.org/10.1016/j.envint.2014.12.010] [PMID: 25603422]

[32] Yuce G, Pinarbasi A, Ozcelik S, Ugurluoglu D. Soil and water pollution derived from anthropogenic activities in the Porsuk River Basin, Turkey. Environ Geol 2006; 49(3): 359-75.
[http://dx.doi.org/10.1007/s00254-005-0072-5]

[33] Geissen V, Ramos FQ, De J Bastidas-Bastidas P, *et al.* Soil and water pollution in a banana production region in tropical Mexico. Bull Environ Contam Toxicol 2010; 85(4): 407-13.
[http://dx.doi.org/10.1007/s00128-010-0077-y] [PMID: 20734023]

[34] Aktar W, Sengupta D, Chowdhury A. Impact of pesticides use in agriculture: Their benefits and hazards. Interdiscip Toxicol 2009; 2(1): 1-12.
[http://dx.doi.org/10.2478/v10102-009-0001-7] [PMID: 21217838]

[35] Akyıl D, Özkara A, Erdoğmuş SF, Eren Y, Konuk M, Sağlam E. Micronucleus assay in human lymphocytes after exposure to alloxydim sodium herbicide *in vitro.* Cytotechnology 2015; 67(6): 1059-66.
[http://dx.doi.org/10.1007/s10616-014-9746-8] [PMID: 25017922]

[36] Köhler HR, Triebskorn R. Wildlife ecotoxicology of pesticides: can we track effects to the population level and beyond? Science 2013; 341(6147): 759-65.
[http://dx.doi.org/10.1126/science.1237591] [PMID: 23950533]

[37] Ostrowski SR, Wilbur S, Chou CHSJ, *et al.* Agency for toxic substances and disease registry's 1997 priority list of hazardous substances. latent effects—carcinogenesis, neurotoxicology, and developmental deficits in humans and animals. Toxicol Ind Health 1999; 15(7): 602-44.
[http://dx.doi.org/10.1177/074823379901500702] [PMID: 10677885]

[38] USEPA. Health effects support document for aldrin/dieldrin, US Environmental Protection Agency, Office of Water (4304T). Washington, DC: Health and Ecological Criteria Division 2003; p. 20460.

[39] Yadav SK. Heavy metals toxicity in plants: An overview on the role of glutathione and phytochelatins in heavy metal stress tolerance of plants. S Afr J Bot 2010; 76(2): 167-79.
[http://dx.doi.org/10.1016/j.sajb.2009.10.007]

[40] Rajapaksha RMCP, Tobor-Kapłon MA, Bååth E. Metal toxicity affects fungal and bacterial activities in soil differently. Appl Environ Microbiol 2004; 70(5): 2966-73.
[http://dx.doi.org/10.1128/AEM.70.5.2966-2973.2004] [PMID: 15128558]

[41] Khan A, Khan S, Khan MA, Qamar Z, Waqas M. The uptake and bioaccumulation of heavy metals by

[42] Friedlová M. The influence of heavy metals on soil biological and chemical properties. Soil Water Res 2010; 5(1): 21-7.
[http://dx.doi.org/10.17221/11/2009-SWR]

[43] Xie Y, Fan J, Zhu W, *et al.* Effect of heavy metals pollution on soil microbial diversity and bermudagrass genetic variation. Front Plant Sci 2016; 7: 755.
[http://dx.doi.org/10.3389/fpls.2016.00755] [PMID: 27303431]

[44] Keunen E, Remans T, Bohler S, Vangronsveld J, Cuypers A. Metal-induced oxidative stress and plant mitochondria. Int J Mol Sci 2011; 12(10): 6894-918.
[http://dx.doi.org/10.3390/ijms12106894] [PMID: 22072926]

[45] Emamverdian A, Ding Y, Mokhberdoran F, Xie Y. Heavy metal stress and some mechanisms of plant defense response. ScientWorldJ 2015; 2015: 1-18.
[http://dx.doi.org/10.1155/2015/756120] [PMID: 25688377]

[47] Jaffré T, Brooks RR, Lee J, Reeves RD. *Sebertia acuminata*: A hyperaccumulator of nickel from new caledonia. Science 1976; 193(4253): 579-80.
[http://dx.doi.org/10.1126/science.193.4253.579] [PMID: 17759588]

[48] Baker AJM, Brooks RR. Terrestrial higher plants which hyperaccumulate metallic elements : A review of their distribution, ecology and phytochemistry. Biorecovery 1989; 1: 81-126.

[49] Singh S, Parihar P, Singh R, Singh VP, Prasad SM. Heavy metal tolerance in plants: Role of transcriptomics, proteomics, metabolomics, and ionomics. Front Plant Sci 2016; 6(1143): 1143.
[http://dx.doi.org/10.3389/fpls.2015.01143] [PMID: 26904030]

[50] Yang X, Feng Y, He Z, Stoffella PJ. Molecular mechanisms of heavy metal hyperaccumulation and phytoremediation. J Trace Elem Med Biol 2005; 18(4): 339-53.
[http://dx.doi.org/10.1016/j.jtemb.2005.02.007] [PMID: 16028496]

[51] Stepanauskas R, Glenn TC, Jagoe CH, Tuckfield RC, Lindell AH, McArthur J. Elevated microbial tolerance to metals and antibiotics in metal-contaminated industrial environments. Environ Sci Technol 2005; 39(10): 3671-8.
[http://dx.doi.org/10.1021/es048468f] [PMID: 15952371]

[52] Wright MS, Peltier GL, Stepanauskas R, McArthur JV. Bacterial tolerances to metals and antibiotics in metal-contaminated and reference streams. FEMS Microbiol Ecol 2006; 58(2): 293-302.
[http://dx.doi.org/10.1111/j.1574-6941.2006.00154.x] [PMID: 17064270]

[53] Gadd GM. Microbial influence on metal mobility and application for bioremediation. Geoderma 2004; 122(2-4): 109-19.
[http://dx.doi.org/10.1016/j.geoderma.2004.01.002]

[54] Mishra D, Rhee YH. Microbial leaching of metals from solid industrial wastes. J Microbiol 2014; 52(1): 1-7.
[http://dx.doi.org/10.1007/s12275-014-3532-3] [PMID: 24390831]

[55] Gholami RM, Borghei SM, Mousavi SM. Bacterial leaching of a spent Mo–Co–Ni refinery catalyst using *Acidithiobacillus ferrooxidans* and *Acidithiobacillus thiooxidans*. Hydrometallurgy 2011; 106(1-2): 26-31.
[http://dx.doi.org/10.1016/j.hydromet.2010.11.011]

[56] Naseri T, Bahaloo-Horeh N, Mousavi SM. Bacterial leaching as a green approach for typical metals recovery from end-of-life coin cells batteries. J Clean Prod 2019; 220: 483-92.
[http://dx.doi.org/10.1016/j.jclepro.2019.02.177]

[57] Hofmann M, Heine T, Malik L, *et al.* Screening for microbial metal-chelating siderophores for the removal of metal ions from solutions. Microorganisms 2021; 9(1): 111.

[http://dx.doi.org/10.3390/microorganisms9010111] [PMID: 33466508]

[58] Nair A, Juwarkar AA, Singh SK. Production and characterization of siderophores and its application in arsenic removal from contaminated soil. Water Air Soil Pollut 2007; 180(1-4): 199-212.
[http://dx.doi.org/10.1007/s11270-006-9263-2]

[59] Gupta VK, Nayak A, Bhushan B, Agarwal S. A critical analysis on the efficiency of activated carbons from low-cost precursors for heavy metals remediation. Crit Rev Environ Sci Technol 2015; 45(6): 613-68.
[http://dx.doi.org/10.1080/10643389.2013.876526]

[60] Xu Y, Liang X, Xu Y, *et al*. Remediation of heavy metal-polluted agricultural soils using clay minerals: A review. Pedosphere 2017; 27(2): 193-204.
[http://dx.doi.org/10.1016/S1002-0160(17)60310-2]

[61] Martin TA, Ruby MV. Review of *in situ* remediation technologies for lead, zinc, and cadmium in soil. Rem J 2004; 14(3): 35-53.
[http://dx.doi.org/10.1002/rem.20011]

[62] Zhang C, Zhu M, Zeng G, *et al*. Active capping technology: A new environmental remediation of contaminated sediment. Environ Sci Pollut Res Int 2016; 23(5): 4370-86.
[http://dx.doi.org/10.1007/s11356-016-6076-8] [PMID: 26762937]

[63] Cho YM, Ghosh U, Kennedy AJ, *et al*. Field application of activated carbon amendment for *in-situ* stabilization of polychlorinated biphenyls in marine sediment. Environ Sci Technol 2009; 43(10): 3815-23.
[http://dx.doi.org/10.1021/es802931c] [PMID: 19544893]

[64] Bonaglia S, Rämö R, Marzocchi U, *et al*. Capping with activated carbon reduces nutrient fluxes, denitrification and meiofauna in contaminated sediments. Water Res 2019; 148: 515-25.
[http://dx.doi.org/10.1016/j.watres.2018.10.083] [PMID: 30408736]

[65] Abramovitch RA, ChangQing L, Hicks E, Sinard J. *In situ* remediation of soils contaminated with toxic metal ions using microwave energy. Chemosphere 2003; 53(9): 1077-85.
[http://dx.doi.org/10.1016/S0045-6535(03)00572-1] [PMID: 14512111]

[66] Schmidt R, Gudbjerg J, Sonnenborg TO, Jensen KH. Removal of NAPLs from the unsaturated zone using steam: Prevention of downward migration by injecting mixtures of steam and air. J Contam Hydrol 2002; 55(3-4): 233-60.
[http://dx.doi.org/10.1016/S0169-7722(01)00192-9] [PMID: 11999631]

[67] Beyke G, Fleming D. *In situ* thermal remediation of DNAPL and LNAPL using electrical resistance heating. Rem J 2005; 15(3): 5-22.
[http://dx.doi.org/10.1002/rem.20047]

[68] Dellisanti F, Rossi PL, Valdrè G. In-field remediation of tons of heavy metal-rich waste by Joule heating vitrification. Int J Miner Process 2009; 93(3-4): 239-45.
[http://dx.doi.org/10.1016/j.minpro.2009.09.002]

[69] Vidonish JE, Zygourakis K, Masiello CA, Gao X, Mathieu J, Alvarez PJJ. Pyrolytic treatment and fertility enhancement of soils contaminated with heavy hydrocarbons. Environ Sci Technol 2016; 50(5): 2498-506.
[http://dx.doi.org/10.1021/acs.est.5b02620] [PMID: 26284736]

[70] Pape A, Switzer C, McCosh N, Knapp CW. Impacts of thermal and smouldering remediation on plant growth and soil ecology. Geoderma 2015; 243-244: 1-9.
[http://dx.doi.org/10.1016/j.geoderma.2014.12.004]

[71] USEPA. How To Evaluate Alternative Clean-up Technologies for Underground Storage Tank Sites, US Environmental Protection Agency, Land and Emergency Management (5401R). Washington, DC: Health and Ecological Criteria Division 1994; p. 20460.

[72] Benschoten JEV, Matsumoto MR, Young WH. Evaluation and analysis of soil washing for seven lead-

contaminated soils. J Environ Eng 1997; 123(3): 217-24.
[http://dx.doi.org/10.1061/(ASCE)0733-9372(1997)123:3(217)]

[73]   Marino MA, Brica RM, Neale CN. Heavy metal soil remediation: The effects of attrition scrubbing on a wet gravity concentration process. Environ Prog 1997; 16(3): 208-14.
[http://dx.doi.org/10.1002/ep.3300160318]

[74]   Cauwenberg P, Verdonckt F, Maes A. Flotation as a remediation technique for heavily polluted dredged material. 1. A feasibility study. Sci Total Environ 1998; 209(2-3): 113-119, 113-119.
[http://dx.doi.org/10.1016/S0048-9697(97)00305-7] [PMID: 9514033]

[75]   Ehsan S, Prasher SO, Marshall WD. A washing procedure to mobilize mixed contaminants from soil: II. Heavy metals. J Environ Qual 2006; 35(6): 2084-91.
[http://dx.doi.org/10.2134/jeq2005.0475] [PMID: 17071877]

[76]   Kuo S, Lai MS, Lin CW. Influence of solution acidity and $CaCl_2$ concentration on the removal of heavy metals from metal-contaminated rice soils. Environ Pollut 2006; 144(3): 918-25.
[http://dx.doi.org/10.1016/j.envpol.2006.02.001] [PMID: 16603295]

[77]   Finžgar N, Leštan D. Multi-step leaching of Pb and Zn contaminated soils with EDTA. Chemosphere 2007; 66(5): 824-32.
[http://dx.doi.org/10.1016/j.chemosphere.2006.06.029] [PMID: 16879858]

[78]   Hubicki Z, Koodynsk D. Selective removal of heavy metal ions from waters and wastewaters using ion exchange methods. Ion Exch Tech 2012.

[79]   Li JR, Wang X, Yuan B, Fu ML. Layered chalcogenide for Cu^{2+} removal by ion-exchange from wastewater. J Mol Liq 2014; 200: 205-12.
[http://dx.doi.org/10.1016/j.molliq.2014.09.008]

[80]   Manos MJ, Malliakas CD, Kanatzidis MG. Heavy-metal-ion capture, ion-exchange, and exceptional acid stability of the open-framework chalcogenide (NH(4))(4)In(12)Se(20). Chemistry 2007; 13(1): 51-8.
[http://dx.doi.org/10.1002/chem.200600892] [PMID: 17120263]

[81]   AlOthman ZA, Alam MM, Naushad M. Heavy toxic metal ion exchange kinetics: Validation of ion exchange process on composite cation exchanger nylon 6,6 Zr(IV) phosphate. J Ind Eng Chem 2013; 19(3): 956-60.
[http://dx.doi.org/10.1016/j.jiec.2012.11.016]

[82]   Pinel-Raffaitin P, Le Hecho I, Amouroux D, Potin-Gautier M. Distribution and fate of inorganic and organic arsenic species in landfill leachates and biogases. Environ Sci Technol 2007; 41(13): 4536-41.
[http://dx.doi.org/10.1021/es0628506] [PMID: 17695893]

[83]   Gounaris V, Anderson PR, Holsen TM. Characteristics and environmental significance of colloids in landfill leachate. Environ Sci Technol 1993; 27(7): 1381-7.
[http://dx.doi.org/10.1021/es00044a013]

[84]   Evanko CR, Dzombak DA. Remediation of metals-contaminated soils and groundwater. Technology Evaluation Report: TE-97-01. 1997.

[85]   Andreottola G, Dallago L, Ferrarese E. Feasibility study for the remediation of groundwater contaminated by organolead compounds. J Hazard Mater 2008; 156(1-3): 488-98.
[http://dx.doi.org/10.1016/j.jhazmat.2007.12.044] [PMID: 18242831]

[86]   Aluthgun Hewage S, Batagoda JH, Meegoda JN. Remediation of contaminated sediments containing both organic and inorganic chemicals using ultrasound and ozone nanobubbles. Environ Pollut 2021; 274: 116538.
[http://dx.doi.org/10.1016/j.envpol.2021.116538] [PMID: 33540254]

[87]   Ince M, Kaplan İnce O. An overview of adsorption technique for heavy metal removal from water/wastewater: A critical review. Int J Pure Appl Sci 2017; 3(2): 10-9.
[http://dx.doi.org/10.29132/ijpas.358199]

[88] Renu MA, Agarwal M, Singh K. Heavy metal removal from wastewater using various adsorbents: A review. J Water Reuse Desalin 2017; 7(4): 387-419.
[http://dx.doi.org/10.2166/wrd.2016.104]

[89] Yang X, Liu L, Tan W, Liu C, Dang Z, Qiu G. Remediation of heavy metal contaminated soils by organic acid extraction and electrochemical adsorption. Environ Pollut 2020; 264: 114745.
[http://dx.doi.org/10.1016/j.envpol.2020.114745] [PMID: 32416427]

[90] Charpentier TVJ, Neville A, Lanigan JL, Barker R, Smith MJ, Richardson T. Preparation of magnetic carboxymethylchitosan nanoparticles for adsorption of heavy metal ions. ACS Omega 2016; 1(1): 77-83.
[http://dx.doi.org/10.1021/acsomega.6b00035] [PMID: 31457118]

[91] Rajendran K, Lin R, Wall DM, Murphy JD. Influential aspects in waste management practices.Taherzadeh MJ. Bolton K, Wong J, & Pandey A. Sustainable Resource Recovery and Zero Waste Approaches 2019; pp. 65-78.

[92] Mohan D, Kumar H, Sarswat A, Alexandre-Franco M, Pittman CU Jr. Cadmium and lead remediation using magnetic oak wood and oak bark fast pyrolysis bio-chars. Chem Eng J 2014; 236: 513-28.
[http://dx.doi.org/10.1016/j.cej.2013.09.057]

[93] Li AY, Deng H, Jiang YH, et al. Super-efficient removal of heavy metals from wastewater by Mg-loaded biochars: Adsorption characteristics and removal mechanisms. Langmuir 2020; 36(31): 9160-74.
[http://dx.doi.org/10.1021/acs.langmuir.0c01454] [PMID: 32644798]

[94] Sonawdekar S. Bioremediation: A boon to hydrocarbon degradation. Int J Environ Sci 2012; 2: 2408-23.

[95] Cristaldi A, Conti GO, Jho EH, et al. Phytoremediation of contaminated soils by heavy metals and PAHs. A brief review. Environ TechnolInnov 2017; 8: 309-26.
[http://dx.doi.org/10.1016/j.eti.2017.08.002]

[96] Verma S, Kuila A. Bioremediation of heavy metals by microbial process. Environ TechnolInnov 2019; 14: 100369.
[http://dx.doi.org/10.1016/j.eti.2019.100369]

[97] Leong YK, Chang JS. Bioremediation of heavy metals using microalgae: Recent advances and mechanisms. Bioresour Technol 2020; 303: 122886.
[http://dx.doi.org/10.1016/j.biortech.2020.122886] [PMID: 32046940]

[98] García-Carmona M, Romero-Freire A, Sierra Aragón M, Martínez Garzón FJ, Martín Peinado FJ. Evaluation of remediation techniques in soils affected by residual contamination with heavy metals and arsenic. J Environ Manage 2017; 191: 228-36.
[http://dx.doi.org/10.1016/j.jenvman.2016.12.041] [PMID: 28110163]

[99] Chemlal R, Abdi N, Lounici H, Drouiche N, Pauss A, Mameri N. Modeling and qualitative study of diesel biodegradation using biopile process in sandy soil. Int Biodeterior Biodegradation 2013; 78: 43-8.
[http://dx.doi.org/10.1016/j.ibiod.2012.12.014]

[100] Akbari A, Ghoshal S. Pilot-scale bioremediation of a petroleum hydrocarbon-contaminated clayey soil from a sub-Arctic site. J Hazard Mater 2014; 280: 595-602.
[http://dx.doi.org/10.1016/j.jhazmat.2014.08.016] [PMID: 25218258]

[101] Azubuike CC, Chikere CB, Okpokwasili GC. Bioremediation techniques–classification based on site of application: Principles, advantages, limitations and prospects. World J Microbiol Biotechnol 2016; 32(11): 180.
[http://dx.doi.org/10.1007/s11274-016-2137-x] [PMID: 27638318]

[102] Coulon F, Al Awadi M, Cowie W, et al. When is a soil remediated? Comparison of biopiled and windrowed soils contaminated with bunker-fuel in a full-scale trial. Environ Pollut 2010; 158(10):

3032-40.
[http://dx.doi.org/10.1016/j.envpol.2010.06.001] [PMID: 20656385]

[103] Dentel SK, Qi Y. Management of sludges, biosolids, and residuals. Biology 2014.
[http://dx.doi.org/10.1016/B978-0-12-382182-9.00049-9]

[104] Nelson MJ, Nakhla G, Zhu J. Fluidized-bed bioreactor applications for biological wastewater treatment: A review of research and developments. Engineering 2017; 3(3): 330-42.
[http://dx.doi.org/10.1016/J.ENG.2017.03.021]

[105] Ibrahim AEDM, Hamdona S, El-Naggar M, et al. Heavy metal removal using a fixed bed bioreactor packed with a solid supporter. Beni Suef Univ J Basic Appl Sci 2019; 8(1): 1.
[http://dx.doi.org/10.1186/s43088-019-0002-3]

[106] Fulekar MH, Sharma J, Tendulkar A. Bioremediation of heavy metals using biostimulation in laboratory bioreactor. Environ Monit Assess 2012; 184(12): 7299-307.
[http://dx.doi.org/10.1007/s10661-011-2499-3] [PMID: 22270588]

[107] Su J, Bai Y, Huang T, Wei L, Gao C, Wen Q. Multifunctional modified polyvinyl alcohol: A powerful biomaterial for enhancing bioreactor performance in nitrate, Mn(II) and Cd(II) removal. Water Res 2020; 168: 115152.
[http://dx.doi.org/10.1016/j.watres.2019.115152] [PMID: 31614240]

[108] Philp JC, Atlas RM. Bioremediation of contaminated soils and aquifers.Bioremediation: applied microbial solutions for real-world environmental cleanup, 139–236 American Society for Microbiology (ASM). Washington: Press 2005.
[http://dx.doi.org/10.1128/9781555817596.ch5]

[109] Johnson PC, Johnson RL, Bruce CL, Leeson A. Advances in *in situ* air sparging/biosparging. Bioremediat J 2001; 5(4): 251-66.
[http://dx.doi.org/10.1080/20018891079311]

[110] Møller J, Winther P, Lund B, Kirkebjerg K, Westermann P. Bioventing of diesel oil-contaminated soil: Comparison of degradation rates in soil based on actual oil concentration and on respirometric data. J Ind Microbiol 1996; 16(2): 110-6.
[http://dx.doi.org/10.1007/BF01570070]

[111] Frutos FJG, Escolano O, García S, Babín M, Fernández MD. Bioventing remediation and ecotoxicity evaluation of phenanthrene-contaminated soil. J Hazard Mater 2010; 183(1-3): 806-13.
[http://dx.doi.org/10.1016/j.jhazmat.2010.07.098] [PMID: 20800967]

[112] Thomé A, Reginatto C, Cecchin I, Colla LM. Bioventing in a residual clayey soil contaminated with a blend of biodiesel and diesel oil. J Environ Eng 2014; 140(11): 06014005.
[http://dx.doi.org/10.1061/(ASCE)EE.1943-7870.0000863]

[113] Kim S, Krajmalnik-Brown R, Kim JO, Chung J. Remediation of petroleum hydrocarbon-contaminated sites by DNA diagnosis-based bioslurping technology. Sci Total Environ 2014; 497-498: 250-9.
[http://dx.doi.org/10.1016/j.scitotenv.2014.08.002] [PMID: 25129160]

[114] Ali H, Khan E, Sajad MA. Phytoremediation of heavy metals—Concepts and applications. Chemosphere 2013; 91(7): 869-81.
[http://dx.doi.org/10.1016/j.chemosphere.2013.01.075] [PMID: 23466085]

[115] Nirola R, Megharaj M, Palanisami T, Aryal R, Venkateswarlu K, Ravi Naidu . Evaluation of metal uptake factors of native trees colonizing an abandoned copper mine : A quest for phytostabilization. J Sustain Min 2015; 14(3): 115-23.
[http://dx.doi.org/10.1016/j.jsm.2015.11.001]

[116] Farraji H, Zaman NQ, Tajuddin R, Faraji H. Advantages and disadvantages of phytoremediation: A concise review. Int J Environ Sci Technol 2016; 2: 69-75.

[117] Muramoto S, Oki Y. Removal of some heavy metals from polluted water by water hyacinth (*Eichhornia crassipes*). Bull Environ Contam Toxicol 1983; 30(1): 170-7.

[http://dx.doi.org/10.1007/BF01610117] [PMID: 6839042]

[118] Akinbile CO, Yusoff MS. Assessing water hyacinth (*Eichhornia crassopes*) and lettuce (*Pistia stratiotes*) effectiveness in aquaculture wastewater treatment. Int J Phytoremediation 2012; 14(3): 201-11.
[http://dx.doi.org/10.1080/15226514.2011.587482] [PMID: 22567705]

[119] Bolan NS, Park JH, Robinson B, Naidu R, Huh KY. Phytostabilization. Adv Agron 2011; 112: 145-204.
[http://dx.doi.org/10.1016/B978-0-12-385538-1.00004-4]

[120] Yan A, Wang Y, Tan SN, Mohd Yusof ML, Ghosh S, Chen Z. Phytoremediation: A promising approach for revegetation of heavy metal-polluted land. Front Plant Sci 2020; 11: 359.
[http://dx.doi.org/10.3389/fpls.2020.00359] [PMID: 32425957]

[121] Hassan SE, Hijri M, St-Arnaud M. Effect of arbuscular mycorrhizal fungi on trace metal uptake by sunflower plants grown on cadmium contaminated soil. N Biotechnol 2013; 30(6): 780-7.
[http://dx.doi.org/10.1016/j.nbt.2013.07.002] [PMID: 23876814]

[122] Li Y, Peng J, Shi P, Zhao B. The effect of Cd on mycorrhizal development and enzyme activity of *Glomus mosseae* and *Glomus intraradices* in *Astragalus sinicus* L. Chemosphere 2009; 75(7): 894-9.
[http://dx.doi.org/10.1016/j.chemosphere.2009.01.046] [PMID: 19232430]

[123] Burken JG, Schnoor JL. Uptake and metabolism of atrazine by poplar trees. Environ Sci Technol 1997; 31(5): 1399-406.
[http://dx.doi.org/10.1021/es960629v]

[124] Sakakibara M, Watanabe A, Inoue M, Sano S, Kaise T. Phytoextraction and phytovolatilization of arsenic from as-contaminated soils by *Pteris vittata*. Proc Annu Int Conf Soils Sediments Water Energy.

[125] Moreno FN, Anderson CWN, Stewart RB, Robinson BH. Mercury volatilisation and phytoextraction from base-metal mine tailings. Environ Pollut 2005; 136(2): 341-52.
[http://dx.doi.org/10.1016/j.envpol.2004.11.020] [PMID: 15840542]

[126] Hokura A, Omuma R, Terada Y, *et al.* Arsenic distribution and speciation in an arsenic hyperaccumulator fern by X-ray spectrometry utilizing a synchrotron radiation source. J Anal At Spectrom 2006; 21(3): 321-8.
[http://dx.doi.org/10.1039/b512792k]

[127] Sheoran V, Sheoran AS, Poonia P. Role of hyperaccumulators in phytoextraction of metals from contaminated mining sites: A review. Crit Rev Environ Sci Technol 2010; 41(2): 168-214.
[http://dx.doi.org/10.1080/10643380902718418]

[128] Rascio N, Navari-Izzo F. Heavy metal hyperaccumulating plants: How and why do they do it? And what makes them so interesting? Plant Sci 2011; 180(2): 169-81.
[http://dx.doi.org/10.1016/j.plantsci.2010.08.016] [PMID: 21421358]

[129] García-Salgado S, García-Casillas D, Quijano-Nieto MA, Bonilla-Simón MM. Arsenic and heavy metal uptake and accumulation in native plant species from soils polluted by mining activities. Water Air Soil Pollut 2012; 223(2): 559-72.
[http://dx.doi.org/10.1007/s11270-011-0882-x]

[130] Zhuang P, Yang QW, Wang HB, Shu WS. Phytoextraction of heavy metals by eight plant species in the field. Water Air Soil Pollut 2007; 184(1-4): 235-42.
[http://dx.doi.org/10.1007/s11270-007-9412-2]

[131] Robinson BH, Lombi E, Zhao FJ, McGrath SP. Uptake and distribution of nickel and other metals in the hyperaccumulator *Berkheya coddii*. New Phytol 2003; 158(2): 279-85.
[http://dx.doi.org/10.1046/j.1469-8137.2003.00743.x]

[132] Meers E, Vandecasteele B, Ruttens A, Vangronsveld J, Tack FMG. Potential of five willow species (*Salix* spp.) for phytoextraction of heavy metals. Environ Exp Bot 2007; 60(1): 57-68.

[http://dx.doi.org/10.1016/j.envexpbot.2006.06.008]

[133] Ahmadpour P, Ahmadpour F, Sadeghi S, Tayefeh FH, Soleimani M, Abdu AB. Evaluation of four plant species for phytoremediation of copper-contaminated soil.Soil Remediation and Plants. Academic Press 2014; pp. 147-205.

[134] Dushenkov V, Kumar PBAN, Motto H, Raskin I. Rhizofiltration: The use of plants to remove heavy metals from aqueous streams. Environ Sci Technol 1995; 29(5): 1239-45.
[http://dx.doi.org/10.1021/es00005a015] [PMID: 22192017]

[135] Abhilash PC, Pandey VC, Srivastava P, *et al.* Phytofiltration of cadmium from water by (L.) Buchenau grown in free-floating culture system. J Hazard Mater 2009; 170(2-3): 791-7.
[http://dx.doi.org/10.1016/j.jhazmat.2009.05.035] [PMID: 19523759]

[136] Sut-Lohmann M, Jonczak J, Raab T. Phytofiltration of chosen metals by aquarium liverwort (*Monosoleum tenerum*). Ecotoxicol Environ Saf 2020; 188: 109844.
[http://dx.doi.org/10.1016/j.ecoenv.2019.109844] [PMID: 31727495]

[137] Gardea-Torresdey JL, Gonzalez JH, Tiemann KJ, Rodriguez O, Gamez G. Phytofiltration of hazardous cadmium, chromium, lead and zinc ions by biomass of *Medicago sativa* (Alfalfa). J Hazard Mater 1998; 57(1-3): 29-39.
[http://dx.doi.org/10.1016/S0304-3894(97)00072-1]

[139] López ML, Peralta-Videa JR, Parsons JG, Benitez T, Gardea-Torresdey JL. Gibberellic acid, kinetin, and the mixture indole-3-acetic acid-kinetin assisted with EDTA-induced lead hyperaccumnulation in alfalfa plants. Environ Sci Technol 2007; 41(23): 8165-70.
[http://dx.doi.org/10.1021/es0714080] [PMID: 18186354]

[140] Mihopoulos P. Vapor phase treatment of PCE in a soil column by lab-scale anaerobic bioventing. Water Res 2000; 34(12): 3231-7.
[http://dx.doi.org/10.1016/S0043-1354(00)00023-3]

[141] Chen X, Chen X, Wan X, Weng B, Huang Q. Water hyacinth (*Eichhornia crassipes*) waste as an adsorbent for phosphorus removal from swine wastewater. Bioresour Technol 2010; 101(23): 9025-30.
[http://dx.doi.org/10.1016/j.biortech.2010.07.013] [PMID: 20674342]

[142] Glass JB, Orphan VJ. Trace metal requirements for microbial enzymes involved in the production and consumption of methane and nitrous oxide. Front Microbiol 2012; 3: 61.
[http://dx.doi.org/10.3389/fmicb.2012.00061] [PMID: 22363333]

[143] Bruins MR, Kapil S, Oehme FW. Microbial resistance to metals in the environment. Ecotoxicol Environ Saf 2000; 45(3): 198-207.
[http://dx.doi.org/10.1006/eesa.1999.1860] [PMID: 10702338]

[144] Chen C, Li L, Huang K, *et al.* Sulfate-reducing bacteria and methanogens are involved in arsenic methylation and demethylation in paddy soils. ISME J 2019; 13(10): 2523-35.
[http://dx.doi.org/10.1038/s41396-019-0451-7] [PMID: 31227814]

[145] Vijayaraghavan K, Yun YS. Bacterial biosorbents and biosorption. Biotechnol Adv 2008; 26(3): 266-91.
[http://dx.doi.org/10.1016/j.biotechadv.2008.02.002] [PMID: 18353595]

[146] Kang SY, Lee JU, Kim KW. Biosorption of Cr(III) and Cr(VI) onto the cell surface of *Pseudomonas aeruginosa*. Biochem Eng J 2007; 36(1): 54-8.
[http://dx.doi.org/10.1016/j.bej.2006.06.005]

[147] Vijayaraghavan K, Yun YS. Utilization of fermentation waste (*Corynebacterium glutamicum*) for biosorption of Reactive Black 5 from aqueous solution. J Hazard Mater 2007; 141(1): 45-52.
[http://dx.doi.org/10.1016/j.jhazmat.2006.06.081] [PMID: 16879915]

[148] Loukidou MX, Matis KA, Zouboulis AI, Liakopoulou-Kyriakidou M. Removal of As(V) from wastewaters by chemically modified fungal biomass. Water Res 2003; 37(18): 4544-52.
[http://dx.doi.org/10.1016/S0043-1354(03)00415-9] [PMID: 14511725]

[149] Giese EC, Jordão CS. Biosorption of lanthanum and samarium by chemically modified free *Bacillus subtilis* cells. Appl Water Sci 2019; 9(8): 182.
[http://dx.doi.org/10.1007/s13201-019-1052-3]

[150] Jeon C, Höll WH. Chemical modification of chitosan and equilibrium study for mercury ion removal. Water Res 2003; 37(19): 4770-80.
[http://dx.doi.org/10.1016/S0043-1354(03)00431-7] [PMID: 14568064]

[151] Jianlong W. Biosorption of copper(II) by chemically modified biomass of *Saccharomyces cerevisiae*. Process Biochem 2002; 37(8): 847-50.
[http://dx.doi.org/10.1016/S0032-9592(01)00284-9]

[152] Coyle P, Philcox JC, Carey LC, Rofe AM. Metallothionein: the multipurpose protein. Cell Mol Life Sci CMLS 2002; 59: 627-47.
[http://dx.doi.org/10.1007/s00018-002-8454-2]

[153] Juárez-Rebollar D, Rios C, Nava-Ruíz C, Méndez-Armenta M. Metallothionein in brain disorders. Oxid Med Cell Longev 2017; 2017: 1-12.
[http://dx.doi.org/10.1155/2017/5828056] [PMID: 29085556]

[154] Huckle JW, Morby AP, Turner JS, Robinson NJ. Isolation of a prokaryotic metallothionein locus and analysis of transcriptional control by trace metal ions. Mol Microbiol 1993; 7(2): 177-87.
[http://dx.doi.org/10.1111/j.1365-2958.1993.tb01109.x] [PMID: 8446025]

[155] Westenberg DJ, Guerinot ML. Regulation of bacterial gene expression by metals.Advances in Genetics. Academic Press 1997; pp. 187-238.

[156] Gold B, Deng H, Bryk R, *et al*. Identification of a copper-binding metallothionein in pathogenic mycobacteria. Nat Chem Biol 2008; 4(10): 609-16.
[http://dx.doi.org/10.1038/nchembio.109] [PMID: 18724363]

[157] Shao Z, Sun F. Intracellular sequestration of manganese and phosphorus in a metal-resistant fungus *Cladosporium cladosporioides* from deep-sea sediment. Extremophiles 2007; 11(3): 435-43.
[http://dx.doi.org/10.1007/s00792-006-0051-0] [PMID: 17265162]

[158] Helbig K, Bleuel C, Krauss GJ, Nies DH. Glutathione and transition-metal homeostasis in *Escherichia coli*. J Bacteriol 2008; 190(15): 5431-8.
[http://dx.doi.org/10.1128/JB.00271-08] [PMID: 18539744]

[159] Lima AIG, Pereira SIA, De Almeida Paula Figueira EM, Caldeira GCN, De Matos Caldeira HDQ. Cadmium detoxification in roots of *Pisum sativum* seedlings: Relationship between toxicity levels, thiol pool alterations and growth. Environ Exp Bot 2006; 55(1-2): 149-62.
[http://dx.doi.org/10.1016/j.envexpbot.2004.10.008]

[160] Igiri BE, Okoduwa SIR, Idoko GO, Akabuogu EP, Adeyi AO, Ejiogu IK. Toxicity and bioremediation of heavy metals contaminated ecosystem from tannery wastewater: A review. J Toxicol 2018; 2018: 1-16.
[http://dx.doi.org/10.1155/2018/2568038] [PMID: 30363677]

[161] Grass G, Rensing C. Genes involved in copper homeostasis in *Escherichia coli*. J Bacteriol 2001; 183(6): 2145-7.
[http://dx.doi.org/10.1128/JB.183.6.2145-2147.2001] [PMID: 11222619]

[162] Thelwell C, Robinson NJ, Turner-Cavet JS. An SmtB-like repressor from *Synechocystis* PCC 6803 regulates a zinc exporter. Proc Natl Acad Sci 1998; 95(18): 10728-33.
[http://dx.doi.org/10.1073/pnas.95.18.10728] [PMID: 9724772]

[163] Diels L, Dong Q, Lelie D, Baeyens W, Mergeay M. The *czc* operon of *Alcaligenes eutrophus* CH34: From resistance mechanism to the removal of heavy metals. J Ind Microbiol 1995; 14(2): 142-53.
[http://dx.doi.org/10.1007/BF01569896] [PMID: 7766206]

[164] Roane TM. Lead resistance in two bacterial isolates from heavy metal-contaminated soils. Microb

Ecol 1999; 37(3): 218-24.
[http://dx.doi.org/10.1007/s002489900145] [PMID: 10227879]

[165] Suh JH, Yun JW, Kim DS. Effect of extracellular polymeric substances (EPS) on Pb^{2+} accumulation by *Aureobasidium pullulans*. Bioprocess Eng 1999; 21(1): 1-4.
[http://dx.doi.org/10.1007/PL00009061]

[166] Hirst CN, Cyr H, Jordan IA. Distribution of exopolymeric substances in the littoral sediments of an oligotrophic lake. Microb Ecol 2003; 46(1): 22-32.
[http://dx.doi.org/10.1007/s00248-002-1064-6] [PMID: 12754660]

[167] Costerton JW, Cheng KJ, Geesey GG, *et al.* Bacterial biofilms in nature and disease. Annu Rev Microbiol 1987; 41(1): 435-64.
[http://dx.doi.org/10.1146/annurev.mi.41.100187.002251] [PMID: 3318676]

[168] Sheng GP, Yu HQ, Yue ZB. Production of extracellular polymeric substances from *Rhodopseudomonas acidophila* in the presence of toxic substances. Appl Microbiol Biotechnol 2005; 69(2): 216-22.
[http://dx.doi.org/10.1007/s00253-005-1990-6] [PMID: 15843928]

[169] Iyer A, Mody K, Jha B. Multifunctional modified polyvinyl alcohol: A powerful biomaterial for enhancing bioreactor perf by an exopolysaccharide producing marine *Enterobacter cloaceae*. Mar Pollut Bull 2004; 49(11-12): 974-7.
[http://dx.doi.org/10.1016/j.marpolbul.2004.06.023] [PMID: 15556183]

[170] Knopf B, König H. Biomethylation of heavy metals in soil and terrestrial invertebrates.Soil Heavy Metals, Soil Biology 19, 315–328. Springer-Verlag 2009.

[171] Tripathi P, Khare P, Barnawal D, *et al.* Bioremediation of arsenic by soil methylating fungi: Role of *Humicola* sp. strain 2WS1 in amelioration of arsenic phytotoxicity in *Bacopa monnieri* L. Sci Total Environ 2020; 716: 136758.
[http://dx.doi.org/10.1016/j.scitotenv.2020.136758] [PMID: 32092818]

[172] Qin J, Rosen BP, Zhang Y, Wang G, Franke S, Rensing C. Arsenic detoxification and evolution of trimethylarsine gas by a microbial arsenite *S*-adenosylmethionine methyltransferase. Proc Natl Acad Sci 2006; 103(7): 2075-80.
[http://dx.doi.org/10.1073/pnas.0506836103] [PMID: 16452170]

[173] Su SM, Zeng XB, Li LF, *et al.* Arsenate reduction and methylation in the cells of *Trichoderma asperellum* SM-12F1, *Penicillium janthinellum* SM-12F4, and *Fusarium oxysporum* CZ-8F1 investigated with X-ray absorption near edge structure. J Hazard Mater 2012; 243: 364-7.
[http://dx.doi.org/10.1016/j.jhazmat.2012.09.061] [PMID: 23122191]

[174] Chen J, Li J, Zhang H, Shi W, Liu Y. Bacterial heavy metal and antibiotic resistance genes in a copper tailing dam area in Northern China. Front Microbiol 2019; 10: 1916.
[http://dx.doi.org/10.3389/fmicb.2019.01916] [PMID: 31481945]

[175] Challenger F. Biological methylation. Chem Rev 1945; 36(3): 315-61.
[http://dx.doi.org/10.1021/cr60115a003]

[176] Wakatsuki T. Metal oxidoreduction by microbial cells. J Ind Microbiol 1995; 14(2): 169-77.
[http://dx.doi.org/10.1007/BF01569900] [PMID: 7766210]

[177] Boyd ES, Barkay T. The mercury resistance operon: From an origin in a geothermal environment to an efficient detoxification machine. Front Microbiol 2012; 3: 349.
[http://dx.doi.org/10.3389/fmicb.2012.00349] [PMID: 23087676]

[178] García C, Moreno DA, Ballester A, Blázquez ML, González F. Bioremediation of an industrial acid mine water by metal-tolerant sulphate-reducing bacteria. Miner Eng 2001; 14(9): 997-1008.
[http://dx.doi.org/10.1016/S0892-6875(01)00107-8]

[179] Miao Z, He H, Tan T, *et al.* Biotreatment of Mn^{2+} and Pb^{2+} with sulfate-reducing bacterium *Desulfuromonas alkenivorans* S-7. J Environ Eng 2018; 144(3): 04017116.

[http://dx.doi.org/10.1061/(ASCE)EE.1943-7870.0001330]

[180] Vaxevanidou K, Papassiopi N, Paspaliaris I. Removal of heavy metals and arsenic from contaminated soils using bioremediation and chelant extraction techniques. Chemosphere 2008; 70(8): 1329-37.
[http://dx.doi.org/10.1016/j.chemosphere.2007.10.025] [PMID: 18037468]

[181] Jafarian V, Ghaffari F. A unique metallothionein-engineered in *Escherichia coli* for biosorption of lead, zinc, and cadmium; absorption or adsorption?. Microbiol 2017; 86: 73-81.
[http://dx.doi.org/10.1134/S0026261717010064]

[182] Silver S, Phung LT. A bacterial view of the periodic table: genes and proteins for toxic inorganic ions. J Ind Microbiol Biotechnol 2005; 32(11-12): 587-605.
[http://dx.doi.org/10.1007/s10295-005-0019-6] [PMID: 16133099]

[183] Hanlon P, Sewalt V. GEMs: Genetically engineered microorganisms and the regulatory oversight of their uses in modern food production. Crit Rev Food Sci Nutr 2021; 61(6): 959-70.
[http://dx.doi.org/10.1080/10408398.2020.1749026] [PMID: 32274948]

[185] Liu S, Zhang F, Chen J, Sun G. Arsenic removal from contaminated soil *via* biovolatilization by genetically engineered bacteria under laboratory conditions. J Environ Sci 2011; 23(9): 1544-50.
[http://dx.doi.org/10.1016/S1001-0742(10)60570-0] [PMID: 22432292]

[186] Bang SW, Clark DS, Keasling JD. Engineering hydrogen sulfide production and cadmium removal by expression of the thiosulfate reductase gene (*phsABC*) from *Salmonella enterica* serovar typhimurium in *Escherichia coli*. Appl Environ Microbiol 2000; 66(9): 3939-44.
[http://dx.doi.org/10.1128/AEM.66.9.3939-3944.2000] [PMID: 10966412]

[187] Sousa C, Kotrba P, Ruml T, Cebolla A, De Lorenzo V. Metalloadsorption by *Escherichia coli* cells displaying yeast and mammalian metallothioneins anchored to the outer membrane protein LamB. J Bacteriol 1998; 180(9): 2280-4.
[http://dx.doi.org/10.1128/JB.180.9.2280-2284.1998] [PMID: 9573175]

[188] Kiyono M, Pan-Hou H. Genetic engineering of bacteria for environmental remediation of mercury. J Health Sci 2006; 52(3): 199-204.
[http://dx.doi.org/10.1248/jhs.52.199]

[189] Nishino K, Nikaido E, Yamaguchi A. Regulation of multidrug efflux systems involved in multidrug and metal resistance of *Salmonella enterica* serovar Typhimurium. J Bacteriol 2007; 189(24): 9066-75.
[http://dx.doi.org/10.1128/JB.01045-07] [PMID: 17933888]

[190] Herrmann L, Schwan D, Garner R, *et al*. Helicobacter pylori cadA encodes an essential Cd(II)-Zn(II)-Co(II) resistance factor influencing urease activity. Mol Microbiol 1999; 33(3): 524-36.
[http://dx.doi.org/10.1046/j.1365-2958.1999.01496.x] [PMID: 10417643]

[191] Deng X, Wilson DB. Bioaccumulation of mercury from wastewater by genetically engineered *Escherichia coli*. Appl Microbiol Biotechnol 2001; 56(1-2): 276-9.
[http://dx.doi.org/10.1007/s002530100620] [PMID: 11499944]

[192] Makarova KS, Haft DH, Barrangou R, *et al*. Evolution and classification of the CRISPR–Cas systems. Nat Rev Microbiol 2011; 9(6): 467-77.
[http://dx.doi.org/10.1038/nrmicro2577] [PMID: 21552286]

[193] Cooper LA, Stringer AM, Wade JT. Determining the specificity of cascade binding, interference, and primed adaptation *in vivo* in the *Escherichia coli* Type I-E CRISPR-Cas System. MBio 2018; 9(2): e02100-17.
[http://dx.doi.org/10.1128/mBio.02100-17] [PMID: 29666291]

[194] Zhang C, Konermann S, Brideau NJ, *et al*. Structural basis for the RNA-guided ribonuclease activity of CRISPR-Cas13d. Cell 2018; 175(1): 212-223.e17.
[http://dx.doi.org/10.1016/j.cell.2018.09.001] [PMID: 30241607]

[195] Ran FA, Hsu PD, Wright J, Agarwala V, Scott DA, Zhang F. Genome engineering using the CRISPR-

Cas9 system. Nat Protoc 2013; 8(11): 2281-308.
[http://dx.doi.org/10.1038/nprot.2013.143] [PMID: 24157548]

[196] Doudna JA, Charpentier E. The new frontier of genome engineering with CRISPR-Cas9. Science 2014; 346(6213): 1258096-6.
[http://dx.doi.org/10.1126/science.1258096] [PMID: 25430774]

[197] Tang X, Lowder LG, Zhang T, *et al.* A CRISPR–Cpf1 system for efficient genome editing and transcriptional repression in plants. Nat Plants 2017; 3(3): 17018.
[http://dx.doi.org/10.1038/nplants.2017.18] [PMID: 28211909]

[198] Canver MC, Joung JK, Pinello L. Impact of genetic variation on CRISPR-Cas targeting. CRISPR J 2018; 1(2): 159-70.
[http://dx.doi.org/10.1089/crispr.2017.0016] [PMID: 31021199]

[199] Comfort N. From controlling elements to transposons: Barbara mcclintock and the nobel prize1. Trends Genet 2001; 17(8): 475-8.
[http://dx.doi.org/10.1016/S0168-9525(01)02383-6] [PMID: 11485821]

[200] Makałowski W, Gotea V, Pande A, Makałowska I. Transposable elements: Classification, identification, and their use as a tool for comparative genomics.Evolutionary Genomics Methods in Molecular Biology. New York, NY: Humana 2019; 1910.
[http://dx.doi.org/10.1007/978-1-4939-9074-0_6]

[201] Zhao S, Jiang E, Chen S, *et al.* PiggyBac transposon vectors: The tools of the human gene encoding. Transl Lung Cancer Res 2016; 5(1): 120-5.
[PMID: 26958506]

[202] Li S, Zhang A, Xue H, Li D, Liu Y. One-Step piggyBac transposon-based CRISPR/Cas9 activation of multiple genes. Mol Ther Nucleic Acids 2017; 8: 64-76.
[http://dx.doi.org/10.1016/j.omtn.2017.06.007] [PMID: 28918057]

[203] Nishizawa-Yokoi A, Toki S. A *piggyBac* -mediated transgenesis system for the temporary expression of CRISPR/Cas9 in rice. Plant Biotechnol J 2021; 19(7): 1386-95.
[http://dx.doi.org/10.1111/pbi.13559] [PMID: 33529430]

[204] Rubin BE, Diamond S, Cress BF, *et al.* Targeted genome editing of bacteria within microbial communities. bioRxiv 2020.
[http://dx.doi.org/10.1101/2020.07.17.209189]

[205] Kalyabina VP, Esimbekova EN, Kopylova KV, Kratasyuk VA. Pesticides: formulants, distribution pathways and effects on human health–a review. Toxicol Rep 2021; 8: 1179-1192.
[http://dx.doi.org/10.1016/j.toxrep.2021.06.004]

[206] Jaishankar M, Tseten T, Anbalagan N, Mathew BB, Beeregowda KN. Toxicity, mechanism and health effects of some heavy metals. Interdiscip Toxicol 2014; 7(2): 60-72.
[http://dx.doi.org/10.2478/intox-2014-0009] [PMID: 26109881]

[207] De Lorenzo V. Systems biology approaches to bioremediation. Curr Opin Biotechnol 2008; 19(6): 579-89.
[http://dx.doi.org/10.1016/j.copbio.2008.10.004] [PMID: 19000761]

[208] Fang L, Cai P, Chen W, Liang W, Hong Z, Huang Q. Impact of cell wall structure on the behavior of bacterial cells in the binding of copper and cadmium. Colloids Surf A Physicochem Eng Asp 2009; 347(1-3): 50-5.
[http://dx.doi.org/10.1016/j.colsurfa.2008.11.041]

[209] Sayler GS, Ripp S. Field applications of genetically engineered microorganisms for bioremediation processes. Curr Opin Biotechnol 2000; 11(3): 286-9.
[http://dx.doi.org/10.1016/S0958-1669(00)00097-5] [PMID: 10851144]

[210] Margoshes M, Vallee BL. A cadmium protein from equine kidney cortex. J Am Chem Soc 1957; 79(17): 4813-4.

[http://dx.doi.org/10.1021/ja01574a064]

[211] Kao CM, Chien HY, Surampalli RY, Sung WP. Application of biopile system for the remediation of petroleum-hydrocarbon contaminated soils. World Environmental and Water Resources Congress 2009.
[http://dx.doi.org/10.1061/41036(342)260]

[212] Kuhlman LR. Windrow composting of agricultural and municipal wastes. Resour Conserv Recycling 1990; 4(1-2): 151-60.
[http://dx.doi.org/10.1016/0921-3449(90)90039-7]

[213] Huang H, Jia Q, Jing W, Dahms HU, Wang L. Screening strains for microbial biosorption technology of cadmium. Chemosphere 2020; 1(251): 126428.
[http://dx.doi.org/10.1016/j.chemosphere.2020.126428]

[214] Bae W, Mehra RK, Mulchandani A, Chen W. Genetic engineering of *Escherichia coli* for enhanced uptake and bioaccumulation of mercury. Appl Environ Microbiol 2001; 67(11): 5335-8.
[http://dx.doi.org/10.1128/AEM.67.11.5335-5338.2001] [PMID: 11679366]

[215] Kotrba P, Ruml T. Surface display of metal fixation motifs of bacterial P1-type ATPases specifically promotes biosorption of Pb^{2+} by *Saccharomyces cerevisiae*. Appl Environ Microbiol 2010; 76(8): 2615-22.
[http://dx.doi.org/10.1128/AEM.01463-09] [PMID: 20173062]

[216] Chen J, Qin J, Zhu YG, De Lorenzo V, Rosen BP. Engineering the soil bacterium *Pseudomonas putida* for arsenic methylation. Appl Environ Microbiol 2013; 79(14): 4493-5.
[http://dx.doi.org/10.1128/AEM.01133-13] [PMID: 23645194]

[217] Ackerley DF, Gonzalez CF, Keyhan M, Blake R II, Matin A. Mechanism of chromate reduction by the *Escherichia coli* protein, NfsA, and the role of different chromate reductases in minimizing oxidative stress during chromate reduction. Environ Microbiol 2004; 6(8): 851-60.
[http://dx.doi.org/10.1111/j.1462-2920.2004.00639.x] [PMID: 15250887]

[218] Almaguer-Cantú V, Morales-Ramos LH, Balderas-Rentería I. Biosorption of lead (II) and cadmium (II) using *Escherichia coli* genetically engineered with mice metallothionein I. Water Sci Technol 2011; 63(8): 1607-13.
[http://dx.doi.org/10.2166/wst.2011.225] [PMID: 21866758]

[219] Dungan RS, Frankenberger WT Jr. Biotransformations of selenium by *Enterobacter cloacae* SLD1a-1: Formation of dimethylselenide. Biogeochemistry 2001; 55(1): 73-86.
[http://dx.doi.org/10.1023/A:1010640307328]

[220] Ranjard L, Prigent-Combaret C, Nazaret S, Cournoyer B. Methylation of inorganic and organic selenium by the bacterial thiopurine methyltransferase. J Bacteriol 2002; 184(11): 3146-9.
[http://dx.doi.org/10.1128/JB.184.11.3146-3149.2002] [PMID: 12003960]

[221] Eswayah AS, Smith TJ, Gardiner PHE. Microbial transformations of selenium species of relevance to bioremediation. Appl Environ Microbiol 2016; 82(16): 4848-59.
[http://dx.doi.org/10.1128/AEM.00877-16] [PMID: 27260359]

CHAPTER 7

Designing the Metabolic Capacities of Environmental Bioprocesses through Genome Editing

Ashish Kumar Singh[1,2,#], **Bhagyashri Poddar**[1,2,#], **Rakesh Kumar Gupta**[1,2,#], **Suraj Prabhakarrao Nakhate**[1,2,#], **Vijay Varghese**[2,#], **Anshuman A. Khardenavis**[1,2,*] and **Hemant J. Purohit**[2]

[1] *Academy of Scientific and Innovative Research (AcSIR), Ghaziabad-201002, India*

[2] *Environmental Biotechnology and Genomics Division, CSIR-NEERI, Nehru Marg, Nagpur-440020, Maharashtra, India*

Abstract: The ubiquity of the CRISPR gene system in bacteria and archaea is characterized by the Cas9 protein, which functions in the repression and activation of several genes. This inherent function of the CRISPR system can find application in bioprocess optimization in environmental and health research. Owing to the complex and dynamic nature of microbial communities catalysing the bioremediation of urban and industrial toxic waste effluents in wastewater treatment plant (WWTP)/common effluent treatment plant (CETP), such sites represent a relatively untapped area for applying the CRISPR technique. DNA editing using CRISPR can enable the site-specific enhancement in process efficiency of bacterial remediation, which under normal conditions is hampered by its non-selectivity and saturation of binding sites with multiple non-targeted pollutants. Similarly, under the second generation bio-refinery concept, CRISPR can serve as a powerful tool in strengthening and improving the anaerobic bio-processes by genome editing in microbes for the heterogeneous expression of various genes associated with anaerobic digestion. Not only has the CRISPR system been used to insert desired genes in the host genome but also to regulate the expression of the host-specific genes. The role of methanotrophic and nitrogen metabolizing bacteria in shaping the atmospheric gaseous composition can also be monitored *via* CRISPR aided manipulation so as to regulate the nutritional exchange between the atmosphere and the soil. Additionally, genome editing of targeted organisms and crops has found extensive applications in various areas ranging from the nutrigenomics, food and pharmaceutical industry, diagnostics and therapeutics, health and disease prevention.

---

* **Corresponding author Anshuman A. Khardenavis:** Academy of Scientific and Innovative Research (AcSIR), Ghaziabad-201002, India; and Environmental Biotechnology and Genomics Division, CSIR-NEERI, Nehru Marg, Nagpur- 440020, Maharashtra, India; E-mail: aa_khardenavis@neeri.res.in

# Equal contribution

Prakash M. Halami & Aravind Sundararaman (Eds.)
All rights reserved-© 2024 Bentham Science Publishers

Keywords: Anaerobic bioprocesses, Bioremediation, CRISPR, Gene editing, Gut microbiome.

INTRODUCTION

Clustered regularly interspaced short palindromic repeats (CRISPR) system is widely distributed among bacteria and archaea with 87% of the archaeal and 45% of the bacterial genomes and plasmids showing the presence of CRISPR motifs and Cas proteins [1]. This system has evolved through the battle between bacteria and phages [2, 3]. The presence or absence of CRISPR sequences in a prokaryote is associated with the defence mechanism against the phages, while the presence of more than one array of CRISPR sequences corresponds to the selective maintenance by an organism under pathogenic environmental stress and depends on the ecological niche of the organism [4]. For example, in water and wastewater and their distribution systems, the pathogenic microorganisms mostly reside within the biofilms where warfare-like complex interaction is frequently prevalent amongst the members of different microbial communities like protozoa predation, bacterial and viral lysis *etc*. In such interactions, the phage DNA can be merged into the bacterial genome by horizontal gene transfer resulting in the formation of prophages [5]. Thus, the bacteria that protect themselves from the phages are often encountered with foreign DNA fragments integrated into the location of routinely dissected short palindromic repeats — CRISPR-associated proteins (CRISPR/Cas) as spacers [6]. The functionality of the CRISPR system depends on the presence of the CRIPSR-associated genes (*Cas1* and *Cas2*) in the spacer region of the arrays. The spacer sequences between the CRISPR sequences of bacteria are associated with the foreign genetic material linked to viruses or other mobile genetic components.

Genome editing is a powerful tool in basic biology encasing novel capabilities in microbial genomes. The prokaryotic immune system has given rise to the emerging technology called CRISPR/Cas, which has encouraged researchers to easily alter unique organisms for different applications. The use of CRISPR/Cas systems in microbiome editing is one of the most promising approaches for controlling the gene expression and regulation of metabolites and protein production. CRISPR has become a gene-editing technique of choice that has been proven and widely used to treat or prevent diseases. This approach has been used as an important tool for stress typing in epidemiology for outbreaks and the identification of sources of infection [7, 8]. The established relationship between the exposed surfaces of our body and the production of metabolites, host-immune response, and gut-brain axis make the gut microbiome a potential target for gene therapies [9].

Besides the medical industry, the CRISPR technique has opened new avenues as a powerful genetic engineering tool in protecting, repairing, and saving the environment from harmful anthropogenic activity. The inherent functions of the CRISPR system can be used for many environmental aspects such as biofuel production, bioplastics, biosensing, pesticide reduction, food waste bioremediation, greenhouse gas emissions, water and wastewater [10]. Briefly, wastewater is a cocktail of many pollutants like organic carbon, nutrients (nitrogen and phosphorus), pathogens, and other contaminants. All these components pose a severe threat to human health and environmental integrity, thus making it imperative to clean the water resource. Treatment of such a contaminated resource by microorganisms is the best way owing to the inexpensive and sustainable approach. In view of strengthening and improving the aerobic and anaerobic bioprocesses, exploitation and augmentation of genetically engineered microbes can be a crucial strategy. Second generation bio-refinery caters to the eco-friendly lignocellulose waste disposal *via* anaerobic digestion, simultaneously providing the benefit of biomethane produced. Microbes engineered for heterologous expression of various genes obtained from source organisms can find valuable application in bioaugmentation for hastening the bioprocess operating in bio-refineries.

CRISPR–based genome editing has also been applied across the field of food science. Genome editing has found applications in the targeted engineering of crops, including corn, rice, and tomatoes, for improving their growth and nutrition potential by inserting traits for drought and disease resistance, resistance to insecticides, and survival under low nutrition/fertilizer conditions. Similarly, genome editing has been demonstrated to improve the yield from animal breeds through desirable alteration and herd genetics selection for disease resistant animals [11].

This chapter discusses the significance of gene editing techniques based on CRISPR/Cas in the optimization of various bioprocesses in environmental research. The application of this tool in improving the efficiency of both aerobic and anaerobic bioprocesses, such as wastewater treatment, anaerobic digestion for biogas and volatile fatty acid production, and landfill gas management, has been highlighted.

CRISPR AND BIOREMEDIATION

The global increase in human population and the industrial revolution are the major turning points in human history which have led to a change in society, economy, politics, and particularly in the environment. The large-scale production and use of chemicals over the past few decades, and the unchecked discharge of toxic waste effluents have resulted in critical issues related to environmental degradation [12]. The persistent nature of hazardous and emerging pollutants and consequent environmental problems have brought the possibility of long-term environmental disasters into the public conscience [13]. To date, more than 200 toxic persistent compounds have been identified, with most of the pollutants having no defined regulatory standard though they can be detrimental to living beings at environmentally admissible concentrations [14]. Pharmaceutical industries, flame retardants, pesticides used in agricultural practices, personal care products, long range transportation, and backyard barrel burning are the major sources of frequently detected environmental pollutants which adversely affect the soil microbiome and its activities with reference to the biogeochemical cycling of elements. These sources emit a host of hazardous organic chemicals like petroleum hydrocarbons, nitro-aromatic compounds, phenolic compounds, halogenated hydrocarbons, polychlorinated biphenyls (PCBs), polycyclic aromatic hydrocarbons (PAHs), xenobiotic compounds, volatile organic compounds (VOCs), and pesticides, along with inorganic compounds such as phosphates, nitrates, salts, and heavy metals, *i.e.*, copper (Cu), arsenic (As), selenium (Se), zinc (Zn), lead (Pb), mercury (Hg), cadmium (Cd), nickel (Ni), chromium (Cr), and silver (Ag) [15].

Conventional Practices for Improvement of Bioremediation Efficiencies

Various strategies have been developed, including the physical and chemical methods for minimizing the deleterious effect of pollutants on aquatic life, human health as well as the environment [16]. However, physio-chemical treatment is not effective in abolishing the pollutants to the required level [17]. Bioremediation is an emerging biological process for the clean-up of toxic pollutants using microbial metabolic pathways to transform (biotransformation), degrade (biodegradation), and adsorb (bioadsorption) a wide variety of contaminants. It offers an eco-friendly and cost-effective solution for the remediation and/or removal of persistent and hazardous contaminants from soil, water, and sediments using the natural ability of microorganisms with a minimum disruption of the natural ecosystem, thus making this process less expensive and amenable for wide-scale public acceptance. However, the wastewater treatment systems represent a genetically diverse microbial population with broad catabolic/metabolic capacities [18]. The microbial community residing in such systems is characterized by diverse metabolic pathways that are strengthened with enzymatic activities encoded by a wide range of novel genes for the catabolic capacity of interest. These variable factors directly or indirectly affect the overall metabolism of a microbial community with reference to their gene expression in the wastewater treatment system, for which an in-depth understanding of their intra-species interaction is needed.

In view of a significant increase in the estimated number of contaminated sites and accumulation of toxic substances, bioremediation as a tool for decreasing the concentration of such contaminants from the environment has been exploited successfully in many countries. In spite of all its advantages, bioremediation also has its own limitations. Microbial bioremediation is a slow process which consequently results in a lower removal of contaminants from the polluted site reducing its applicability and efficacy for the remediation of all types of organic and inorganic pollutants [19]. Moreover, the biodegradation of compounds mediated by microbial enzymes is often carried out partially, thus releasing intermediates into the environment, which may be more toxic compared to the parent compound [15]. Another major challenge with microbial remediation is the non-selectivity of the process, which in the scenario of the presence of multiple pollutants, results in the saturation of the binding site with non-targeted pollutants [20]. These non-targeted pollutants may also decrease the remediation rate of the pollutants of concern regardless of the number of free binding sites available on the microbial cell wall. In order to meet the bioremediation challenge, all the active microbial communities need to be more tolerant to non-targeted pollutants, should be selectively available for the targeted pollutants, and should possess extremely intensive metabolic enzymes potent enough to remove significant levels of contaminants.

Recent Advances in Gene Editing for Enhanced Bioremediation

Modern genomic engineering approaches for manipulating the epigenome through metabolomic and transcriptomic tools have enabled researchers to make changes in the genomic sequences *via* nucleotide mutation. The considerable remedial power of microorganisms can be explored by partial or complete genome sequencing of model microbes employed in the bioremediation processes. There are several studies where the genome sequencing of the model bioremediators such as *Shewanella oneidensis, Microbacterium oleivorans, Rhodococcus ruber, Bacillus subtilis*, and *Pseudomonas putida* have been carried out to outline the genomic modification of targeted gene. The sequencing step helps to identify and characterize the target gene which forms an important step towards enhancing the bioremediation capability of the microbial enzymes for the degradation and detoxification of a number of pollutants. Linking the genomic editing tools with bioprocess optimization could aid in the decision-making process during treatment in WWTPs or CETPs, thereby enabling to achieve maximum efficiency of pollutant removal [18]. All these approaches are dependent on the presence of selective markers such as genes for antimetabolite, herbicide, and antibiotic resistance, which give rise to serious environmental risks [21].

The practice of genetic engineering in bioremediation boosts the pollutant removal efficiency of microorganisms with a higher selectivity for the pollutant of interest. First and foremost, genome engineering enables alteration in the expression of specific genes and exploits a new route under the genome editing mediated enhanced microbial remediation (Fig. 1) [22]. The recent developments in genetic engineering have achieved a breakthrough by the discovery of CRISPR-assisted bioremediation [23]. Programmable DNA editing using CRISPR is the most promising tool in functional genomics that enables the site-specific and efficient enhancement of targeted genes in bacteria. Owing to the high complexity, diversity, and variability of the pollutants, and the complex and dynamic nature of microbial communities residing in water and wastewater environment, WWTPs/CETPs are considered as a relatively untapped area for application of CRISPR techniques. CRISPR based bioremediation can help in mitigating the ill effects of pollutants present in the environment by removing the toxic components and purifying soil, water, and sediment [24]. Thus, the above variability in the stressed environments may pose serious challenges before the specific microorganisms important in environmental pollution control, with reference to their efficient and robust functioning [21].

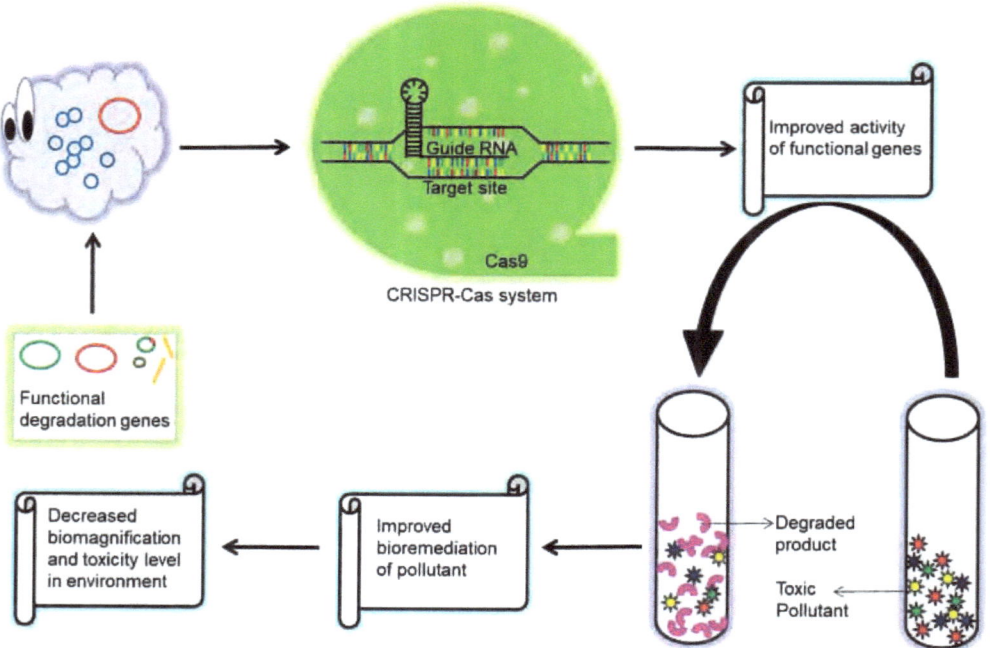

Fig. (1). CRISPR/Cas genome editing of potent bacterial species for enhanced bioremediation.

To date, the CRISPR system has been applied in water and wastewater, particularly associated with identifying the pathogenic microorganisms and multidrug resistance in many WWTPs with an aim for providing clean water and protecting the environment. CRISPR editing tool could be employed to enhance the activity of the metabolic pathways at the genetic level in microbes that are good candidates for pollutant removal and treatment of wastewater.

The genome-editing tool boxes rely on the sequence-specific CRISPR/Cas systems that may overcome these limitations. CRISPR/Cas genome editing provides a precise and proficient molecular mechanism for targeted genome engineering that has found widespread application in bioremediation *via* adsorption, transformation, and degradation of the contaminants [25]. Future execution of these molecular tools in microorganisms is expected to improve our understanding of genomic engineering and designing microbial bioconversion processes in the aquatic environment [26].

Metabolic enzymes that play an important role in removing pollutants can be targeted for CRISPR editing so as to achieve newer and higher levels of enhancement in the gene/enzyme functions. One such study was performed for accelerating the bioremediation of pollutants by *P. putida* and *P. fluorescens* wherein CRISPR was applied to develop a successful gene repression tool for these model organisms [27]. On a similar note, CRISPR was also applied to enhance the expression of functional genes (*mpd, opd, vgb, gfp, pnpA, linC, pnpB, linB, linD* and *linA, etc.*) in *P. putida* for the bioremediation of broad spectrum of pollutants [23]. In a recent study, alkane degradation from crude oil was assessed in a rhamnolipid (RL) biosurfactant producing *P. aeruginosa,* a strategy which allowed the isolate to emulsify oil, thereby enabling its survival in the high alkane containing environment. Analysis by quorum sensing revealed that the high RL yield was dependent upon the higher production of AHL (N-acyl-homoserine lactone) signal molecules. In addition, four genes (*alkB, almA*, P450, and *ladA*) encoding alkane hydroxylases (AHs) were also found to be responsible for alkane degradation in crude oil. The AH encoding gene and AHL synthesis are the two most important factors which can be targeted by CRISPR/Cas genome editing technology to achieve enhanced crude oil recovery or RL production [28].

S. oneidensis, a potent metal reducing bacterium is important in view of decreasing the load of sites heavily contaminated with metals. Corts *et al.* [29] recombineered a single-stranded DNA oligonucleotide and coupled it with CRISPR/Cas9-mediated counter-selection that allowed the selection of a highly versatile recombinase independent mutant of *S. oneidensis* with an average efficiency greater (>90%) than the efficiency by recombineering (~5%) alone. Further studies on *S. oneidensis* MR-1 by Li *et al.* [30] revealed the application of

CRISPR technology in enhancing waste management and bioremediation by promoting the extracellular electron flux (EET). Tuning the EET by CRISPR-ddAsCpf1 conformity enabled the repression of wild type gene, increased the transcription of riboflavin and L-lactate metabolic genes, and hence enhanced the indirect and direct electron flux which favoured more effective and efficient environmental remediation of organic pollutants, methyl orange, and chromium (typical heavy metal) by *S. oneidensis* MR-1.

Liang *et al.* [31] successfully developed a CRISPR/Cas9 genome editing system for *Rhodococcus ruber*, an organic solvent-tolerant strain with diverse metabolic pathways for bioremediation, biotransformation, and bioconversion [32]. With the introduction of the Che9c61, Che9c60, and bacteriophage recombinases, *R. ruber* showed a reduction in the byproduct formation from 2.54 g/L to 0.5 g/L with enhancement in biocatalytic activity [31]. Thus, the successful establishment of CRISPR in wild type *Rhodococcus* can produce the desirable mutants/traits that make a promising platform for bioremediation. Another study performed using CRISPR on *Dehalococcoides mccartyi*, a hydrogen-utilizing, obligate organohalide-respiring bacterium, reported that the integrated genomic island containing dehalogenase (VcrA) encoding vcrABC operon was circularized. This enzyme catalysed the transformation of vinyl chloride to non-toxic ethane [33].

A decade ago, Coutte *et al.* [34] replaced the native promoter P_{srfA} in *B. subtilis* with the constitutive promoter P_{repU} after functional *sfp* gene integration and pps (plipastatin synthasase) operon disruption. The higher expression of P_{repU} mutants showed a 5-fold increase in surfactin production that enhanced the growth, and haemolytic and antifungal activities of *B. subtilis* hence playing an important role in bioremediation. To overcome this aspect, *B. subtilis* was subjected to CRISPR-derived genome editing for the deletion of pps operon using plamid delivered *S. pyogenes* spCas9, sgRNA (target specific), and a donor DNA template. The resulting SpCas9/sgRNA endonuclease system was shown to disrupt the plipastatin operon and increase the surfactin production by mutant strain leading to an advancement in the bioremediation efficiency [35]. To extend the application of *B. subtilis* in bioremediation, a CRISPR-Cpf1 toolkit has been recently developed for deletion/insertion of single gene/gene cluster such as *sacA, aprE, ganA, ligV, ligD*, and bac operon with high efficiency [36].

CRISPR Based Genetic Manipulation in Nitrogen Metabolizing Bacteria

In addition to the bioremediation of organic pollutants, microorganisms are significantly involved in nutrient cycles like nitrogen and phosphorus cycles through multiple elements which control the pollutant transformation and transportation in water bodies. The nitrogen cycle is amongst the most complex chemical cycles in the environment. As the most abundant gas in the atmosphere and one of the most essential elements for life, atmospheric nitrogen is fixed by nitrogen fixing bacteria in a form which can be utilized to produce proteins, nucleic acids, amino acids, ammonia, and other molecules required by living organisms. The inter-conversion of nitrogen in various molecular variants in the nitrogen cycle is performed by nitrogen metabolizing bacteria along with other life forms. Hence, modifying or enhancing such bacteria can play a key role in environmental restoration as well can hold importance in pharmaceuticals, increasing agricultural productivity, and other products beneficial to humans. CRISPR technologies are being used in these nitrogen metabolizing bacteria to achieve such objectives. In a study involving the metabolic engineering of *Corynebacterium glutamicum*, RecT was used for the incorporation of synthetic single-stranded oligodeoxyribonucleotides in the genome, whereas CRISPR/Cas9 system was used for counter-selection of the negative mutants [37]. The recombineered plasmids carrying CRISPR/Cas9 and RecT, respectively, targeted multiple gene knockouts producing plasmid-cured mutants. The plasmids used for carrying RecT based on pCC1 and pEKts-Cas9 derived plasmids were used for carrying Cas9-sgRNA with Cas9 gene cloned from *S. pyogenes* [37]. The deletion efficiency of the mentioned combination was further demonstrated by separate deletion of three different genes for enhanced production of γ-aminobutyric acid (GABA). CRISPR technologies have also been applied in the metabolic engineering of cyanobacterium *Anabaena* sp. PCC 7120, where a nuclease-deficient Cas9 (dCas9) and a sgRNA were used for repression studies of the target gene [38]. The repression of the glutamine synthetase gene (*glnA*) responsible for nitrogen assimilation in the cell was studied for the production of ammonium. The dCas9 expression was regulated using TetR (tetracycline resistance gene repressor protein) induction system and the sgRNA expression aided by heterocyst-specific *nifB* (Nuclear factor 1 B-type) promoter was found to be effectively inhibiting the *glnA* transcription resulting in ammonium secretion from the cells.

WWTPs can be considered as a point of control for the environmental spread of antimicrobial resistance (AMR), where the mobility of virulence genetic elements shared among species was more common. In a study, comparative genomics of multidrug-resistance in *Enterococcus* spp. detected a maximum of 16 virulence genes in *E. faecium*, while in *E. faecalis*, *E. casseliflavus*, and *E. gallinarum*, only 6, 5, and 3 virulence genes were detected, respectively [39]. This result showed that the virulence genes were more common in *E. faecalis* and *E. faecium*, than in *E. casseliflavus* and *E. gallinarum*. Analysis of the functional CRISPR/Cas arrays with multi-drug resistance was only detected in *E. faecalis* genomes, with one genome containing a prophage that was lacking in *E. faecium* despite multi-drug resistance. This study suggested that genes associated with phage and CRISPR/Cas arrays could potentially serve as environmental biomarkers.

CRISPR AND METHANOTROPHS

Methanotrophs are bacteria which can oxidize methane and utilize it as a sole carbon and energy source. Methane oxidation in the environment is necessary to control the global uploading of greenhouse gas (GHG) in the atmosphere. The oxidation of highly stable methane by methanotrophs is attributed to the presence of the enzyme methane monooxygenase, which depending on its location in the cell, is classified into two types: soluble methane monooxygenase and particulate methane monooxygenase. The two methane monooxygenases oxidize methane to yield methanol which is then utilized by the cell for carbon and energy needs, a reaction catalysed by the enzyme methanol dehydrogenase/alcohol dehydrogenase. Methanotrophs have also found industrial applications for their capability to produce methanol, an industrial and laboratory solvent [40].

Transformation Efficiencies in Methanotrophs

The highest transformation efficiency achieved to date in a methanotrophic bacterium is through the electroporation method in type I methanotrophic strain derived from *Methylomicrobium buryatense* 5GB1C, where plasmids like pAWP89 (*Kmr*) and pFC25 (*Hmr*) infused with *mmoD, sMMO, glgA1* and *Ppmo-xylE* genes were used to reach the efficiency of approximately 5×10^5 CFUs per μg DNA [41]. As type I methanotrophs generally lack soluble methane monooxygenase, introducing genes for this enzyme or additional promoters for particulate methane monooxygenase can improve the methane oxidation potency in such microbes [42]. Similar electroporation methods were used for the introduction of succinate producing genes in strains derived from type I methanotrophic bacterium *Methylomonas* sp. DH-1 using plasmids derived from pAWP89, pCM184 and pCM351 [43]. The highest efficiency achieved was 3.2×10^5 CFUs per μg DNA, and the highest yield achieved for succinate was 0.295mmol/g DCW/d. These studies indicate that the maximum transformation efficiencies achieved in methanotrophs *via* aforesaid methods were lower than those required for the effective transformation by the CRISPR-based methods. Thus, to utilise the full potential of transformation through CRISPR technologies, the transformation efficiencies needed to be higher than those discussed above [44].

CRISPR Based Genetic Manipulation in Methanotrophs

The sequenced genome of *Methylobacillus flagellatus* was found to have a CRISPR region of 32 nucleotides in 93 copies interspaced by 33 to 39 nucleotide non-identical sequences [45]. Whereas, in another study, different strains of the methanotroph, *Methylobacterium* have been found to carry more than one array of CRISPR sequences; *M. terrae* 17Sr1-28, *M. currus* PR1016A, *M. platani* PMB02, *M. aquaticum* DSM 16371, *M. tarhaniae* N4211, *M. frigidaeris* IER25-16, and *M. organophilum* JCM 2833 had 3, 2, 5, 3, 5, 3, and 0 arrays of CRISPR sequences, respectively [46]. Contrary to this, *Methylovulum psychrotolerans* strain HV10-M2 genome sequence was found to have no CRISPR sequence at all [47]. Another methanotrophic bacterium, *Methylocystis heyer* iH2, was found to have at least one CRISPR locus and a set of Cas genes in its genome [48].

CRISPR/Cas9 systems application in methanotrophs has been studied in the well-known methanotroph *Methylococcus capsulatus* [49]. The expression vector pCAH01 having a tetracycline induced promoter/operator- PtetA was used to enhance the expression of GFP (green fluorescent protein) of the pCAH01 in the host *M. capsulatus*, similar to its expression in an earlier experiment with another methanotrophic strain *M. buryatense* 5GB1 [50]. The fluorescence was found to be increased by 10-folds in the presence of anhydrotetracycline inducer for *M. capsulatus* cells encompassing pCAH01-GFP. The *Cas9* and *Cas9D10A* (a Cas9 variant which produces single strand nicks) genes were cloned in the pCAH01 vector downstream of PtetA to create pCas9 and pCas9D10A, respectively. The promoter for methanol dehydrogenase gene *mxaF* (Pmxa) was cloned alongside the guide RNA (gRNA) in addition to a 1-kb DNA repair template in a pBBR1 plasmid to produce pgRNA. The promoter was selected among the promoters of the sMMO hydroxylase component *mmoX* (PmmoX), the gamma subunit of particulate methane monooxygenase 1 (PpmoC1), particulate methane monooxygenase 2 (PpmoC2) and the methanol dehydrogenase gene *mxaF* (Pmxa) based on their lowest expression efficiencies of GFP in *E. coli*. When the nucleases, Cas9 and Cas9D10A, were expressed without the expression of gRNA in *M. capsulatus*, bacterial death and, thus, growth was found to be unaffected. However, when the Cas9 was expressed with 4 targeted gRNA, the transformed *M. capsulatus* cells showed around 99% death. On the other hand, when the Cas9D10A and mmoX targeted gRNA were co-expressed in *M. capsulatus*, there was no observed cell death. This indicated that the ssDNA nicks did not affect the cell death as the *M. capsulatus* repair system removed the single strand nicks and the expression of gRNA is required for the action of Cas nucleases.

CRISPR AND ANAEROBIC DIGESTION

CRISPR Based Genetic Manipulation of Hydrolytic Bacteria

Among the four steps of anaerobic digestion (AD), the first step of hydrolysis that crumbs the polymeric substrate into a form utilizable by microbes is the rate limiting step ultimately controlling the downstream steps. Lignocellulosic biomass constitutes the recalcitrant natural polysaccharides imparting inherent rigidity, thus resisting rapid solubilisation to monomeric or dimeric sugars. Potent lignocellulose degraders that survive in natural habitats fail to adapt to an artificial environment in an anaerobic digester. Such conditions compel the development and exploitation of genetically engineered microbes that can survive in the anaerobic digester and perform efficient lignocellulose degradation.

The natural degraders of rigid plant cell wall polymers are white-rot fungi that express a catalogue of potent enzymes like the lytic polysaccharide monooxygenases (LPMO). The phytopathogenic fungus *Heterrobasidion irregulare* is a causative agent of wood decay in conifers [51]. This fungus expresses 10 Auxiliary Activity family 9 (AA9) LPMOs, two of which, *viz.*, HiLPMO9H and HiLPMO9I, have been introduced and heterologously expressed in *Pichia pastoris* employing CRISPR/Cas [26]. Both the CBM containing LPMOs were endoglucanases that possessed a distinct active site; the active site of HiLPMO9H was located at the C1 carbon of the pyranose ring at β-1,4-glycosidic linkages, whereas the oxidation and cleavage of the internal bonds at the C4 carbon of glucose units were catalysed by HiLPMO9I that was also active against glucomannan.

CRISPR Based Genetic Manipulation of Acidogenic Bacteria

The second step in the AD process is characterized by the activity of acidogens which convert the mono/dimeric components produced during the first step to organic acids, including volatile fatty acids (VFAs). The whole-genome sequence of one such organism, *Clostridium bornimense* strain M2/40T, obtained from a two-stage anaerobic digester operating under mesophilic condition, and fed with the lignocellulosic substrate maize silage or wheat straw identified it as an acidogen. In-depth bio-informatic analysis revealed the existence of three prophage regions and two CRISPR/Cas systems. Additionally, the glycosyl hydrolases (GH), CelK (GH9), and CelA (GH48) encoded by a putative cellulosomal gene cluster on the second replicon (chromid) were shown to be active on the substrates xylan and xyloglucan whereas XghA (GH74) was highly active on xyloglucan. This strain was classified to be an acidogenic bacterium on the basis of fermentation pathways having all the enzymes for hydrogen, acetate, formate, lactate, butyrate, and ethanol production [52].

CRISPR Based Genetic Manipulation of Methanogenic Bacteria

The next step in AD consists of the conversion of the VFAs, especially acetic acid, into methane and carbon dioxide, a reaction carried out by methanogens belonging to the phylum Archaea. Members of this group are fastidious growth-environment requiring microbes which makes the genome editing job difficult. Nayak and Metcalf [53] reported the first-ever Cas9 mediated genome editing in the slow-growing methanogenic archaeon *Methanosarcina acetivorans* saving the construction of a complex strain by the introduction of double mutation. CRISPR/Cas9 represented a versatile toolbox that aided in the programming of cellular processes without genome editing, which paved the way for developing a deeper understanding of the gene function and behaviour of microbial cells in response to various environmental challenges [54]. Considering the CRISPR interference studies in methanogens, a brilliant work was reported by Dhamad and Lessner [55] wherein the authors achieved a precise and targeted repression of genes involved in nitrogen fixation in the archaebacterium, *M. acetivorans*. To construct the CRISPRi-dCas9 system for targeted gene repression, the CRISPR/Cas9 system of *M. acetivorans* was restructured by replacing the functional *Cas9* with the *dCas9* (lacking the endonuclease activity). Repression of nif operon (nifHI1I2DKEN) that encoded molybdenum nitrogenase and *nifD* (Nitrogenase molybdenum-iron protein) gene showed that the abundance of nif transcripts was >90% reduced in the archaeal strains expressing the gRNAs thereby greatly affecting the cell growth, while NifD was not detected in the cell lysate. Similarly, when NifB (nitrogenase cofactor biogenesis) was targeted, the transcription of the *nifB* gene declined to >85% but did not affect the growth of the strains [55].

Gene knockout leads to the deletion of a particular targeted gene from the genome, which is ultimately visible in the altered phenotype. Archaebacterial cells are surrounded by the proteinaceous slime layer (S-layer), which serves as an anchor for the archaeaphage, maintains the cell size/shape, and is necessary for reproduction. The absence of the S-layer prevents viral infection (Virus SSV1), suggesting its importance in the various cellular processes [56]. The authors reported a study utilizing the endogenous CRISPR type III complex for silencing of *slaB* responsible for the S-layer membrane anchor in hyperthermophilic archaeon *Sulfolobus solfataricus*. Gene knockout resulted in reduced or absence of S-layer lattice, distorted cell shape, and decreased surface glycosylation. The importance of the S-layer was reflected in the ceased reproduction and widening of the cell diameter and nucleic acid content. Further, the altered phenotype was restored on the re-introduction of missing or deleted genes.

CRISPR/Cas9 Assisted Strain Built-Up and Introduction of Novel Capabilities By Parallel Metabolic Pathway Engineering (PMPE)

Naturally, 'glucose,' the most preferred substrate for metabolic machinery when available in proximity, represses the utilization of other carbohydrates, a process called catabolite repression. In the case of substrates consisting of heterogeneous composition of carbohydrates, *viz.*, lignocellulosic plant biomass, glucose interferes with the simultaneous utilization of xylose, the second most abundant carbohydrate on the earth. Therefore, abolishing the catabolite repression *via* genome engineering enables microbial factories to readily uptake and metabolize more than one preferred sugar resulting in a more cost-friendly bioprocess. Increased diversity in sugar uptake and metabolism results in depreciation in dependency on energy harvest from the preferred sugar substrate. Removal of catabolite repression results in energy harvest from other sugar substrates expending it for microbial cell growth, while the primary substrate glucose is targeted for enhanced acid or sugar fermentation. Fujiwara *et al.* [57] engineered a complex strain of *Escherichia coli* with the aid of CRISPR/Cas two-plasmid system by introducing a xylose catabolic pathway that directly flows into the TCA cycle when cultivated under aerobic conditions resulting in 4.09 g/L cis, cis-muconic acid production with a high yield (0.31 g/g of glucose) and L-tyrosine production at 64% of the theoretical yield. Under micro-aerobic conditions, products like organic acids and ethanol were produced as byproducts. Such genetically engineered strains of microbes could prove to be a milestone in the anaerobic digestion of pretreated lignocellulosic agricultural biomass to methane.

Strategies For Improving The Efficiency of CRISPR/Cas9

The exploitation of the CRISPR/Cas system for genome editing apparently suffers from the difficulty in achieving the required transformation efficiency and survival of the transformed organisms. Hence various strategies have been suggested to ensure improved transformation efficiency and survival rate of transformants. *Clostridium thermocellum,* a potent cellulose solubilizing anaerobe, has been a subject of interest in biofuel industries [58]. Walker *et al.* [59] studied and characterized a native Type I–B and heterologous Type II CRISPR/Cas associated system from *C. thermocellum*. The study pointed out that GEOcas9 obtained from *Geobacillus stearothermophilus*, the thermophilic Cas9 variant (TypeII) was functional over the other two thermophilic Cas9 variants in *C. thermocellum*. A recent experimental application of CRISPR/Cas mediated genome editing in the case of Orotidine 5'-phosphate decarboxylase (pyrF) gene found obstacles in the homologous recombination between repair template and the flanking ends of the genome concerning both the systems [59].

Walker *et al.* [59] reported the successful introduction of thermophillic Binases (*Bacillus intemedius* Ribonuclease) from *Acidithiobacillus caldus* and by virtue of that, the authors were able to achieve increased recombination efficiency. Results showed a doubling of the genome editing efficiency at the *pyrF* locus in the case of the Type I-B system from 40% to 71%, whereas in Type II GeoCas9 system, when aided with recombineering machinery, efficiency improved strikingly from

12.5% to 94%. A similar approach for restructuring the CRISPR/Cas9 system to introduce multiple gene silencing in a single attempt was reported by Wang *et al.* [60]. Herein, the restructured CRISPR/Cas9 system was assisted with chromosome-borne Cas9-RecET that potentially enhanced the homologous recombination efficiency and gene deletion. The method is so potent that it could delete genomic regions of 10-20 kb and, on the other hand, could introduce very site-specific mutations by the insertion of an expression cassette of size 2.5-, 5.7- and 7.5-kb. The success of the above-suggested method was reflected from the complex genome editing like the deletion of regulatory elements *argR* and *farR* of L-arginine operon as well as the introduction of site-directed mutation of *argB* and *pgi* genes, generating an L-arginine accumulating mutant. The genetically engineered strain of *C. glutamicum* produced the highest ever reported titre of 1,2-propanediol (6.75 ± 0.46 g/L). The above studies demonstrate that restructuring the CRISPR/Cas9 gene editing machinery with other recombinases could increase its applicability in better engineering of various microbial strains for both increased lignocellulose degradation and biofuel production.

CRISPR and Volatile Fatty Acid (VFA) Production

Water soluble short chain fatty acids (SCFAs) containing up to six carbon atoms which can be distilled at atmospheric pressure are known as volatile fatty acids (VFAs). Acetic, butyric, and propionic acid are the significant VFAs for the industrial production of chemicals (like acetic anhydride, solvents, calcium butyrate, esters *etc.*), biofuels (like biodiesel, biogas, biohydrogen *etc.*), bioplastics, [polyhydroxyalkanoates (PHAs)] *etc.* [61, 62]. Traditional production of VFAs is achieved *via* chemical routes, which relies on petrochemical sources [63]. Since petroleum reserves are on the decline, scientists are exploring other modes of VFA production which are environment-friendly and sustainable. One of the methods being investigated is microbial production, a route considered to be beneficial due to its renewable and pollution-free nature, which can lead to a reduction in waste. The VFAs are important products synthesized during the acidogenesis and acetogenesis stages of anaerobic digestion. However, low VFA yields and the high cost of raw materials make the process unviable for industrial production. In view of the above, recombinant DNA technology and strategies based on CRISPR technology are being explored to enhance its production.

CRISPR Based Genetic Manipulation of VFA Producing Bacteria

Genome sequence analysis of multiple VFA producing bacteria has confirmed the presence of a CRISPR system [64 - 67]. However, very few researchers have attempted the application of CRISPR/Cas9 to enhance VFA production, the reason being a lack of genomic tools. Introducing DNA molecules of interest into the acetogenic bacteria is difficult due to their thick cell walls, as they are mostly Gram positive. Another reason is the low recombination efficiency of acetogens. Nevertheless, scientists have developed tools based on CRISPR to circumvent the above restrictions. Most of these tools have been aimed at *Clostridium ljungdahlii*, a model acetogenic bacterium [68]. Both genome editing and gene expression regulation studies using CRISPR based tools have been carried out successfully to improve the yield of VFA.

With reference to the genome editing studies for maximizing VFA production, among all the CRISPR types, type II CRISPR/Cas9 has been the most studied and the most popular one used for metabolic engineering applications. Seo *et al.* [69] examined the addition of acid tolerance genes *via* CRISPR/Cas9 to the genome of *E. coli* BL21. The heptanoic acid tolerance genes, *dsrA* and *rcsB* were obtained from another *E. coli* strain K-12, and the native transcription unit (TU) in the case of *dsrA* along with a synthetic TU in the case of *rcsB* were incorporated in the *E. coli* BL21 genome. The expression level of the *dsrA* was comparatively lower than in *E. coli* K-12, however, increased expression levels of *rcsB* were observed in comparison to that in *E. coli* K-12. These aspects accrued the transformed strain with higher survival and growth rates under heptanoic acid stress as compared to the host *E. coli* BL21. In another study by Huang *et al.* [70], CRISPR/Cas9 was used to incorporate butyric acid production pathway (*thl, crt-bcd, etfB-etfA-hbd, ptb-buk* genes and gene clusters) from *C. acetobutylicum* into the genome of *C. ljungdahlii*. The resultant strain yielded 1.01 g/L butyric acid in 3 days using synthesis gas and displayed good genetic stability resulting in continuous butyric acid production even after repeated sub-culturing. In a similar study targeting the improved production of fatty acids, 8 genes (*FAA1, FAA4, POX1, ARE2, PAH1, LPP1, DPP1,* and *ARE1*) were knocked out concurrently in *Saccharomyces cerevisiae* using GTRCRISPR. A 30-fold rise in free fatty acid accumulation was observed in the engineered strain (559.52 mg/L) as compared to the wild type (19.93 mg/L) with only two rounds of gene deletion [71].

With reference to gene expression based on CRISPRi (CRISPR interference) and CRISPRa (CRISPR activation), very few studies have been carried out on the regulation of gene expression for enhancing VFA production. For instance, Tan *et al.* [72] utilised CRISPR to delete *ackA, fadE, fumAC* genes while simultaneously overexpressing the *fabZ* gene in *E. coli*. The resultant engineered strain produced 442 mg/L octanoic acid, which was 61% higher than the wild-type strain. VFA accumulation further rose to a titre of 1 g/L upon optimisation in fed-batch mode.

While only a few researchers have applied CRISPR for enhancing VFA production, there are studies which have used CRISPR for the production of other metabolites from acetogens. For instance, Wang *et al.* [73] used CRISPR to delete *pta* and *buk* genes (which are important for acetate and butyrate production, respectively) in *C. saccharoperbutylacetonicum* N1-4, a high butanol producing strain. While acetate and butyrate production was not fully eliminated, higher butanol production of 19 g/L by *pta-buk* double deletion mutant in the P2 medium was observed. A study by Zhang *et al.* [74] utilised the endogenous CRISPR/Cas system for multiplex genome editing in *C. tyrobutyricum,* a butyric acid producing bacterium and engineered the strain for high-level butanol production (26.2 g/L). In yet another study, CRISPRi was used to engineer a strain of *C. ljungdahlii* for 3 hydroxybutyrate (3HB) production with down-regulation of *pta* gene leading to the doubling of the 3HB titre [75].

LIMITATIONS OF CRISPR AND STRATEGIES TO OVERCOME THE DRAWBACKS

In spite of the significant number of applications of CRISPR/Cas9 in multiple areas, it is plagued by the drawback of recognizing only the GC-rich Protospacer Adjacent Motif (PAM) sequences which is problematic in the case of genomes with low GC content [68]. Thus, the sequences that can be targeted by CRISPR are dependent upon the presence/availability of PAM sequences in the genome [1] since the absence of PAM sequences would prevent its use. Furthermore, biases within the PAM sequences for G+C content would also limit its use. The restriction of the CRISPR/Cas9 system in recognizing only the GC-rich PAM sequences can be overcome with the use of other Cas nucleases, such as CRISPR/Cas12a, which identified AT-rich PAM sequences. Another benefit of Cas12a was that when several genes were targeted, only a single promoter and terminator were needed for the expression of numerous repeats and spacers, unlike Cas9, which would require several promoters and terminators in such a case [76]. Thus, Cas12a held the potential in simplifying the process of multiplex genome editing. Zhao *et al.* [78] used CRISPR/Cas12a to delete the genes *pyrE, pta, adhE1,* and *ctf* in butyric acid producing *C. ljungdahlii*. The researchers further created a CRISPRi (CRISPR interference) system using DNase-deactivated FnCas12a (ddCas12a) to repress the selected genes achieving a high repression rate of 80% for the majority of the binding sites. The authors reported a 20-40% decline in ethanol production (a by-product) and a subsequent increase in butyric acid synthesis.

Another option for overcoming the aforesaid deficiencies of the CRISPR/Ca9 system has been suggested to be the use of genetically modified nucleases which do not have this bias [77]. However, multiple studies have shown that genome editing using Class 2 CRISPR systems is not always effective, especially among prokaryotes. The intrinsic toxicity of large heterologous nucleases is presumed to be one of the reasons [74]. Vento *et al.* [78] reviewed the limitations of using SpCas9 for genome editing and found that overexpression of SpCas9 could be cytotoxic in some bacteria, with few to no colonies being observed after transformation with SpCas9. Potential alternatives such as Cas12a and Cas9n were suggested to be comparatively less toxic. Another option was to utilize the endogenous nucleases of the host for editing, which required the introduction of only gRNA and recombineering template in such a case. For instance, Pyne *et al.* [79] observed that genome editing in *C. pasteurianum* using a heterologous Type 2 CRISPR/Cas9 system yielded an editing efficiency of 25% as compared to editing by Type 1-B CRISPR/Cas system, which generated 100% editing efficiency. The endogenous Type 1 CRISPR system has multiple advantages. Since most proteins required for genome editing are already naturally present in the cells except nuclease, heterologous helpers are not needed for the production of Crna guides, while only synthetic CRISPR arrays are required, thereby making the process of multiplex gene uncomplicated [80]. Another benefit is that the Class 1 systems are more common in Bacteria and Archaea than Class 2 systems, thus highlighting their immense potential for diverse applications [6]. Most studies on fatty acid production currently utilise the Class 2 systems for genome editing, thus highlighting the tremendous scope of exploiting the endogenous CRISPR system for enhancing VFA production in future studies. CRISPR/Cas9 system is very specific, however in practice, that is not the case, with multiple studies indicating the common occurrence of off-target mutations. These unwanted modifications at random sites could be detrimental to applications like gene therapy. To reduce the off-target mutations, several computational methods based on sequence similarity have been developed in addition to the methods using Cas9 nickase or fCas9. While each of these techniques reduces the off-target effects, they come with their own set of limitations [81].

FUTURE PERSPECTIVES

In view of strengthening and improving the bio-processes like anaerobic digestion (AD), exploitation and augmentation of genetically engineered microbes can be a crucial strategy. Toxicity caused by volatile fatty acids (VFAs) can be withstood by actively introducing VFA metabolizing genetically engineered acetogens. It can introduce resistance against various environmental stress conditions to methanogens, thus improving their utility in the environmental rejuvenation process and, at the same time, can improve economic aspects of anaerobic digestion. More research and future work will broaden the range of applications of CRISPR in the diagnosis and control of infectious diseases, non-infectious diseases, and metabolic disorders related to gut dysbiosis. Accurate, speedy, and reliable diagnosis will help in precise and flexible editing of the gut microbiome using CRISPR for correcting gut dysbiosis. Also, applications of engineered probiotics, prebiotics as well as synbiotics using CRISPR would be very beneficial for the treatment of gut related disorders and in other health aspects.

CONCLUSION

CRISPR/Cas as a genome editing tool provides an opportunity to construct more robust recombinant microbial cells by introducing site-directed mutations. The studies on CRISPR/Cas in bioremediation could be extended to other environmental pollutants such as xenobiotics, petroleum hydrocarbon, PAHs, and PCBs. As seen from the reviewed literature, many researchers have reported genes involved in the bioremediation of a wide range of pollutants in bacteria. These genes have demonstrated tremendous potential for future experimental studies using CRISPR/Cas mediated enhancement of bacterial metabolic enzymes, and their increased efficiency towards the detoxification and the removal of environmental contaminants. In addition, CRISPR could also be applied for the development of more competent bacterial strains with enhanced capacities for the degradation of pollutants. Strategies adopted for modifying the gene of interest, their entire catabolic pathway, homeostasis and expression that promote growth, hyper-accumulation, tolerance and degradation, *etc.*, can prove to be revolutionary in environmental clean-up *via* bacteria. Altogether, bacteria and CRISPR-derived upgrades could take bioremediation to the next level, allowing the successful reclamation of polluted soils and waters.

ACKNOWLEDGEMENTS

All the authors have contributed equally to the manuscript.

The authors would like to acknowledge Director, CSIR-NEERI and AcSIR-NEERI for providing essential resources for the research work. The manuscript has been internally checked for plagiarism through iThenticate software and assigned the KRC No. CSIR-NEERI/KRC/2021/MARCH/EBGD/2. SPN and RKG are thankful to UGC, while AKS is thankful to CSIR for Senior Research Fellowship for carrying out this research. Funds from the Science and Engineering Research Board (SERB), New Delhi for the project-EMR/2016/006589 are gratefully acknowledged.

REFERENCES

[1] Chylinski K, Makarova KS, Charpentier E, Koonin EV. Classification and evolution of type II CRISPR-Cas systems. Nucleic Acids Res 2014; 42(10): 6091-105.
[http://dx.doi.org/10.1093/nar/gku241] [PMID: 24728998]

[2] Deveau H, Garneau JE, Moineau S. CRISPR/Cas system and its role in phage-bacteria interactions. Annu Rev Microbiol 2010; 64(1): 475-93.
[http://dx.doi.org/10.1146/annurev.micro.112408.134123] [PMID: 20528693]

[3] Sun CL, Thomas BC, Barrangou R, Banfield JF. Metagenomic reconstructions of bacterial CRISPR loci constrain population histories. ISME J 2016; 10(4): 858-70.
[http://dx.doi.org/10.1038/ismej.2015.162] [PMID: 26394009]

[4] Weissman J, Fagan WF, Johnson PLF. Selective maintenance of multiple CRISPR arrays across prokaryotes. CRISPR J 2018; 1(6): 405-13.
[http://dx.doi.org/10.1089/crispr.2018.0034] [PMID: 31021246]

[5] Bobay LM, Touchon M, Rocha EPC. Pervasive domestication of defective prophages by bacteria. Proc Natl Acad Sci 2014; 111(33): 12127-32.
[http://dx.doi.org/10.1073/pnas.1405336111] [PMID: 25092302]

[6] Makarova KS, Wolf YI, Alkhnbashi OS, et al. An updated evolutionary classification of CRISPR–Cas systems. Nat Rev Microbiol 2015; 13(11): 722-36.
[http://dx.doi.org/10.1038/nrmicro3569] [PMID: 26411297]

[7] Horváth A, Patonay A, Bánhegyi D, et al. The first case of human alveolar echinococcosis in Hungary. Orv Hetil 2008; 149(17): 795-9.
[http://dx.doi.org/10.1556/OH.2008.28281] [PMID: 18426761]

[8] Shariat N, Dudley EG. CRISPRs: molecular signatures used for pathogen subtyping. Appl Environ Microbiol 2014; 80(2): 430-9.
[http://dx.doi.org/10.1128/AEM.02790-13] [PMID: 24162568]

[9] Gilbert JA, Blaser MJ, Caporaso JG, Jansson JK, Lynch SV, Knight R. Current understanding of the human microbiome. Nat Med 2018; 24(4): 392-400.
[http://dx.doi.org/10.1038/nm.4517] [PMID: 29634682]

[10] Srivastav S, Fatima M, Ahmad MH, Mondal AC. Impact of CRISPR-based gene editing in environmental biotechnology.Emerging Trends in Environmental Biotechnology. CRC Press 2022; pp. 131-42.
[http://dx.doi.org/10.1201/9781003186304-11]

[11] Joshi RK, Bharat SS, Mishra R. Engineering drought tolerance in plants through CRISPR/Cas genome editing. 3 Biotech 2020; 10(9): 1-14.

[12] Paul D, Pandey G, Pandey J, Jain RK. Accessing microbial diversity for bioremediation and environmental restoration. Trends Biotechnol 2005; 23(3): 135-42.
[http://dx.doi.org/10.1016/j.tibtech.2005.01.001] [PMID: 15734556]

[13] Pandey G, Jain RK. Bacterial chemotaxis toward environmental pollutants: Role in bioremediation. Appl Environ Microbiol 2002; 68(12): 5789-95.
[http://dx.doi.org/10.1128/AEM.68.12.5789-5795.2002] [PMID: 12450797]

[14] Barber LB, Keefe SH, Brown GK, et al. Persistence and potential effects of complex organic contaminant mixtures in wastewater-impacted streams. Environ Sci Technol 2013; 47(5): 2177-88.
[http://dx.doi.org/10.1021/es303720g] [PMID: 23398602]

[15] Kumar V, Shahi SK, Singh S. Bioremediation: An eco-sustainable approach for restoration of contaminated sites. In: Singh J, Sharma D, Kumar G, Sharma N, Eds. Microbial Bioprospecting for Sustainable Development. Singapore: Springer 2018; pp. 115-36.
[http://dx.doi.org/10.1007/978-981-13-0053-0_6]

[16] Dermont G, Bergeron M, Mercier G, Richer-Laflèche M. Soil washing for metal removal: A review of physical/chemical technologies and field applications. J Hazard Mater 2008; 152(1): 1-31.
[http://dx.doi.org/10.1016/j.jhazmat.2007.10.043] [PMID: 18036735]

[17] Ayangbenro A, Babalola O. Metal (loid) bioremediation: Strategies employed by microbial polymers. Sustainability 2018; 10(9): 3028.
[http://dx.doi.org/10.3390/su10093028]

[18] Purohit HJ, Kapley A, Khardenavis A, Qureshi A, Dafale NA. Insights in waste management

bioprocesses using genomic tools. Adv Appl Microbiol 2016; 97: 121-70.
[http://dx.doi.org/10.1016/bs.aambs.2016.09.002] [PMID: 27926430]

[19] Sangwan S, Dukare A. Microbe-mediated bioremediation: An eco-friendly sustainable approach for environmental clean-up.advances in Soil Microbiology: Recent Trends and Future Prospects: Volume 1: Soil-Microbe Interaction. Singapore: Springer 2018; pp. 145-63.
[http://dx.doi.org/10.1007/978-981-10-6178-3_8]

[20] Sutherland DL, Ralph PJ. Microalgal bioremediation of emerging contaminants : Opportunities and challenges. Water Res 2019; 164: 114921.
[http://dx.doi.org/10.1016/j.watres.2019.114921] [PMID: 31382151]

[21] Liu DF, Li WW. Genome editing techniques promise new breakthroughs in water environmental microbial Biotechnologies. ACS ES&T Water 2021; 1(4): 745-7.
[http://dx.doi.org/10.1021/acsestwater.0c00276]

[22] Wolt JD, Wang K, Yang B. The regulatory status of genome–edited crops. Plant Biotechnol J 2016; 14(2): 510-8.
[http://dx.doi.org/10.1111/pbi.12444] [PMID: 26251102]

[23] Jaiswal S, Singh DK, Shukla P. Gene editing and systems biology tools for pesticide bioremediation: A review. Front Microbiol 2019; 10: 87.
[http://dx.doi.org/10.3389/fmicb.2019.00087] [PMID: 30853940]

[24] Thijs S, Sillen W, Rineau F, Weyens N, Vangronsveld J. Towards an enhanced understanding of plant–microbiome interactions to improve phytoremediation: Engineering the metaorganism. Front Microbiol 2016; 7: 341.
[http://dx.doi.org/10.3389/fmicb.2016.00341] [PMID: 27014254]

[25] Dangi AK, Sharma B, Hill RT, Shukla P. Bioremediation through microbes: Systems biology and metabolic engineering approach. Crit Rev Biotechnol 2019; 39(1): 79-98.
[http://dx.doi.org/10.1080/07388551.2018.1500997] [PMID: 30198342]

[26] Liu B, Olson Å, Wu M, Broberg A, Sandgren M. Biochemical studies of two lytic polysaccharide monooxygenases from the white-rot fungus *Heterobasidion irregulare* and their roles in lignocellulose degradation. PLoS One 2017; 12(12): e0189479.
[http://dx.doi.org/10.1371/journal.pone.0189479] [PMID: 29228039]

[27] Tan SZ, Reisch CR, Prather KLJ. A robust CRISPR interference gene repression system in Pseudomonas. J Bacteriol 2018; 200(7): e00575-17.
[http://dx.doi.org/10.1128/JB.00575-17] [PMID: 29311279]

[28] Xu A, Wang D, Ding Y, *et al.* Integrated comparative genomic analysis and phenotypic profiling of *Pseudomonas aeruginosa* isolates from crude oil. Front Microbiol 2020; 11: 519.
[http://dx.doi.org/10.3389/fmicb.2020.00519] [PMID: 32300337]

[29] Corts AD, Thomason LC, Gill RT, Gralnick JA. Efficient and precise genome editing in Shewanella with recombineering and CRISPR/Cas9-mediated counter-selection. ACS Synth Biol 2019; 8(8): 1877-89.
[http://dx.doi.org/10.1021/acssynbio.9b00188] [PMID: 31277550]

[30] Li J, Tang Q, Li Y, *et al.* Rediverting electron flux with an engineered CRISPR-ddAsCpf1 system to enhance the pollutant degradation capacity of *Shewanella oneidensis*. Environ Sci Technol 2020; 54(6): 3599-608.
[http://dx.doi.org/10.1021/acs.est.9b06378] [PMID: 32062962]

[31] Liang Y, Jiao S, Wang M, Yu H, Shen Z. A CRISPR/Cas9-based genome editing system for *Rhodococcus ruber* TH. Metab Eng 2020; 57: 13-22.
[http://dx.doi.org/10.1016/j.ymben.2019.10.003] [PMID: 31610242]

[32] Bell KS, Philp JC, Aw DW, Christofi N. The genus Rhodococcus. J Appl Microbiol 1998; 85(2): 195-210.

[http://dx.doi.org/10.1046/j.1365-2672.1998.00525.x] [PMID: 9750292]

[33] Molenda O, Tang S, Lomheim L, *et al.* Extrachromosomal circular elements targeted by CRISPR-Cas in *Dehalococcoides mccartyi* are linked to mobilization of reductive dehalogenase genes. ISME J 2019; 13(1): 24-38.
[http://dx.doi.org/10.1038/s41396-018-0254-2] [PMID: 30104577]

[34] Coutte F, Leclère V, Béchet M, *et al.* Effect of *pps* disruption and constitutive expression of *srfA* on surfactin productivity, spreading and antagonistic properties of *Bacillus subtilis* 168 derivatives. J Appl Microbiol 2010; 109(2): 480-91.
[http://dx.doi.org/10.1111/j.1365-2672.2010.04683.x] [PMID: 20148996]

[35] So Y, Park SY, Park EH, *et al.* A highly efficient CRISPR-Cas9-mediated large genomic deletion in *Bacillus subtilis*. Front Microbiol 2017; 8: 1167.
[http://dx.doi.org/10.3389/fmicb.2017.01167] [PMID: 28690606]

[36] Hao W, Suo F, Lin Q, *et al.* Design and construction of portable CRISPR-Cpf1-mediated genome editing in bacillus subtilis 168 oriented toward multiple utilities. Front Bioeng Biotechnol 2020; 8: 524676.
[http://dx.doi.org/10.3389/fbioe.2020.524676] [PMID: 32984297]

[37] Cho JS, Choi KR, Prabowo CPS, *et al.* CRISPR/Cas9-coupled recombineering for metabolic engineering of *Corynebacterium glutamicum*. Metab Eng 2017; 42: 157-67.
[http://dx.doi.org/10.1016/j.ymben.2017.06.010] [PMID: 28649005]

[38] Higo A, Ehira S. Spatiotemporal gene repression system in the heterocyst-forming multicellular cyanobacterium *Anabaena* sp. PCC 7120. ACS Synth Biol 2019; 8(4): 641-6.
[http://dx.doi.org/10.1021/acssynbio.8b00496] [PMID: 30865823]

[39] Sanderson H, Ortega-Polo R, Zaheer R, *et al.* Comparative genomics of multidrug-resistant Enterococcus spp. isolated from wastewater treatment plants. BMC Microbiol 2020; 20(1): 20.
[http://dx.doi.org/10.1186/s12866-019-1683-4] [PMID: 31980014]

[40] Patel SKS, Mardina P, Kim SY, Lee JK, Kim IW. Biological methanol production by a type II methanotroph *Methylocystis bryophila*. J Microbiol Biotechnol 2016; 26(4): 717-24.
[http://dx.doi.org/10.4014/jmb.1601.01013] [PMID: 26838340]

[41] Yan X, Chu F, Puri AW, Fu Y, Lidstrom ME. Electroporation-based genetic manipulation in type I methanotrophs. Appl Environ Microbiol 2016; 82(7): 2062-9.
[http://dx.doi.org/10.1128/AEM.03724-15] [PMID: 26801578]

[42] Hanson RS, Hanson TE. Methanotrophic bacteria. Microbiol Rev 1996; 60(2): 439-71.
[http://dx.doi.org/10.1128/mr.60.2.439-471.1996] [PMID: 8801441]

[43] Nguyen DTN, Lee OK, Hadiyati S, Affifah AN, Kim MS, Lee EY. Metabolic engineering of the type I methanotroph *Methylomonas* sp. DH-1 for production of succinate from methane. Metab Eng 2019; 54: 170-9.
[http://dx.doi.org/10.1016/j.ymben.2019.03.013] [PMID: 30986511]

[44] Henard CA, Guarnieri MT. Metabolic engineering of methanotrophic bacteria for industrial biomanufacturing.Methane biocatalysis: Paving the way to sustainability. Cham: Springer 2018; pp. 117-32.
[http://dx.doi.org/10.1007/978-3-319-74866-5_8]

[45] Chistoserdova L, Lapidus A, Han C, *et al.* Genome of *Methylobacillus flagellatus*, molecular basis for obligate methylotrophy, and polyphyletic origin of methylotrophy. J Bacteriol 2007; 189(11): 4020-7.
[http://dx.doi.org/10.1128/JB.00045-07] [PMID: 17416667]

[46] Kim J, Chhetri G, Kim I, Kim H, Kim MK, Seo T. *Methylobacterium terrae* sp. nov., a radiation-resistant bacterium isolated from gamma ray-irradiated soil. J Microbiol 2019; 57(11): 959-66.
[http://dx.doi.org/10.1007/s12275-019-9007-9] [PMID: 31463788]

[47] Mateos-Rivera A, Islam T, Marshall IPG, Schreiber L, Øvreås L. High-quality draft genome of the

methanotroph *Methylovulum psychrotolerans* Str. HV10-M2 isolated from plant material at a high-altitude environment. Stand Genomic Sci 2018; 13(1): 10.
[http://dx.doi.org/10.1186/s40793-018-0314-2] [PMID: 29686747]

[48] Oshkin IY, Miroshnikov KK, Grouzdev DS, Dedysh SN. Pan-genome-based analysis as a framework for demarcating two closely related methanotroph genera Methylocystis and Methylosinus. Microorganisms 2020; 8(5): 768.
[http://dx.doi.org/10.3390/microorganisms8050768] [PMID: 32443820]

[49] Tapscott T, Guarnieri MT, Henard CA. Development of a CRISPR/Cas9 system for *Methylococcus capsulatus in vivo* gene editing. Appl Environ Microbiol 2019; 85(11): e00340-19.
[http://dx.doi.org/10.1128/AEM.00340-19] [PMID: 30926729]

[50] Garg S, Clomburg JM, Gonzalez R. A modular approach for high-flux lactic acid production from methane in an industrial medium using engineered *Methylomicrobium buryatense* 5GB1. J Ind Microbiol Biotechnol 2018; 45(6): 379-91.
[http://dx.doi.org/10.1007/s10295-018-2035-3] [PMID: 29675615]

[51] Mgbeahuruike EE, Yrjönen T, Vuorela H, Holm Y. Bioactive compounds from medicinal plants: Focus on Piper species. S Afr J Bot 2017; 112: 54-69.
[http://dx.doi.org/10.1016/j.sajb.2017.05.007]

[52] Tomazetto G, Hahnke S, Wibberg D, Pühler A, Klocke M, Schlüter A. *Proteiniphilum saccharofermentans* str. M3/6^T isolated from a laboratory biogas reactor is versatile in polysaccharide and oligopeptide utilization as deduced from genome-based metabolic reconstructions. Biotechnol Rep 2018; 18: e00254.
[http://dx.doi.org/10.1016/j.btre.2018.e00254] [PMID: 29892569]

[53] Nayak DD, Metcalf WW. Cas9-mediated genome editing in the methanogenic archaeon *Methanosarcina acetivorans*. Proc Natl Acad Sci 2017; 114(11): 2976-81.
[http://dx.doi.org/10.1073/pnas.1618596114] [PMID: 28265068]

[54] Xu X, Qi LS. A CRISPR–dCas toolbox for genetic engineering and synthetic biology. J Mol Biol 2019; 431(1): 34-47.
[http://dx.doi.org/10.1016/j.jmb.2018.06.037] [PMID: 29958882]

[55] Dhamad AE, Lessner DJ. CRISPRi-dCas9 system for archaea and its use to examine gene function during nitrogen fixation by *Methanosarcina acetivorans*. Appl Environ Microbiol 2020; 86(21): e01402-20.
[http://dx.doi.org/10.1128/AEM.01402-20] [PMID: 32826220]

[56] Zink IA, Pfeifer K, Wimmer E, Sleytr UB, Schuster B, Schleper C. CRISPR-mediated gene silencing reveals involvement of the archaeal S-layer in cell division and virus infection. Nat Commun 2019; 10(1): 4797.
[http://dx.doi.org/10.1038/s41467-019-12745-x] [PMID: 31641111]

[57] Fujiwara R, Noda S, Tanaka T, Kondo A. Metabolic engineering of *Escherichia coli* for shikimate pathway derivative production from glucose–xylose co-substrate. Nat Commun 2020; 11(1): 1-12.
[PMID: 31911652]

[58] Tian L, Cervenka ND, Low AM, Olson DG, Lynd LR. A mutation in the AdhE alcohol dehydrogenase of *Clostridium thermocellum* increases tolerance to several primary alcohols, including isobutanol, n-butanol and ethanol. Sci Rep 2019; 9(1): 1736.
[http://dx.doi.org/10.1038/s41598-018-37979-5] [PMID: 30741948]

[59] Walker JE, Lanahan AA, Zheng T, *et al.* Development of both type I–B and type II CRISPR/Cas genome editing systems in the cellulolytic bacterium *Clostridium thermocellum*. Metab Eng Commun 2020; 10: e00116.
[http://dx.doi.org/10.1016/j.mec.2019.e00116] [PMID: 31890588]

[60] Wang B, Hu Q, Zhang Y, *et al.* A RecET-assisted CRISPR–Cas9 genome editing in *Corynebacterium glutamicum*. Microb Cell Fact 2018; 17(1): 63.

[http://dx.doi.org/10.1186/s12934-018-0910-2] [PMID: 29685154]

[61] Lee WS, Chua ASM, Yeoh HK, Ngoh GC. A review of the production and applications of waste-derived volatile fatty acids. Chem Eng J 2014; 235: 83-99.
[http://dx.doi.org/10.1016/j.cej.2013.09.002]

[62] Kim NJ, Lim SJ, Chang HN. Volatile fatty acid platform: Concept and application. In: Chang HN. (Ed.) Emerging Areas in Bioengineering. 2018; Chapter 10, (pp 173-90), Wiley☐VCH Verlag GmbH & Co. KGaA.

[63] Zacharof MP, Lovitt RW. Complex effluent streams as a potential source of volatile fatty acids. Waste Biomass Valoriz 2013; 4(3): 557-81.
[http://dx.doi.org/10.1007/s12649-013-9202-6]

[64] Brown SD, Nagaraju S, Utturkar S, *et al.* Comparison of single-molecule sequencing and hybrid approaches for finishing the genome of *Clostridium autoethanogenum* and analysis of CRISPR systems in industrial relevant Clostridia. Biotechnol Biofuels 2014; 7(1): 40.
[http://dx.doi.org/10.1186/1754-6834-7-40] [PMID: 24655715]

[65] Wu Q, Liu T, Zhu L, Huang H, Jiang L. Insights from the complete genome sequence of *Clostridium tyrobutyricum* provide a platform for biotechnological and industrial applications. J Ind Microbiol Biotechnol 2017; 44(8): 1245-60.
[http://dx.doi.org/10.1007/s10295-017-1956-6] [PMID: 28536840]

[66] Nethery MA, Henriksen ED, Daughtry KV, Johanningsmeier SD, Barrangou R. Comparative genomics of eight Lactobacillus buchneri strains isolated from food spoilage. BMC Genomics 2019; 20(1): 902.
[http://dx.doi.org/10.1186/s12864-019-6274-0] [PMID: 31775607]

[67] Piwowarek K, Lipińska E, Hać-Szymańczuk E, Kieliszek M, Kot AM. Sequencing and analysis of the genome of *Propionibacterium freudenreichii* T82 strain: Importance for industry. Biomolecules 2020; 10(2): 348.
[http://dx.doi.org/10.3390/biom10020348] [PMID: 32102319]

[68] Jin S, Bae J, Song Y, *et al.* Synthetic biology on acetogenic bacteria for highly efficient conversion of c1 gases to biochemicals. Int J Mol Sci 2020; 21(20): 7639.
[http://dx.doi.org/10.3390/ijms21207639] [PMID: 33076477]

[69] Seo JH, Baek SW, Lee J, Park JB. Engineering *Escherichia coli* BL21 genome to improve the heptanoic acid tolerance by using CRISPR-Cas9 system. Biotechnol Bioprocess Eng 2017; 22(3): 231-8.
[http://dx.doi.org/10.1007/s12257-017-0158-4]

[70] Huang H, Chai C, Yang S, Jiang W, Gu Y. Phage serine integrase-mediated genome engineering for efficient expression of chemical biosynthetic pathway in gas-fermenting *Clostridium ljungdahlii*. Metab Eng 2019; 52: 293-302.
[http://dx.doi.org/10.1016/j.ymben.2019.01.005] [PMID: 30633974]

[71] Zhang Y, Wang J, Wang Z, *et al.* A gRNA-tRNA array for CRISPR-Cas9 based rapid multiplexed genome editing in *Saccharomyces cerevisiae*. Nat Commun 2019; 10(1): 1053.
[http://dx.doi.org/10.1038/s41467-019-09005-3] [PMID: 30837474]

[72] Tan Z, Yoon JM, Chowdhury A, *et al.* Engineering of *E. coli* inherent fatty acid biosynthesis capacity to increase octanoic acid production. Biotechnol Biofuels 2018; 11(1): 87.
[http://dx.doi.org/10.1186/s13068-018-1078-z] [PMID: 29619083]

[73] Wang S, Dong S, Wang P, Tao Y, Wang Y. Genome editing in *Clostridium saccharoperbutylacetonicum* N1-4 with the CRISPR-Cas9 system. Appl Environ Microbiol 2017; 83(10): e00233-17.
[http://dx.doi.org/10.1128/AEM.00233-17] [PMID: 28258147]

[74] Zhang J, Zong W, Hong W, Zhang ZT, Wang Y. Exploiting endogenous CRISPR-Cas system for

multiplex genome editing in *Clostridium tyrobutyricum* and engineer the strain for high-level butanol production. Metab Eng 2018; 47: 49-59.
[http://dx.doi.org/10.1016/j.ymben.2018.03.007] [PMID: 29530750]

[75] Woolston BM, Emerson DF, Currie DH, Stephanopoulos G. Rediverting carbon flux in *Clostridium ljungdahlii* using CRISPR interference (CRISPRi). Metab Eng 2018; 48: 243-53.
[http://dx.doi.org/10.1016/j.ymben.2018.06.006] [PMID: 29906505]

[76] Ao X, Yao Y, Li T, *et al*. A multiplex genome editing method for *Escherichia coli* based on CRISPR-Cas12a. Front Microbiol 2018; 9: 2307.
[http://dx.doi.org/10.3389/fmicb.2018.02307] [PMID: 30356638]

[77] Donohoue PD, Barrangou R, May AP. Advances in industrial biotechnology using CRISPR-Cas systems. Trends Biotechnol 2018; 36(2): 134-46.
[http://dx.doi.org/10.1016/j.tibtech.2017.07.007] [PMID: 28778606]

[78] Vento JM, Crook N, Beisel CL. Barriers to genome editing with CRISPR in bacteria. J Ind Microbiol Biotechnol 2019; 46(9-10): 1327-41.
[http://dx.doi.org/10.1007/s10295-019-02195-1] [PMID: 31165970]

[79] Pyne ME, Bruder MR, Moo-Young M, Chung DA, Chou CP. Harnessing heterologous and endogenous CRISPR-Cas machineries for efficient markerless genome editing in Clostridium. Sci Rep 2016; 6(1): 25666.
[http://dx.doi.org/10.1038/srep25666] [PMID: 27157668]

[80] Zheng Y, Li J, Wang B, *et al*. Endogenous type I CRISPR-Cas: From foreign DNA defense to prokaryotic engineering. Front Bioeng Biotechnol 2020; 8: 62.
[http://dx.doi.org/10.3389/fbioe.2020.00062] [PMID: 32195227]

[81] Peng R, Lin G, Li J. Potential pitfalls of CRISPR/Cas9-mediated genome editing. FEBS J 2016; 283(7): 1218-31.
[http://dx.doi.org/10.1111/febs.13586] [PMID: 26535798]

CHAPTER 8

Genetic Engineering of Methanotrophs: Methods and Recent Advancements

Eleni N. Moutsoglou[1,2] and **Rajesh K. Sani**[1,2,*]

[1] Department of Chemical and Biological Engineering, South Dakota School of Mines and Technology, Rapid City, SD 57701, USA

[2] BuG ReMeDEE Consortium, Rapid City, SD 57701, USA

Abstract: Methanotrophs use methane gas as their carbon and energy source, but their industrial use has not yet fully been realized due to undiscovered genetic engineering methods that could amend their slow growth rate and economically inefficient product yield. This chapter informs upon genetic engineering approaches taken on methanotrophs so far to enable their widespread use in industry, as well as the reasoning behind these interests. Specific examples of successful engineering performed so far, including conjugation and electroporation methods, CRISPR, genome-scale metabolic modeling, and specific vectors reported as successful, are presented. In addition, the reading provides insights into existing knowledge gaps in the field of methanotrophic engineering and future prospects for optimizing growth and product yield from methanotrophs.

Keywords: Conjugation, Electroporation, Genetic engineering, Methanotroph, Methane monooxygenase (MMO) gene.

INTRODUCTION

Methanotrophs

Methanotrophs use C1 compounds as their sole carbon and energy source. Methanotrophs can be found in many diverse environments and are classified as aerobic or anaerobic, depending on their electron acceptor. Methanotrophs are further classified depending on the pathway they use for formaldehyde assimilation. Type I methanotrophs use the ribulose monophosphate pathway for assimilation of formaldehyde into cellular carbon, Type II methanotrophs use the serine pathway, and Type X methanotrophs have hybrid properties of Type I and

---

[*] **Corresponding author Rajesh K. Sani:** Department of Chemical and Biological Engineering, South Dakota School of Mines and Technology, Rapid City, SD 57701, USA; & BuG ReMeDEE Consortium, Rapid City, SD 57701, USA; E-mail: Rajesh.Sani@sdsmt.edu

Prakash M. Halami & Aravind Sundararaman (Eds.)
All rights reserved-© 2024 Bentham Science Publishers

Type II methanotrophs [1]. Key aspects of methanotrophy include methane assimilation, copper accumulation, and metal-dependent gene expression. Since 1970, genetic engineering approaches have focused on well-studied methanotrophs like *Methylococcus capsulatus* Bath, *Methylosinus trichosporium* OB3b, and *Methylocystis parvus* OBBP [2]. Haloalkaliphilic *Methylotuvimicrobium* bacteria like *buryatense* 5GB1C or *alcaliphilum* 20Z have also been of interest for use in the industry because of their fast growth rate [2].

Industrial Use of Methanotrophs

Methanotrophs are of industrial interest because they can produce value-added products from methane, which is a cheap feedstock as well as a greenhouse gas. The most well-studied derived products from methane are methanol, polyhydroxyalkanoates (PHAs), and single-cell protein. Many other chemicals, such as precursors to biofuels, could become industrially relevant through genetic engineering. For example, fatty acids from bacteria are precursors for the production of liquid biofuels, and Type I methanotrophs could be microbial catalysts that substitute methane for sugars as a carbon source.

Why Genetically Engineered Methanotrophs?

For a methanotroph to be as industrially relevant as *E. coli* or yeast, it must be well characterized and tools for its genetic manipulation must be determined. Economically effective methane bioconversion is rare and has only occurred with single-cell protein and PHA products. The slow growth rate is the main problem for industrial use of methanotrophs. Recently spiked interest in methanotrophy has led to the production of new protocols for the genetic engineering of bacteria that are simple and efficient. The high degree of reduction of methane by methanotrophs provides more free electrons for the production of products, and methane bioconversion has a higher carbon conversion efficiency compared to chemical processes (75% compared to 20-50%, respectively) [3]. The ability of Type II methanotrophs to direct high carbon fluxes towards acetoacetyl-CoA under nutrient limited conditions makes them promising for metabolic engineering [4]. As an example, the increased and stable expression of the methane monooxygenase gene (MMO) and methanol dehydrogenase through protein engineering could provide methanotrophic strains with more stable phenotypes, and expression of the MMO in a heterologous host with a higher growth rate could lead to new biotechnologies.

METHODS OF GENETIC ENGINEERING

In General

High-efficiency genome editing requires making a targeted DNA double-strand break in the DNA sequence of interest. Three classes of nucleases can be designed to make this double-strand break at any target. These are zinc-finger nucleases, transcription activator-like effector nucleases, and CRISPR-Cas [5]. A zinc finger nuclease is a hybrid of a DNA cleavage domain from a bacterial protein and a set of zinc fingers that were identified in sequence-specific eukaryotic transcription factors. Transcription activator-like effector nucleases have the same bacterial cleavage domain, but it is instead linked to a DNA recognition module of transcription factors from plant pathogenic bacteria [5]. CRISPR-Cas is a prokaryotic system derived from acquired immunity to invading nucleic acids.

Everything that happens after the targeted double-strand break depends on cellular DNA repair machinery, either by homology-dependent repair or non-homologous end joining [5]. To increase the efficiency of homology-dependent repair machinery, the donor DNA can be designed and linked to the guide RNA, and specific mechanisms that mediate sequence insertions must be considered. Gene-editing systems can enable rapid and high-throughput methanotrophic genetics.

In Methanotrophs

Genome sequences have been published for almost all genera of methanotrophs as either draft or complete sequences [6]. The availability of these sequences provides comparative analyses and allows for regulatory metabolism reconstruction and modelling, as discussed later when genome scale metabolic models (GSMM) are described.

Metabolic engineering in methanotrophs is used to increase metabolic flux and production of end products, as well as to enhance stress tolerance and substrate utilization. The more recent availability of broad-host range plasmids has provided for greater development of genetic techniques for methanotrophs. Protocols for efficient extraction of DNA from methanotrophs as well as plasmid construction and their transfer to *E. colivia* conjugation have been reported [7]. Some vectors, like pCM184, contain antibiotic markers, but unmarked mutants can be created with counterselection systems [6]. Sucrose counterselection using sacB has been successful in *Methylococcus capsulatus* Bath, *Methylomonas* sp. strain 16a, and *Methylomicrobium buryatense*. Insertions and deletions are performed using marker exchange *via* homologous recombination using flanking regions of 500 base pairs of genomic sequence [6].

Conjugation

Conjugation, or the transfer of DNA from one bacterial cell to another, is the method most used for the transformation of methanotrophs [8]. Genetic data is integrated into methanotrophs primarily by conjugation with an *E. coli* donor. Vectors used for conjugation often contain the RK2/RP4 origin of transfer. Conjugation and vector construction are time consuming, so more streamlined approaches are needed. Triparental matings can be used with the donor strain and another strain with a helper plasmid like pRK2013 to mobilize the vector [6]. Matings are done on agar plates with NMS media and supplemented with yeast extract for the donor. Conjugations are done for several days before selection, and efficiencies vary between species. Naladixic acid and rifamycin or rafamycin can be used for selection, as many methanotrophs are resistant to them [6].

Electroporation

Electroporation is a method of transformation less commonly used for methanotrophs that employ electric current to disrupt the cell membrane to allow for DNA uptake. Electroporation techniques have been developed for Type I and Type II methanotrophs, and these techniques have been validated with gene knockouts, deletion, and insertion of foreign fragments into the chromosome. The method of electroporation is simpler than conjugation for genetic engineering in general. Linear DNA fragments can be transferred with electroporation, creating double-crossover deletion and insertion events and eliminating the need for counterselection. Additionally, electroporation allows for direct transfer of genes that might be toxic in other hosts, like *E. coli*, and avoids altering plasmids amid conjugation. Some instances of electroporation with methanotrophs reported include deletion of the pmo operon in *Methylocystis* sp. strain SC2, and electroporation of linear DNA fragments was used to mutate *M. silverstris* BL2 [6].

CRISPR

The procedure of genome editing with CRISPR requires two steps: finding the most effective guide RNA and analyzing clones for the presence of the desired mutation [9]. Since the recognizing fragment of gRNA is only 20 bp long, the CRISPR system can be easily altered to find a desired sequence (compared to zinc finger nucleases and transcription activator-like effector nucleases) [9]. Types of CRISPR mutations include 1) a reading frame shift causing gene knockout, 2) nucleotide substitutions or 3) insertion/deletion using a homology-directed repair template. Nuclease efficacy evaluation is essential for successful genome editing, and methods to assess this efficacy differ in cost and time required. The CRISPR/

Cas9 system could be used in methanotrophs, but this method relies on vector construction, is inefficient, and has off-target effects [2].

Tapscott *et al.* [10] developed a CRISPR/Cas9 system for gene editing *in vivo* for the methanotroph *Methylococcus capsulatus* Bath, which is described below. Literature on CRISPR gene editing in methanotrophs is currently scarce.

Vectors Used in Methanotrophs

Several vectors facilitating allele exchange and allowing insertion and deletion of genes have been applied to methanotrophs. Lindstrom and Wopat [1] found that out of 10 methanotrophic strains studied, only Bath did not contain plasmid DNA. All *M. trichosporium* strains contained three plasmids of sizes 186, 159, and 145 kb. *Methylomonas*, *Methylobacter*, and *Methylocystis* contained plasmids ranging from 50-190 kb in size.

Replicating vectors of the group IncP-1 have been used for heterologous expression in methanotrophs, along with the use of vectors pBBR and IncQ vectors [6]. Replicating plasmids with reporter genes are beneficial for efficient promoter probing. Reporter genes, including GFP, dTomato, and beta-galactosidase have all been successfully used. Table **1** shows plasmids corresponding to replication and deletion/insertion vectors used for genetically engineering methanotrophs.

Table 1. Vectors used for genetic engineering in methanotrophs.

Plasmid	Selectable Marker	Species	Source
Replicating Vectors			
pVK100	Tetracycline, Kanamycin	*M. albus*, *Methylocystis*, *Methylosinus* 6, *M. trichosporium*, *Metylomicrobium album* BG8	Lindstrom *et al.* [11], Murrell [1]
pULB113	Kanamycin	*M. trichosporium*, *M. capsulatus*	Al-Taho *et al.* [12]
pSRK-Km	Kanamycin	*M. capsulatus* Bath	Kalyuzhnaya [6]
pAWP78	Kanamycin	*Methylomicrobium buryatense*	Puri *et al.* [13]
PMHA200_Phps	Kanamycin	*Methylomicrobium alcaliphilum* 20Z	Mustakhimov *et al.* [14]
PMHA203	Kanamycin	*Methylocella silvestris* BL2	Hwang *et al.* [3]
Deletion/Insertion Vectors			
pK18mob	Kanamycin	*M. trichosporium* OB3b	Kalyuzhnaya [6]
pCM184	Tetracycline, Ampicillin, Kanamycin	*M. capsulatus* Bath, *Methylocella silvestris*, *Methylomicrobium alcaliphilum* 20Z	Kalyuzhnaya [6]

(Table 1) cont.....

Plasmid	Selectable Marker	Species	Source
pSUKSM	Kanamycin	*Methylomonas* sp. st. 16a	Ye *et al.* [15]
pAWP78	Kanamycin	*Methylomicrobium buryatense*	Puri *et al.* [13]

Genome-Scale Metabolic Models

Complete genome-scale metabolic models (GSMM) provide insights into metabolic genetic engineering strategies and help to identify the biotechnological potential of a microbe. A GSMM is a collection of annotated metabolic reactions within an organism and allows for simulated metabolic behavior under different conditions. The first genome scale metabolic model of a methanotroph was of *Methylomicrobium buryatense* [4]. Annotated genome sequences, along with literature reviews and biochemical databases, are used to complete the reaction network for a GSMM. The models can be used to engineer microbes to channel metabolic fluxes through certain pathways to optimize product yields. As an example, Gupta *et al.* constructed a complete GSMM of Bath. The model had 535 genes, 899 reactions, and 865 metabolites present [16].

SPECIFIC EXAMPLES OF GENETIC ENGINEERING IN METHANOTROPHS

The following section of this book chapter will focus on specific examples of genetic engineering in methanotrophs, as summarized in Table **2**.

Table 2. Genetic engineering and results reported in methanotrophs.

Species	Technique	Result	Source
Methylosinus trichosporium OB3b	Shuttle and replicative plasmids added to the transformation protocol	Improved gene delivery	Ro *et al.* [17]
Methylococcus capsulatus (Bath)	Constructed CRISPR/Cas9 gene editing system	Disrupted sMMO function	Tapscott *et al.* [10]
Methylomicrobium buryatense 5GB1C, *Methylomonas* sp. LW13, and *Methylobacter tundripaludum* 21/22	Developed the first electroporation technique for Type I methanotroph	Higher transfer efficiency	Yan *et al.* [8]
Bath	Mutated sacB gene for efficient counterselection	Efficient production of mutants	Ishikawa *et al.* [18]
Methylomicrobium buryatense	Deleted farE regulon	Increased fatty acid methyl ester production	Demidenko *et al.* [19]

(Table 2) cont.....

Species	Technique	Result	Source
Methylomicrobium alcaliphilum 20Z	Combined genes from other organisms *via* electroporation	Engineered 2,3-butanediol production	Nguyen *et al.* [20]
OB3b	Knocked out mbnT using marker exchange mutagenesis	Determined a methanobactin uptake gene	Gu *et al.* [21]
M. buryatense 5GB1C	Used site-specific recombination for marker-free genome modification	Allowed for efficient markerless genomic deletion	Liu *et al.* [2]
OB3b	Expressed recombinant sMMO in mutated OB3b with sMMO operon deleted		Lock *et al.* [22]
Methylomonas sp. 16a	Integrated hemoglobin genes into the chromosome	Improved astaxanthin production	Tao *et al.* [23]
M. buryatense 5GB1C	Diverted carbon flux through acetyl-CoA pathway with reverse β-oxidation pathway genes	Produced crotonic acid non-native to the host	Garg *et al.* [24]
Methylomonas sp. DH-1	Modified enzymes in the TCA cycle	Increased succinate production tenfold	Nguyen *et al.* [25]
Methylococcus capsulatus	Heterologously expressed lactate dehydrogenases from three species into *M. capsulatus*	Lactic acid was produced at a 20x higher level	Subbian *et al.* [26]
Methanosarcina acetovorans	Induced reverse methanogenesis	Produced acetate	Soo *et al.* [29]
M. buryatense	Engineered bacteria w/ varying promoter and ribosome-binding site combinations	Produced 14x higher L-lactate	Garg *et al.* [30]

Ro *et al.* [17] modified existing transformation protocols for *Methylosinus trichosporium* OB3b. Genetic tools previously developed for OB3b were used to mutate genes encoding important proteins. More recent studies have focused on the sMMO gene, proteins in the pmo operon, methanobactin transport proteins, and genes involved with methanol dehydrogenase [17]. They added additional shuttle and replicative plasmids to existing transformation protocols for improvements in gene delivery and genotyping. For the selection of OB3b mutants, kanamycin and gentamicin were used. They also provided specific transformation protocols for OB3b.

Tapscott *et al.* used a series of broad-host-range expression plasmids to create a CRISPR/Cas9 gene-editing system in *Methylococcus capsulatus* (Bath) [10]. Broad-host-range replicative plasmids containing RP4/RK2, RSF1010, and pBBR1 replicons have been found to be functional in Bath and have enabled the use of promotor-probe vectors and heterologous gene expression in the bacteria

[17]. Chromosomal insertions and genetic mutations using allelic-exchange vectors and sucrose counterselection have been reported in Bath. They also successfully introduced a premature stop codon into the sMMO hydroxylase component-encoding mmoX gene with Cas9 nickase, successfully upsetting the sMMO function.

Yan *et al.* developed the first electroporation technique for use on a Type I methanotroph. The electroporation technique had a high transfer efficiency for three different Type I methanotrophs: *Methylomicrobium buryatense* 5GB1C, *Methylomonas* sp. LW13, and *Methylobacter tundripaludum* 21/22 [8].

Ishikawa *et al.* developed a new genetic tool enabling efficient counterselection in Bath [18]. Bath and OB3b have been studied extensively for the production of single-cell proteins, polyhydroxybutyrates, and methanol production, but the productivity of these processes is still low. In the study, the sacB gene, which is commonly used as a counterselectable marker for disruption of a target gene in gram-negative bacteria, was mutated with a two-point mutation for use in Bath. This allowed for the successful creation of gene disrupted and integrated mutants of Bath. The method developed could be applicable to other methanotrophs besides Bath, as the alignment shows two residues for the construction of a counterselectable marker in other methanotrophs [18]. Depending on the level of sequence similarity, the gene constructed could be useful for a range of Type 1 methanotrophs.

Regulating synthesis pathways can increase lipid production, and most genes homologous to Type II fatty acid biosynthesis pathways can be annotated with bioinformatics. Demidenko *et al.* identified a candidate gene for fatty acid biosynthesis regulation, farE [19]. The farE regulon was investigated with RNA sequence analysis of gene expression in knockout mutants and overexpressing strains. The deletion of the farE regulon caused changes to the fatty acid profile in *M. buryatense* and increased fatty acid methyl ester production. In short, for unmarked gene deletions, a plasmid with sequences flanking the genes to be deleted (pCM433kanT) was introduced, and after conjugation, single-crossover kanamycin resistant *M. buryatense* clones were plated on counterselection plates. The double-crossover clones were then plated on 2.5% sucrose, and colonies were PCR-genotyped. All genes were cloned under that tac-promotor.

The study discovered a novel set of enzymes and regulators involved in fatty acid biosynthesis and supported the theory that metabolic fluxes upstream of fatty acid biosynthesis reduce fatty acid synthesis in methanotrophs. The team also studied whether diverting carbon flux into acyl-CoA and malonyl-CoA would increase fatty acid production. To do this, they deleted the acetate kinase gene and overexpressed acetyl-CoA carboxylase. A 20% increase in fatty acid accumulation resulted. Altering fatty acid synthesis elongation is a viable strategy to alter fatty acid content in methanotrophs.

Nguyen *et al.* metabolically engineered *Methylomicrobium alcaliphilum* 20Z for 2,3-butanediol production from methane [20]. 2,3-BDO gene clusters from native producers were screened, and promoter selection and combination with genes from other organisms were used to optimize their expression. In short, plasmid vectors pAWP89 and pCM433KanT were linearized by inverse PCR using primer pairs for both plasmids. 2,3-BDO synthesis genes from *K. pneumoniae*, *E. aerogenes*, and *B. subtilis* were amplified from genomic DNA and assembled with the linearized pAWP89. Sequences flanking the target genes were amplified from *M. alcaliphilum* 20Z and ligated with the linearized pCM433KanT for unmarked gene deletions, resulting in an allelic exchange vector. Introducing foreign DNA into methanotrophs *via* electroporation is advantageous over conjugation in that fewer steps are involved, and linear DNA can be transformed into the methanotroph.

Gu *et al.* performed a knockout of mbnT in *Methylosinus trichosporium* OB3b using marker exchange mutagenesis to determine if the gene was responsible for methanobactin uptake [21]. Methanobactin is important because it is involved in the copper switch mechanism in MMO. The knockout and marker exchange mutagenesis in the study was performed as follows. The 3' and 5' regions of mbnT were amplified by PCR and inserted into the pK18mobsacB plasmid, and the gentamicin resistance gene was excised from the plasmid. The inserted plasmid was transformed into *E. coli.*, and *E. coli.* was conjugated with OB3b. Transconjugants were selected by plating on NMS with gentamicin. Successful knockouts were confirmed by screening for kanamycin and sucrose resistance, by PCR, and by sequencing. The resulting mutant was able to produce methanobactin but not internalize it.

Liu *et al.* used an Flp/FRT site-specific recombination system and PheS with T251A and A294G substitutions as a counterselectable marker for chromosomal modification in *Methylotuvimicrobium* [2]. PheS, compared to the counterselectable marker sacB, which leaves an FRT site "scar" at the replacement locus, allowed for marker-free genome modification. The work is summarized as follows: First, multiple DNA fragments were fused with overlap PCR, then, PheS was mutated with site-directed mutagenesis. Expression cassettes were inserted at chromosome, and the assembled product was transformed into *M. buryatense* 5GB1C by electroporation. A PZ cassette was constructed using overlap PCR. Deletion mutations to remove glgA1 were performed to generate the deletion amplicon, and it was directly transformed into

5GB1C by electroporation. Finally, the smmo operon was deleted using site-directed mutagenesis, and the deletion amplicon was again transformed into 5GB1C. A PZ* cassette was constructed for *M. alcaliphilum* 20Z and deletion mutants were constructed to remove glgA1 using the same method as for 5GB1C. The method enabled efficient markerless genome deletion in 5GB1C and 20Z [2]. This was advantageous over other methods. No "scar" (which can interfere with subsequent modifications in the same host) was left after modification, and counterselection had a positive rate of more than 92%.

Lock *et al.* engineered sMMO to increase its catalytic activity by six times [22]. sMMO can oxygenate over 100 hydrocarbons and is one of the most catalytically diverse oxidizers. sMMO expression in *E. coli* has not been successful. The team used homologous expression systems where sMMO was expressed in the sMMO-deleted, methane-oxidizing bacteria OB3b. Recombinant sMMO was expressed in the mutated OB3b with the six-gene operon encoding sMMO deleted. The plasmid pT2ML, which was modified from pTJS175 to allow for the cloning of mutant genes in a single step, was used. pT2ML is an *E.coli-M. trichosporium* shuttle vector allowing for cloning of mutated mmoX genes in a single directed cloning step. The recombinant genes were expressed from the bacteria's natural promoters. The system allowed recombinant genes to be introduced into the host *via* conjugation, and the results were sMMO expression equivalent to that in the wild-type organism.

Tao *et al.* improved astaxanthin production in *Methylomonas sp.* 16a [23]. The genome of the methanotroph contains three hemoglobin genes, so a cassette with these hemoglobin genes, along with astaxanthin genes from *Brevundimonas vesicularis*, was integrated into the chromosome. This allowed for the production of a plasmid-free *Methylomonas* strain with increased production capability of the carotenoid. The increased level of oxygen provided by the hemoglobin significantly increased the activity of the astaxanthin synthesis enzymes.

Garg *et al.* produced crotonic acid non-native to the host from methane in *M. buryatense* 5GB1C [24]. This was done by diverting carbon flux through the acetyl-CoA metabolic pathway using reverse β-oxidation pathway genes. The presence of the Embden-Meyerhof-Parnas, Entner-Doudoroff, Pentose Phosphate, and Phosphoketolase pathways in *M. buryatense* indicates that engineering strategies established already in yeast and *E. coli* could be implemented in the methanotroph as "drop-ins". Therefore, they used this strategy to produce crotonic acid in *M. buryatense* 5GB1C by expressing a modified β-oxidation pathway previously demonstrated in *E. coli*.

Nguyen *et al.* engineered *Methylomonas* sp. DH-1 to accumulate succinate, which is a top building block chemical for the agricultural and pharmaceutical industries [25]. The TCA cycle and the enzymes diverting carbon to acetate and formate production were modified or deleted to increase succinate production. Deleting succinate dehydrogenase in the TCA cycle increased succinate production tenfold compared to the wild type strain.

Subbian heterologously expressed lactate dehydrogenases from *E. coli*, *Pseudomonas aeruginosa* MTCC 424, and *Pectobacterium carotovorum* MTCC 1428 under the control of methanol dehydrogenase and sigma 70 promoters in *Methylococcus capsulatus* [26]. Subbian expressed the succinate pathway genes malate dehydrogenase, pyruvate carboxylase, and phosphoenolpyruvate carboxylase in *M. capsulatus*, resulting in 20 times higher levels of lactic acid produced [27].

Production of high levels of methanol from methanotrophs requires inhibiting methanol dehydrogenase [28]. Soo *et al.* [29]. engineered *Methanosarcina acetovorans* to produce acetate (a biofuel precursor) by introducing methyl-CoA reductase and inducing reverse methanogenesis.

Garg *et al.* [24] engineered *M. buryatense* with different promoter and ribosome-binding site combinations. In the synthetic biology approach taken, three sets of plasmids carrying varying promoters (two inducible and one constitutive) were engineered, and the new strain produced L-lactate from methane at 14 times higher than the native strain. The study reported that ribosome binding sites have a large impact on product synthesis, and ribosome binding sites perform differently with different promoters. Modular assembly of genetics, along with traditional metabolic engineering, can be used to optimize methanotrophic strains for industrial production.

EXISTING KNOWLEDGE GAPS

Knowledge gaps in methanotrophy must be resolved to realize the industrial production of value-added products from methane gas. Unsolved properties in methanotrophs include the identity of the pMMO electron donor, the electron transfer components in methane oxidation, and how carbon flux is regulated [6]. The pathways downstream from primary methane assimilation are not well known, and how methanotrophs shift metabolism based on environmental changes is also not well known.

A comparison of several genomes has shown that core metabolic pathways are similar, but even close relatives can have a large divergence in their metabolic pathways [6]. Gaps in knowledge make it difficult to construct metabolic models and, further make it difficult to genetically engineer the metabolism of methanotrophs. Regarding the unknown pMMO electron donor, predictive metabolic models cannot be created without an understanding of the NAD(P)H/ATP ratio within the cell. Kalyuzhnaya *et al.* reported that ubiquinol is the most likely electron donor to pMMO, but the source of electrons to reduce ubiquinone to ubiquinol is unknown [6]. It is possible that pMMO has multiple sources of electrons depending on environmental conditions.

Another knowledge gap is the uncertainty in the validity of hypothesized synthesis pathways within methanotrophs. For example, cell yield is higher on methane than methanol in methanotrophs, going against the result from the deciphered energy-using step of methane oxidation. This suggests methane oxidation does not operate as currently predicted. One hypothesis is that pMMO interacts directly with methanol dehydrogenase for efficient energy coupling. A recent study of this interaction suggests the presence of an unknown additional protein [6]. The lack of understanding of the regulation of sMMO by copper, methane, and other compounds slows progress in expressing sMMO in heterologous hosts and applicable biotechnology.

PROSPECTS

Despite the successful outcomes in the genetic engineering of methanotrophs mentioned previously, methane bioconversion industrial processes are still largely focused on natural methanotrophic strains [7]. A large knowledge base of the physiology of methanotrophs exists, but molecular biology and genetic studies have slackened. Developing better and more electroporation protocols would accelerate metabolic engineering.

sMMO has been successfully transformed into a heterologous host only in a small number of reports, including in *E. coli*, *Pseudomonas mendocina*, and *P. putida*. Heterologous expression of the enzyme is difficult due to unique MMO regulatory sequences, and the sMMO enzyme is less stable in transformed hosts compared to the native organism [7]. Successful unmarked allelic exchange with sucrose counterselection has been successful in *Methylomonas* sp. 16a and Bath, but the methods have not been validated or used in most methanotrophs [13]. One interesting feature of methane oxidation is that 40% of the energy of a methane molecule is lost in the first oxidation step from methanol to methane. Improving the efficiency of methane oxidation could be realized by engineering methane monooxygenase into a di-oxygenase to double the efficiency of the process [28].

The use of *E. coli* for methane bioconversion could be promising because of its high growth rate and extensive knowledge of its physiology. The presence of the Embden-Meyerhof-Parnas, Entner-Doudoroff, Pentose Phosphate, and

Phosphoketolase pathways in *M. buryatense* means that engineering strategies established already in industrial strains like yeast and *E. coli* could be implemented in the methanotroph, as well as others, as "drop-ins" [24]. However, only the β-subunit of pMMO has been successfully expressed in *E. coli*, and at a very low activity. This lack of expression of MMO is the main obstacle for the success of synthetic methanotrophy [3].

CONCLUSION

Genetic engineering of methanotrophs is necessary for their widespread industrial use. Low methane solubility in liquid and the inherently slow growth rate of methanotrophs preclude their economic viability. As listed in Table **1**, several vectors already exist to engineer methanotrophic bacteria. With a deeper understanding of their metabolic pathways, their industrial potential can be fully realized.

ACKNOWLEDGMENTS

We acknowledge support from the National Science Foundation (Award #1736255, #1849206, and #1920954) and the Department of Chemical and Biological Engineering at the South Dakota Mines.

REFERENCES

[1] Murrell JC. Genetics and molecular biology of methanotrophs. FEMS Microbiol Lett 1992; 88(3-4): 233-48.
[http://dx.doi.org/10.1111/j.1574-6968.1992.tb04990.x] [PMID: 1515161]

[2] Liu Y, He X, Zhu P, Cheng M, Hong Q, Yan X. $pheS^{AG}$ based rapid and efficient markerless mutagenesis in *Methylotuvimicrobium*. Front Microbiol 2020; 11: 441.
[http://dx.doi.org/10.3389/fmicb.2020.00441] [PMID: 32296398]

[3] Hwang IY, Nguyen AD, Nguyen TT, Nguyen LT, Lee OK, Lee EY. Biological conversion of methane to chemicals and fuels: Technical challenges and issues. Appl Microbiol Biotechnol 2018; 102(7): 3071-80.
[http://dx.doi.org/10.1007/s00253-018-8842-7] [PMID: 29492639]

[4] Bordel S, Rodríguez Y, Hakobyan A, Rodríguez E, Lebrero R, Muñoz R. Genome scale metabolic modeling reveals the metabolic potential of three Type II methanotrophs of the genus *Methylocystis*. Metab Eng 2019; 54: 191-9.
[http://dx.doi.org/10.1016/j.ymben.2019.04.001] [PMID: 30999053]

[5] Carroll D. Genome editing: Past, present, and future. Yale J Biol Med 2017; 90(4): 653-9.
[PMID: 29259529]

[6] Kalyuzhnaya MG, Puri AW, Lidstrom ME. Metabolic engineering in methanotrophic bacteria. Metab Eng 2015; 29: 142-52.
[http://dx.doi.org/10.1016/j.ymben.2015.03.010] [PMID: 25825038]

[7] Khmelenina VN, Rozova ON, But CY, *et al.* Biosynthesis of secondary metabolites in methanotrophs: Biochemical and genetic aspects (review). Prikl Biokhim Mikrobiol 2015; 51(2): 140-50.
[http://dx.doi.org/10.7868/S0555109915020087] [PMID: 26027349]

[8]   Yan X, Chu F, Puri AW, Fu Y, Lidstrom ME. Electroporation-based genetic manipulation in type I methanotrophs. Appl Environ Microbiol 2016; 82(7): 2062-9.
[http://dx.doi.org/10.1128/AEM.03724-15] [PMID: 26801578]

[9]   Lomov NA, Viushkov VS, Petrenko AP, Syrkina MS, Rubtsov MA. Methods of evaluating the efficiency of CRISPR/Cas genome editing. Mol Biol 2019; 53(6): 982-97.
[PMID: 31876277]

[10]   Tapscott T, Guarnieri MT, Henard CA. Development of a CRISPR/Cas9 system for *Methylococcus capsulatus in vivo* gene editing. Appl Environ Microbiol 2019; 85(11): e00340-19.
[http://dx.doi.org/10.1128/AEM.00340-19] [PMID: 30926729]

[11]   Lindstrom M, Wopat A, Nunn D, *et al.* Manipulation of methanotrophs.Genetic Control of Environmental Pollutants. Basic Life Sciences book series 1984; 28: p. 319.

[12]   Al-Taho N, Warner P. Restoration of phenotype in *Escherichia coli* auxotrophs by PULB113-mediated mobilisation from methylotrophic bacteria. FEMS Microbiol Let 1987; 43(2): 235-9.

[13]   Puri AW, Owen S, Chu F, *et al.* Genetic tools for the industrially promising methanotroph *Methylomicrobium buryatense*. Appl Environ Microbiol 2015; 81(5): 1775-81.
[http://dx.doi.org/10.1128/AEM.03795-14] [PMID: 25548049]

[14]   Mustakhimov II, But SY, Reshetnikov AS, Khmelenina VN, Trotsenko YA. Homo- and heterologous reporter proteins for evaluation of promoter activity in *Methylomicrobium alcaliphilum* 20Z. Prikl Biokhim Mikrobiol 2016; 52(3): 279-86.
[PMID: 29509383]

[15]   Ye R, Yao H, Stead K, *et al.* Construction of the astaxanthin biosynthetic pathway in a methanotrophic bacterium *Methylomonas* sp. strain 16a. Ind Microbiol Biotech. 2007.
[http://dx.doi.org/10.1007/s10295-006-0197-x]

[16]   Gupta A, Ahmad A. Genome-scale metabolic reconstruction and metabolic versatility of an obligate methanotroph *Methylococcus capsulatus* str. Bath PeerJ 2019.

[17]   Ro SY, Rosenzweig AC. Recent advances in the genetic manipulation of *Methylosinus trichosporium* OB3b. Methods Enzymol 2018; 605: 335-49.
[http://dx.doi.org/10.1016/bs.mie.2018.02.011] [PMID: 29909832]

[18]   Ishikawa M, Yokoe S, Kato S, Hori K. Efficient counterselection for *Methylococcus capsulatus* (Bath) by using a mutated *pheS* Gene. Appl Environ Microbiol 2018; 84(23): e01875-18.
[http://dx.doi.org/10.1128/AEM.01875-18] [PMID: 30266726]

[19]   Demidenko A, Akberdin IR, Allemann M, Allen EE, Kalyuzhnaya MG. Fatty acid biosynthesis pathways in *Methylomicrobium buryatense* 5G(B1). Front Microbiol 2017; 7: 2167.
[http://dx.doi.org/10.3389/fmicb.2016.02167] [PMID: 28119683]

[20]   Nguyen AD, Hwang IY, Lee OK, *et al.* Systematic metabolic engineering of *Methylomicrobium alcaliphilum* 20Z for 2,3-butanediol production from methane. Metab Eng 2018; 47: 323-33.
[http://dx.doi.org/10.1016/j.ymben.2018.04.010] [PMID: 29673960]

[21]   Gu W, Farhan Ul Haque M, Baral BS, *et al.* A TonB-dependent transporter is responsible for methanobactin uptake by *Methylosinus trichosporium* OB3b. Appl Environ Microbiol 2016; 82(6): 1917-23.
[http://dx.doi.org/10.1128/AEM.03884-15] [PMID: 26773085]

[22]   Lock M, Nichol T, Murrell JC, Smith TJ. Mutagenesis and expression of methane monooxygenase to alter regioselectivity with aromatic substrates. FEMS Microbiol Lett 2017; 364(13).
[http://dx.doi.org/10.1093/femsle/fnx137] [PMID: 28854685]

[23]   Tao L, Sedkova N, Yao H, Ye RW, Sharpe PL, Cheng Q. Expression of bacterial hemoglobin genes to improve astaxanthin production in a methanotrophic bacterium *Methylomonas* sp. Appl Microbiol Biotechnol 2007; 74(3): 625-33.

[http://dx.doi.org/10.1007/s00253-006-0708-8] [PMID: 17103157]

[24] Garg S, Wu H, Clomburg JM, Bennett GN. Bioconversion of methane to C-4 carboxylic acids using carbon flux through acetyl-CoA in engineered *Methylomicrobium buryatense* 5GB1C. Metab Eng 2018; 48: 175-83.
[http://dx.doi.org/10.1016/j.ymben.2018.06.001] [PMID: 29883803]

[25] Nguyen DTN, Lee OK, Hadiyati S, Affifah AN, Kim MS, Lee EY. Metabolic engineering of the type I methanotroph *Methylomonas* sp. DH-1 for production of succinate from methane. Metab Eng 2019; 54: 170-9.
[http://dx.doi.org/10.1016/j.ymben.2019.03.013] [PMID: 30986511]

[26] Subbian E. Production of lactic acid from organic waste or biogas or methane using recombinant methanotrophic bacteria. US 15303188, 2015.

[27] Subbian E. Production of succinic acid from organic waste or biogas or methane using recombinant methanotrophic bacterium. US 15303184, 2015.

[28] Pieja AJ, Morse MC, Cal AJ. Methane to bioproducts: The future of the bioeconomy? Curr Opin Chem Biol 2017; 41: 123-31.
[http://dx.doi.org/10.1016/j.cbpa.2017.10.024] [PMID: 29197255]

[29] Soo VWC, McAnulty MJ, Tripathi A, *et al.* Reversing methanogenesis to capture methane for liquid biofuel precursors. Microb Cell Fact 2016; 15(1): 11.
[http://dx.doi.org/10.1186/s12934-015-0397-z] [PMID: 26767617]

[30] Garg S, Clomburg JM, Gonzalez R. A modular approach for high-flux lactic acid production from methane in an industrial medium using engineered *Methylomicrobium buryatense* 5GB1. J Ind Microbiol Biotechnol 2018; 45(6): 379-91.
[http://dx.doi.org/10.1007/s10295-018-2035-3] [PMID: 29675615]

CHAPTER 9

Genome Editing in *Cyanobacteria*

Bathula Srinivas[1] and **Prakash M. Halami**[2,3,*]

[1] *Department of Biotechnology, School of Herbal Studies and Naturo Sciences, Dravidian University, Kuppam-517426, India*

[2] *Department of Microbiology and Fermentation Technology, CSIR-Central Food Technological Research Institute, Mysuru-570020, India*

[3] *Academy of Scientific and Innovative Research (AcSIR), Ghaziabad, Uttar Pradesh, India*

Abstract: *Cyanobacteria* are potential organisms being exploited for a wide range of biotechnological applications. They are photosynthetic bacteria and grow in a carbon-free medium and become attractive hosts for biotechnology industries. *Cyanobacteria* can utilize solar energy and atmospheric CO_2 for the growth and synthesis of biomolecules. It is used in many large-scale preparations of various bioproducts such as pharmaceuticals, biofuels, *etc*. *Cyanobacteria* become target organisms for the next generation of biofactories for producing desired products with a low-cost technology. The problem in the metabolic engineering of *Cyanobacteria* is due to ploidy. It has multiple copies of chromosomes ranging from 3-218 copies. There are 12 copies of the genome in *Synechocystis* PCC 6803 and 3 copies in *Synechococcus* PCC 7942. Segregation analysis in the conventional genetic approaches of *Cyanobacteria* becomes laborious due to its polyploidy. Modern genome editing tools such as CRISPR-Cas9 and 12 are available to perform genome editing. CRISPR-Cas9 has been used in a wide range of *Cyanobacteria* such as *Synechococcus elongates* UTEX 2973, *Synechocystis* sp. PCC 6803. To avoid toxic effects caused by Cas-9, a low-level expression system is adopted in *Cyanobacteria*. Cas-9 base genome editing was applied in *Synechococcus* and produced succinate 11-fold higher than the normal. Cas-9 is used to cure plasmids in *Synechocystis* sp. PCC 6803 to develop a shuttle vector for heterologous expression. Another variant of genome editing tool is CRISPR-Cas12a, which is successfully used in *Synechocystis* sp.

Keywords: *Cyanobacteria*, CRISPR-Cas9, Metabolic engineering, *Synechococcus elongates*.

INTRODUCTION

Cyanobacteria are photosynthetic microorganisms living in both marine and freshwater systems [1, 2]. Oxygenic photosynthesis was released approximately

---

* **Corresponding author Prakash M. Halami:** Department Microbiology and Fermentation Technology, CSIR-Central Food Technological Research Institute, Mysuru-570020, India; & Academy of Scientific and Innovative Research (AcSIR), Ghaziabad, Uttar Pradesh, India E-mail: prakashalami@cftri.res.in

Prakash M. Halami & Aravind Sundararaman (Eds.)
All rights reserved-© 2024 Bentham Science Publishers

2.5 billion years ago in primitive *Cyanobacteria* [3]. *Cyanobacteria* are the most promising microorganisms for the sustainable production of a wide range of biotechnological products for the humankind. The free oxygen (O_2) is increased in the atmosphere, that helps in establishing several microorganisms [4, 5]. In present days, *Cyanobacteria* contributes 20%–30% of atmospheric carbon dioxide (CO_2) [6]. *Cyanobacteria* are highly efficient in transforming carbon into a wide range of biomaterials, such as biofuels and commercially important enzymes [7]. *Cyanobacteria* are superior to plants due to many desired features like being capable of synthesizing carbohydrates through photosynthesis [8, 9]. They are capable of growing in unfavourable conditions such as increased temperature, pH and salt concentrations [1, 2]. It can be easily grown on infertile land with the least number of nutrients [10] and is relatively rapid and less expensive in the production of mutants [11]. In addition, chloroplasts are diminished in the internalized *Cyanobacteria* [12] with unique biochemical and physiological features and become an excellent host for producing plant-derived products [13 - 15].

The general features of *Cyanobacteria*, which are desirable for culturing and genetic modification, are listed in Table **1**. In the present years, there is an increasing demand for modifying *Cyanobacteria* for designing markerless selection systems, strong promoters for better expression, reporter proteins and ribosome binding sites. The above modifications made the *Cyanobacteria* a suitable host to transfer genes from desired hosts to make valuable bioproducts. One of the problems with *Cyanobacteria* is the limitation in the growth rate that includes model strains such as *Synechocystis* sp. PCC 6803 (PCC 6803) and *S. elongates* PCC 7942 (PCC 7942). The growth rates of some of them are ca.7 h for PCC6803 [20], compared to 20 min for *Escherichia coli*. In this chapter, we describe many recent developments in the genome engineering of *Cyanobacteria* and their applications in biotechnology. The second problem is the limited number of molecular tools to modify the genome of the Cyanobacterial species.

Table 1. General Features for Growing Genetically Modified *Cyanobacteria*.

S. No.	Desired Features	References
1.	Capable of forming individual colonies on a simple medium.	-
2.	Capable of receiving foreign DNA, either to take native DNA or *via* conjugation or lectroporation.	[28, 29]
3.	Sensitivity to antibiotics for easy selection of transformants.	[30]
4.	Absence of endonucleases that digest foreign DNA. If endonucleases are present, they can be inactivated to improve transformation efficiency.	[31]
5.	Alternatively, specific methylases, restriction inhibitors and liposomes are used during the transfer of DNA.	[32 - 35]

(Table 1) cont.....

S. No.	Desired Features	References
6.	Broad-host-range self-replicating plasmids like RSF1010 can be introduced easily.	-
7.	Existence of homologous recombination (HR) to introduce genetic alterations such as inserting expression cassettes and gene knockouts.	-
8.	Unmarked mutants are important for industrial use and it can be produced by negative selection markers such as sacB and also by CRISPR/Cas.	[11, 36, 37].

There is a rapid progress in the development of modern techniques for *Cyanobacteria*, such as CRISPR/Cas-based tools [16, 17]. It gives an opportunity to design genes with suitable promoters, ribosome binding sites (RBS), coding sequences and terminators [18, 19]. At present, approximately 85 complete genome sequences are available in the form of database such as the CyanoBase database (http://genome.microbedb.jp/cyanobase) [20]. Such databases for *Cyanobacteria* give the opportunity for the production of genome-scale models (GSMs) for a wide range of species, including industrial strains such as *Arthrospira* (*Spirulina platensis* NIES-39) [21, 22]. Metabolic engineering is a process of optimizing the genetic and regulatory functions of the cell to improve the production of various metabolic products [23]. This technique has been successfully applied in *Cyanobacteria* to improve production efficiency [24, 25]. Conventional methods for the improvement of the production of a desired compound are based on random mutations, which take more time to make successful mutations in specific genes [20]. Systems metabolic engineering has given an opportunity to solve the problems associated with random mutagenesis [26]. This systems metabolic engineering involves mathematic models to simulate and predict the output and it is used frequently for the development of a wide range of biomolecules from various microorganisms [27].

Cyanobacteria as a Host for the Heterologous Expression

Cyanobacteria are one of the best hosts for the production of several biomolecules [38]. *Cyanobacteria* are photosynthetic organisms that can fix carbon dioxide to form carbon products which are further converted into valuable compounds [39]. It has become an attractive host for sustainable biotechnological processes, which are in more demand currently [40]. It possesses several advantages over algae and plants, such as easy identification of transformants [41, 42], faster growth and utilizing solar energy for conversion into valuable products [43, 44]. *Cyanobacteria* can be easily cultivated in the absence of arable landmass and potable water [45], and it can also be used to remediate contaminated water, such as removing aromatic hydrocarbons [46, 47]. Cyanobacterial strains have not been much focused to modify to develop for large scale cultivation. *Cyanobacteria* can also synthesize polyhydroxyalkanoates, which are biodegradable plastics [48]. They also produce secondary metabolites like lipopeptides, macrolides, amino acids, pigments, vitamins, fatty acids and amides [42]. *Cyanobacteria* are estimated to produce approximately 1,100 secondary metabolites [49, 50]. Successful efforts are made to produce metabolites such as sugars, alcohols, fatty alcohols, olefins, fatty acids, hydrocarbons and organic acids in large preparations [51, 52].

Cyanobacteria have slow growth compared to other organisms such as *E. coli* and yeast. Some species of *Cyanobacteria* have faster growth that is comparable with industrial yeasts [53, 54]. The availability of genetic tools for modifying the genome is also limited when compared to other heterotrophic model organisms. Another problem with *Cyanobacteria* is polyploid [55], which makes it difficult to get desired mutations [56]. Growing *Cyanobacteria* on a large scale and developing low-cost bioreactors are not well established [52]. Genetic instability is another problem in *Cyanobacteria*, which reduces bioproduction [57], and it is due to repeated DNA motifs that result in homologous recombination [58]. Several efforts have been made in recent years to solve the limitations of *Cyanobacteria*.

Shuttle Vectors Used in *Cyanobacteria*

A shuttle vector is the one which can replicate in two different organisms, and it has ori sites for two different hosts [59]. These vectors will facilitate to study the cloning and expression studies in different hosts. *E. coli* is one among the microorganisms which are well established for multiplication of cloning and expression vectors and successfully transferred into the target organism. Several shuttle vectors are constructed for *Cyanobacteria* and seven vectors are derivatives of RSF1010, which can replicate in *Synechocystis* as a host [60]. These plasmid vectors require antibiotic pressure for their stability in the host [60, 61]. Plasmid pPMQAK1 is a shuttle vector that can replicate in both *E. coli* and *Synechocystis* [62]. This will facilitate to assemble the vector constructs in *E. coli* and finally transfer to *Synechocystis*. The MobA protein encoding gene is present on both the vectors of RSF1010 and PMQAK1, which is necessary for conjugative DNA transfer [63].

Markerless Selection for the Analysis of Transformants in *Cyanobacteria*

The availability of antibiotic resistance markers and the generation of numerous deletions during strain improvement is limited in *Cyanobacteria*. Hence, there is a need to develop markerless selection during genome editing in *Cyanobacteria*. Streptomycin-sensitive rps12 mutation is used for the first time in *S. elongatus* PCC 7942 for markerless selection [64]. This technique involves an expression cassette with two genes, namely kanamycin resistance gene and negative selection marker, rps12, which gives a dominant streptomycin sensitive phenotype. In the second transformation, streptomycin-resistant and kanamycin-sensitive markerless mutants will be produced [65]. The disadvantage of this method requires more time and needs two different suicide vectors.

Alternatively, a new method of counter-selection is developed for *Synechococcus* PCC 7002 and it works based on organic acid toxicity [66]. This method involves two genes, such as the acsA gene and an acetyl-CoA ligase. A defect in the acsA gene is needed in the counter selection method. Another alternative technique was established for *Synechocystis* PCwC 6803, where nickel inducible promoter is used to express toxin gene, namely mazF from *E. coli* [67]. The MazF encodes an endoribonuclease which acts as an inhibitor for protein synthesis in the cell. It acts on the ACA triplet sequence of the Mrna, making into fragments. A simplified version of the markerless gene is established for *Synechocystis* PCC 6803, and it requires a nptI-sacB double selection cassette [68]. This method requires a single vector. The sacB gene is toxic to *Cyanobacteria* when it grows in the presence of sucrose-containing media and becomes resistant to kanamycin. Counter selection by using sacB is not functional in *Synechococcus* PCC 7002 [69]. A few more markerless gene deletion systems have been developed in many Cyanobacterial strains such as *Synechocystis* PCC 6803 and *Synechococcus* PCC 7002 [70]. The principle behind this method is to create knockouts and produce mutants for tolerance against free fatty acids.

Currently, systems without markers involve the CRISPR-based technique that will not require counter selection genes [71]. In principle, CRISPR-Cas clustered regularly interspaced short palindromic repeats (CRISPR)/CRISPR-associated protein (Cas) system is based on the ability to find a nuclease enzyme to a specific site in the genome. This was achieved by single-guide RNA (sgRNA), which is complementary to the target site in the genome of the desired organism. This method is applicable to create single base mutations and knock-out in the desired locations of the genome in several Cyanobacterial species such as *Synechocystis* PCC 6803 [72, 73], *S. elongates* UTEX 2973 [74], *S. elongatus* 7942 [75, 76], and *Nostoc* 7120 [77].

The Role of CRISPR/Cas in Editing the Genome of the *Cyanobacteria*

CRISPR/Cas refers to clustered regularly interspaced short palindromic repeats (CRISPR)/CRISPR-associated protein (Cas) system. This technology is used to specifically alter sites of interest in the genome of a wide range of organisms [78 - 80]. In principle, CRISPR/Cas based tools involve various types of endonucleases, such as types II, V and VI [81, 82]. The endonuclease isolated from *Streptococcus pyogenes* (SpCas9) is a first report and it is used for RNA-guided cleavage of target DNA [83]. This technique has been well applied in many *Cyanobacteria*, including UTEX 2973, PCC 6803, PCC 7942, and the filamentous strain *Nostoc* PCC 7120. In the case of Cas9, it creates a double-stranded break in the DNA with blunt ends (Fig. **1A**), whereas Cas12a produces a double-stranded break with sticky ends (Fig. **1B**) [84]. The desired gene is modified in CRISPR/Cas with the help of single-guide RNA. In contrast to Cas12a, Cas9 utilizes a combination of a crRNA and a tracrRNA. The crRNA is specifically used for identifying a desired site on the genome, and the tracrRNA is used to establish Cas9. In the case of Cas12a enzymes, it possesses higher activity of RNase that provides auto processing of gRNAs, resulting in forming spacer arrays (Fig. **1C**).

These spacer arrays contain gRNAs, and are separated by a direct repeat (DR). Cas12a utilizes DR and produces mature gRNAs. All Cas variants require target DNA in the length of 2-6 nucleotides called a protospacer-adjacent motif (PAM) and produce DSB. The non-homologous end joining (NHEJ) pathway is absent in *Cyanobacteria*, and CRISPR/Cas can effectively create mutation by HR [86 - 89]. Certain Cyanobacterial strains are not transformable by natural methods such as conjugation or electroporation. *Cyanobacteria* such as PCC 6803 subjected to CRISPR/Cas has the following advantages, 1) mutation created in a single event, 2) Multiple sites mutated at the same time, 3) CRISPR/Cas systems used for the strains which are not transformed by natural methods. The total process of inhibition by CRISPRi.DNase inactive Cas enzymes is shown in Fig. (**1D**). These Cas systems are targeted desired site in the DNA by gRNA, which removes RNA polymerase and stops synthesizing mRNA. The total process of transcriptional activation by CRISPRa is shown in Fig. (**1E**). Some of the Cyanobacterial species, which are subjected to CRISPR/Cas are shown in Table **2**.

Table 2. CRISPR/Cas used for editing genes in some of the Cyanobacterial species.

Name of the Species	Cas Variant Used	Type of Expression System	References
Synechocystis PCC 6803	SpCas9	Chromosomal	[88]
S. elongatus UTEX 2973	SpCas9	Episomal	[86]
Nostoc PCC 7120	FnCas12a	Episomal	[89]
S. elongatus PCC 7942	FnCas12a	Episomal	[90]

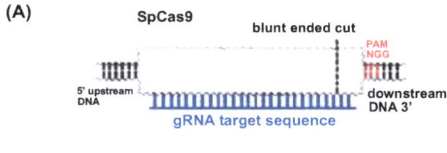

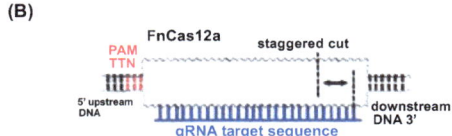

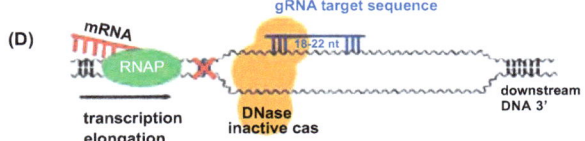

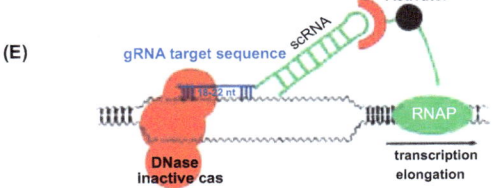

Fig. (1). Basic Mechanism Involved in CRISPR/Cas-based Tools in *Cyanobacteria*. (**A**) CRISPR/Cas system consisting of SpCas9 that binds with a guide RNA(gRNA in green), and it is located in the upstream of an NGG protospacer, a nearest motif (PAM in red). SpCas9 produces a blunt-ended double-stranded break in DNA shown as a black dashed line. (**B**) CRISPR/Cas system consisting of FnCas12a that binds with gRNA. Cas12a produces sticky ends with five nucleotides in the DNA. (**C**) Representation of an expression cassette with three spacers (amber) and direct repeat regions (DR in pink). Cas12a recognizes and cuts the spacers and produces a mature gRNA from each spacer. (**D**) Transcriptional inactivation by CRISPRi. The enzymes which are involved in inactivating DNase are Cas9 and Cas12a, focused on DNA, and it is facilitated by gRNA (18-22 bases). Finally, this complex blocks the RNA polymerase and inhibits transcription. (**E**) Involvement of CRISPRa to activate the transcription. The Cas enzymes are targeted by a gRNA, and it is fused to scRNA. The RNA binding protein (RBP in coral) joins with the transcriptional activator, which results in the activation of RNAP [85].

Applications of CRISPR-Cas9/Cas12a Engineering Tools in *Cyanobacteria*

A wide range of bioproducts like biofuels, pharmaceuticals, and nutrients have been successfully produced from *Cyanobacteria* [91 - 96]. These organisms become major cell factories for metabolic engineering. The problem with *Cyanobacteria* is due to its ploidy. It has multiple copies of 3-218 [100]. In case of *Synechocystis* PCC 6803, it has 12 copies, and it has 3 copies in *Synechococcus* PCC 7942. If conventional methods of mutagenesis are used, it takes more time and performs several experiments to get completely segregated mutant lines. Hence, there is a need to go for single-step gene-editing techniques in *Cyanobacteria* such as CRISPR-Cas systems [97 - 100]. CRISPR-Cas9 is successfully applied to bring specific mutations in *Synechococcus* species *such as S. elongatus* UTEX 2973 [97], *S. elongatus* PCC 7942 [100], and *Synechocystis* sp. PCC 6803 [92]. To avoid the toxic effect produced by the higher expression of Cas9, a reduced level of expression is adopted to make success in gene editing. The succinate production is enhanced by the Cas9 variant, which creates knock-in and knock-out to produce succinate 11-fold higher compared to normal [100]. Endogenous plasmids cure *Synechocystis* sp. PCC 6803 and develop shuttle vectors with *E. coli* for heterologous expression of foreign genes [99]. Due to the less toxic effect of Cas12a, it is applied in various hosts like *Synechocystis* sp. PCC 6803, *S. elongatus* 7942, and *Anabaena* sp. PCC 7120, and it is effective 90% in *Synechococcus*, 63% in *Anabaena* and 44% in *Synechocystis*. Some of the successful achievements made in *Cyanobacteria* by using Cas9 and 12a systems are shown in Table 3.

Table 3. Applications of CRISPR-Cas9/Cas12a in *Cyanobacteria*.

Cyanobacterial Species	Cas Protein Used	Gene(s) for Editing	Achievements	References
S.elongatus UTEX 2973	*Streptomyces* codon-optimized Cas9	*nblA*	First markerless knockout	[101]
Synechocystis sp. PCC 6803, *Anabaena* sp. PCC 7120	*Francisella novicida* Cas12a (Cpf1)	*nblA* or *nifH* k	Direct gene replacement	[102]
S. elongatus PCC 7942	*S. pyogenes* dCas9	CRISPRi repression *eyfp, glgC, sdhA,* and *sdhB*	Succinate production	[103]
Synechocystis sp. PCC 6803	*S. pyogenes* dCas9	CRISPRi repression Ado, Aar, PlsX (Slr1510), Slr2060, and PlsC	Gene repression for fatty acid metabolism	[104]

CONCLUDING REMARKS

Cyanobacteria have become an excellent host for biotechnological production. These organisms have both negative and positive features, and several efforts made by the researchers lead to produce high-value bioproducts from *Cyanobacteria*. Markerless selection and development of shuttle vectors have stimulated the process of gene editing. The conventional strategies of mutagenesis have problems with several copies of ploidy in *Cyanobacteria*, which have been overcome by CRISPR/Cas based technologies for precise genetic manipulations in a single step. Some of the bioproducts produced by using CRISPR/Cas technologies are succinate and free fatty acid. In the future, many more genes of *Cyanobacteria* will be edited to produce commercially valuable products in biotechnology.

REFERENCES

[1] Seckbach J, Ed. Algae and Cyanobacteria in extreme environments. Springer Science & Business Media 2007.
[http://dx.doi.org/10.1007/978-1-4020-6112-7]

[2] Puente-Sánchez F, Arce-Rodríguez A, Oggerin M, *et al.* Viable *Cyanobacteria* in the deep continental subsurface. Proc Natl Acad Sci 2018; 115(42): 10702-7.
[http://dx.doi.org/10.1073/pnas.1808176115] [PMID: 30275328]

[3] Schirrmeister BE, Gugger M, Donoghue PCJ. *Cyanobacteria* and the great oxidation event: Evidence from genes and fossils. Palaeontology 2015; 58(5): 769-85.
[http://dx.doi.org/10.1111/pala.12178] [PMID: 26924853]

[4] Knoll AH. Paleobiological perspectives on early microbial evolution. Cold Spring Harb Perspect Biol 2015; 7(7): a018093.
[http://dx.doi.org/10.1101/cshperspect.a018093] [PMID: 26134315]

[5] Nutman AP, Bennett VC, Friend CRL, Van Kranendonk MJ, Chivas AR. Rapid emergence of life shown by discovery of 3,700-million-year-old microbial structures. Nature 2016; 537(7621): 535-8.
[http://dx.doi.org/10.1038/nature19355] [PMID: 27580034]

[6] Pisciotta JM, Zou Y, Baskakov IV. Light-dependent electrogenic activity of *Cyanobacteria*. PLoS One 2010; 5(5): e10821.
[http://dx.doi.org/10.1371/journal.pone.0010821] [PMID: 20520829]

[7] Lea-Smith DJ, Howe C. The use of Cyanobacteria for biofuel production.Biofuels and Bioenergy. Chichester, UK: JohnWiley & Sons, Ltd. 2017; pp. 143-55.
[http://dx.doi.org/10.1002/9781118350553.ch9]

[8] Dismukes GC, Carrieri D, Bennette N, Ananyev GM, Posewitz MC. Aquatic phototrophs: Efficient alternatives to land-based crops for biofuels. Curr Opin Biotechnol 2008; 19(3): 235-40.
[http://dx.doi.org/10.1016/j.copbio.2008.05.007] [PMID: 18539450]

[9] Long BM, Hee WY, Sharwood RE, *et al.* Carboxysome encapsulation of the CO_2-fixing enzyme Rubisco in tobacco chloroplasts. Nat Commun 2018; 9(1): 3570.
[http://dx.doi.org/10.1038/s41467-018-06044-0] [PMID: 30177711]

[10] Lau NS, Matsui M, Abdullah AAA. *Cyanobacteria*: Photoautotrophic microbial factories for the sustainable synthesis of industrial products. BioMed Res Int 2015; 2015: 1-9.
[http://dx.doi.org/10.1155/2015/754934] [PMID: 26199945]

[11] Lea-Smith DJ, Vasudevan R, Howe CJ. Generation of marked and markerless mutants in model Cyanobacterial species. J Vis Exp 2016; 29(111): e54001.
[PMID: 27286310]

[12] Keeling PJ. Diversity and evolutionary history of plastids and their hosts. Am J Bot 2004; 91(10): 1481-93.
[http://dx.doi.org/10.3732/ajb.91.10.1481] [PMID: 21652304]

[13] Nielsen AZ, Mellor SB, Vavitsas K, *et al.* Extending the biosynthetic repertoires of *Cyanobacteria* and chloroplasts. Plant J 2016; 87(1): 87-102.
[http://dx.doi.org/10.1111/tpj.13173] [PMID: 27005523]

[14] Lin PC, Pakrasi HB. Engineering *Cyanobacteria* for production of terpenoids. Planta 2019; 249(1): 145-54.
[http://dx.doi.org/10.1007/s00425-018-3047-y] [PMID: 30465115]

[15] Yu J, Liberton M, Cliften PF, *et al. Synechococcus elongatus* UTEX 2973, a fast growing Cyanobacterial chassis for biosynthesis using light and CO_2. Sci Rep 2015; 5(1): 8132.
[http://dx.doi.org/10.1038/srep08132] [PMID: 25633131]

[16] Behler J, Vijay D, Hess WR, Akhtar MK. CRISPR-based technologies for metabolic engineering in *Cyanobacteria*. Trends Biotechnol 2018; 36(10): 996-1010.
[http://dx.doi.org/10.1016/j.tibtech.2018.05.011] [PMID: 29937051]

[17] Santos-Merino M, Singh AK, Ducat DC. New applications of synthetic biology tools for Cyanobacterial metabolic engineering. Front Bioeng Biotechnol 2019; 7: 33.
[http://dx.doi.org/10.3389/fbioe.2019.00033] [PMID: 30873404]

[18] Ferreira EA, Pacheco CC, Pinto F, *et al.* Expanding the toolbox for *Synechocystis* sp. PCC 6803: Validation of replicative vectors and characterization of a novel set of promoters. Synth Biol 2018; 3(1): ysy014.
[http://dx.doi.org/10.1093/synbio/ysy014] [PMID: 32995522]

[19] Li S, Sun T, Xu C, Chen L, Zhang W. Development and optimization of genetic toolboxes for a fast-growing cyanobacterium *Synechococcus elongatus* UTEX 2973. Metab Eng 2018; 48: 163-74.
[http://dx.doi.org/10.1016/j.ymben.2018.06.002] [PMID: 29883802]

[20] Fujisawa T, Narikawa R, Maeda S, *et al.* CyanoBase: A large-scale update on its 20th anniversary. Nucleic Acids Res 2017; 45(D1): D551-4.
[http://dx.doi.org/10.1093/nar/gkw1131] [PMID: 27899668]

[21] Yoshikawa K, Aikawa S, Kojima Y, *et al.* Construction of a genome-scale metabolic model of *Arthrospira platensis* NIES-39 and metabolic design for Cyanobacterial bioproduction. PLoS One 2015; 10(12): e0144430.
[http://dx.doi.org/10.1371/journal.pone.0144430] [PMID: 26640947]

[22] Gopalakrishnan S, Pakrasi HB, Maranas CD. Elucidation of photoautotrophic carbon flux topology in *Synechocystis* PCC 6803 using genome-scale carbon mapping models. Metab Eng 2018; 47: 190-9.
[http://dx.doi.org/10.1016/j.ymben.2018.03.008] [PMID: 29526818]

[23] Kumar RR, Prasad S. Metabolic engineering of bacteria. Indian J Microbiol 2011; 51(3): 403-9.
[http://dx.doi.org/10.1007/s12088-011-0172-8] [PMID: 22754024]

[24] Angermayr SA, Gorchs Rovira A, Hellingwerf KJ. Metabolic engineering of *Cyanobacteria* for the synthesis of commodity products. Trends Biotechnol 2015; 33(6): 352-61.
[http://dx.doi.org/10.1016/j.tibtech.2015.03.009] [PMID: 25908503]

[25] Carroll AL, Case AE, Zhang A, Atsumi S. Metabolic engineering tools in model *Cyanobacteria*. Metab Eng 2018; 50: 47-56.
[http://dx.doi.org/10.1016/j.ymben.2018.03.014] [PMID: 29588234]

[26] Nogales J, Gudmundsson S, Thiele I. Toward systems metabolic engineering in *Cyanobacteria*.

Bioengineered 2013; 4(3): 158-63.
[http://dx.doi.org/10.4161/bioe.22792] [PMID: 23138691]

[27] Lee JW, Kim TY, Jang YS, Choi S, Lee SY. Systems metabolic engineering for chemicals and materials. Trends Biotechnol 2011; 29(8): 370-8.
[http://dx.doi.org/10.1016/j.tibtech.2011.04.001] [PMID: 21561673]

[28] Li S, Sun T, Xu C, Chen L, Zhang W. Development and optimization of genetic toolboxes for a fast-growing cyanobacterium *Synechococcus elongatus* UTEX 2973. Metab Eng 2018; 48: 163-74.
[http://dx.doi.org/10.1016/j.ymben.2018.06.002] [PMID: 29883802]

[29] Wendt KE, Pakrasi HB. Genomics approaches to deciphering natural transformation in *Cyanobacteria*. Front Microbiol 2019; 10: 1259.
[http://dx.doi.org/10.3389/fmicb.2019.01259] [PMID: 31231343]

[30] Taton A, Unglaub F, Wright NE, *et al*. Broad-host-range vector system for synthetic biology and biotechnology in *Cyanobacteria*. Nucleic Acids Res 2014; 42(17): e136.
[http://dx.doi.org/10.1093/nar/gku673] [PMID: 25074377]

[31] Elhai J, Wolk CP. Conjugal transfer of DNA to Cyanobacteria.Methods in enzymology. Academic Press 1988; 167: pp. 747-54.
[http://dx.doi.org/10.1016/0076-6879(88)67086-8]

[32] Masukawa H, Inoue K, Sakurai H, Wolk CP, Hausinger RP. Site-directed mutagenesis of the *Anabaena* sp. strain PCC 7120 nitrogenase active site to increase photobiological hydrogen production. Appl Environ Microbiol 2010; 76(20): 6741-50.
[http://dx.doi.org/10.1128/AEM.01056-10] [PMID: 20709836]

[33] Mandakovic D, Trigo C, Andrade D, *et al*. A conserved and novel filamentous Cyanobacterial cell division protein involved in septum localization. Front Microbiol 2016; 7: 94.
[http://dx.doi.org/10.3389/fmicb.2016.00094] [PMID: 26903973]

[34] Jeamton W, Dulsawat S, Tanticharoen M, Vonshak A, Cheevadhanarak S. Overcoming intrinsic restriction enzyme barriers enhances transformation efficiency in *Arthrospira platensis* C1. Plant Cell Physiol 2017; 58(4): 822-30.
[http://dx.doi.org/10.1093/pcp/pcx016] [PMID: 28158667]

[35] Lea-Smith DJ, Vasudevan R, Howe CJ. Generation of marked and markerless mutants in model Cyanobacterial species. J Vis Exp 2016; (111): e54001.
[PMID: 27286310]

[36] Ried JL, Collmer A. An nptI-sacB-sacR cartridge for constructing directed, unmarked mutations in Gram-negative bacteria by marker exchange-eviction mutagenesis. Gene 1987; 57(2-3): 239-46.
[http://dx.doi.org/10.1016/0378-1119(87)90127-2] [PMID: 3319780]

[37] Behler J, Vijay D, Hess WR, Akhtar MK. CRISPR-based technologies for metabolic engineering in *Cyanobacteria*. Trends Biotechnol 2018; 36(10): 996-1010.
[http://dx.doi.org/10.1016/j.tibtech.2018.05.011] [PMID: 29937051]

[38] Knoot CJ, Ungerer J, Wangikar PP, Pakrasi HB. *Cyanobacteria*: Promising biocatalysts for sustainable chemical production. J Biol Chem 2018; 293(14): 5044-52.
[http://dx.doi.org/10.1074/jbc.R117.815886] [PMID: 28972147]

[39] Lau NS, Matsui M, Abdullah AAA. *Cyanobacteria*: Photoautotrophic microbial factories for the sustainable synthesis of industrial products. BioMed Res Int 2015; 2015: 1-9.
[http://dx.doi.org/10.1155/2015/754934] [PMID: 26199945]

[40] Ruffing AM. Engineered *Cyanobacteria*: Teaching an old bug new tricks. Bioeng Bugs 2011; 2(3): 136-49.
[http://dx.doi.org/10.4161/bbug.2.3.15285] [PMID: 21637004]

[41] Parmar A, Singh NK, Pandey A, Gnansounou E, Madamwar D. *Cyanobacteria* and microalgae: A positive prospect for biofuels. Bioresour Technol 2011; 102(22): 10163-72.

[http://dx.doi.org/10.1016/j.biortech.2011.08.030] [PMID: 21924898]

[42] Lau NS, Matsui M, Abdullah AAA. *Cyanobacteria*: Photoautotrophic microbial factories for the sustainable synthesis of industrial products. BioMed Res Int 2015; 2015: 1-9.
[http://dx.doi.org/10.1155/2015/754934] [PMID: 26199945]

[43] Dismukes GC, Carrieri D, Bennette N, Ananyev GM, Posewitz MC. Aquatic phototrophs: Efficient alternatives to land-based crops for biofuels. Curr Opin Biotechnol 2008; 19(3): 235-40.
[http://dx.doi.org/10.1016/j.copbio.2008.05.007] [PMID: 18539450]

[44] Allen DT. Gaseous carbon waste streams utilization status and research needs national academy of sciences. Washington, DC, United States: National Academies of Sciences, Engineering, and Medicine 2018.
[http://dx.doi.org/10.2172/1733359]

[45] Nozzi NE, Oliver JWK, Atsumi S. *Cyanobacteria* as a platform for biofuel production. Front Bioeng Biotechnol 2013; 1: 7.
[http://dx.doi.org/10.3389/fbioe.2013.00007] [PMID: 25022311]

[46] Megharaj M, Venkateswarlu K, Rao AS. Metabolism of monocrotophos and quinalphos by algae isolated from soil. Bull Environ Contam Toxicol 1987; 39(2): 251-6.
[http://dx.doi.org/10.1007/BF01689414] [PMID: 3663978]

[47] Kuritz T, Wolk CP. Use of filamentous *Cyanobacteria* for biodegradation of organic pollutants. Appl Environ Microbiol 1995; 61(1): 234-8.
[http://dx.doi.org/10.1128/aem.61.1.234-238.1995] [PMID: 7534052]

[48] Quintana N, Van der Kooy F, Van de Rhee MD, Voshol GP, Verpoorte R. Renewable energy from *Cyanobacteria*: Energy production optimization by metabolic pathway engineering. Appl Microbiol Biotechnol 2011; 91(3): 471-90.
[http://dx.doi.org/10.1007/s00253-011-3394-0] [PMID: 21691792]

[49] Dittmann E, Gugger M, Sivonen K, Fewer DP. Natural product biosynthetic diversity and comparative genomics of the *Cyanobacteria*. Trends Microbiol 2015; 23(10): 642-52.
[http://dx.doi.org/10.1016/j.tim.2015.07.008] [PMID: 26433696]

[50] Salvador-Reyes LA, Luesch H. Biological targets and mechanisms of action of natural products from marine *Cyanobacteria*. Nat Prod Rep 2015; 32(3): 478-503.
[http://dx.doi.org/10.1039/C4NP00104D] [PMID: 25571978]

[51] Lai M, Lan E. Advances in metabolic engineering of *Cyanobacteria* for photosynthetic biochemical production. Metabolites 2015; 5(4): 636-58.
[http://dx.doi.org/10.3390/metabo5040636] [PMID: 26516923]

[52] Knoot CJ, Ungerer J, Wangikar PP, Pakrasi HB. *Cyanobacteria*: Promising biocatalysts for sustainable chemical production. J Biol Chem 2018; 293(14): 5044-52.
[http://dx.doi.org/10.1074/jbc.R117.815886] [PMID: 28972147]

[53] Yu J, Liberton M, Cliften PF, *et al.* Synechococcus elongatus UTEX 2973, a fast growing Cyanobacterial chassis for biosynthesis using light and CO_2. Sci Rep 2015; 5(1): 8132.
[http://dx.doi.org/10.1038/srep08132] [PMID: 25633131]

[54] Jaiswal D, Sengupta A, Sohoni S, *et al.* Genome features and biochemical characteristics of a robust, fast growing and naturally transformable cyanobacterium *Synechococcus elongatus* PCC 11801 isolated from India. Sci Rep 2018; 8(1): 16632.
[http://dx.doi.org/10.1038/s41598-018-34872-z] [PMID: 30413737]

[55] Griese M, Lange C, Soppa J. Ploidy in *Cyanobacteria*. FEMS Microbiol Lett 2011; 323(2): 124-31.
[http://dx.doi.org/10.1111/j.1574-6968.2011.02368.x] [PMID: 22092711]

[56] Kelly CL, Taylor GM, Hitchcock A, Torres-Méndez A, Heap JT. A rhamnose-inducible system for precise and temporal control of gene expression in *Cyanobacteria*. ACS Synth Biol 2018; 7(4): 1056-66.

[http://dx.doi.org/10.1021/acssynbio.7b00435] [PMID: 29544054]

[57] Jones PR. Genetic instability in *Cyanobacteria* : An elephant in the room? Front Bioeng Biotechnol 2014; 2: 12.
[http://dx.doi.org/10.3389/fbioe.2014.00012] [PMID: 25152885]

[58] Cassier-Chauvat C, Veaudor T, Chauvat F. Comparative genomics of DNA recombination and repair in *Cyanobacteria*: Biotechnological implications. Front Microbiol 2016; 7: 1809.
[http://dx.doi.org/10.3389/fmicb.2016.01809] [PMID: 27881980]

[59] Clark DP, Pazdernik N. Biotechnology. Newnes 2015.

[60] Wang B, Wang J, Zhang W, Meldrum DR. Application of synthetic biology in *Cyanobacteria* and algae. Front Microbiol 2012; 3: 344.
[http://dx.doi.org/10.3389/fmicb.2012.00344] [PMID: 23049529]

[61] Yu Y, You L, Liu D, Hollinshead W, Tang Y, Zhang F. Development of *Synechocystis* sp. PCC 6803 as a phototrophic cell factory. Mar Drugs 2013; 11(8): 2894-916.
[http://dx.doi.org/10.3390/md11082894] [PMID: 23945601]

[62] Huang HH, Camsund D, Lindblad P, Heidorn T. Design and characterization of molecular tools for a Synthetic Biology approach towards developing Cyanobacterial biotechnology. Nucleic Acids Res 2010; 38(8): 2577-93.
[http://dx.doi.org/10.1093/nar/gkq164] [PMID: 20236988]

[63] Frey J, Bagdasarian MM, Bagdasarian M. Replication and copy number control of the broad-host--range plasmid RSF1010. Gene 1992; 113(1): 101-6.
[http://dx.doi.org/10.1016/0378-1119(92)90675-F] [PMID: 1563624]

[64] Takahama K, Matsuoka M, Nagahama K, Ogawa T. High-frequency gene replacement in *Cyanobacteria* using a heterologous rps12 gene. Plant Cell Physiol 2004; 45(3): 333-9.
[http://dx.doi.org/10.1093/pcp/pch041] [PMID: 15047882]

[65] Takahama K, Matsuoka M, Nagahama K, Ogawa T. High-frequency gene replacement in *Cyanobacteria* using a heterologous rps12 gene. Plant Cell Physiol 2004; 45(3): 333-9.
[http://dx.doi.org/10.1093/pcp/pch041] [PMID: 15047882]

[66] Begemann MB, Zess EK, Walters EM, Schmitt EF, Markley AL, Pfleger BF. An organic acid based counter selection system for *Cyanobacteria*. PLoS One 2013; 8(10): e76594.
[http://dx.doi.org/10.1371/journal.pone.0076594] [PMID: 24098537]

[67] Cheah YE, Albers SC, Peebles CAM. A novel counter-selection method for markerless genetic modification in *Synechocystis* sp. PCC 6803. Biotechnol Prog 2013; 29(1): 23-30.
[http://dx.doi.org/10.1002/btpr.1661] [PMID: 23124993]

[68] Viola S, Rühle T, Leister D. A single vector-based strategy for marker-less gene replacement in *Synechocystis* sp. PCC 6803. Microb Cell Fact 2014; 13(1): 4.
[http://dx.doi.org/10.1186/1475-2859-13-4] [PMID: 24401024]

[69] Zhang W, Song X, Eds. Synthetic Biology of Cyanobacteria. Singapore 2018.
[http://dx.doi.org/10.1007/978-981-13-0854-3]

[70] Kojima K, Keta S, Uesaka K, *et al.* A simple method for isolation and construction of markerless Cyanobacterial mutants defective in acyl-acyl carrier protein synthetase. Appl Microbiol Biotechnol 2016; 100(23): 10107-13.
[http://dx.doi.org/10.1007/s00253-016-7850-8] [PMID: 27704180]

[71] Behler J, Vijay D, Hess WR, Akhtar MK. CRISPR-based technologies for metabolic engineering in *Cyanobacteria*. Trends Biotechnol 2018; 36(10): 996-1010.
[http://dx.doi.org/10.1016/j.tibtech.2018.05.011] [PMID: 29937051]

[72] Xiao Y, Wang S, Rommelfanger S, *et al.* Developing a Cas9–based tool to engineer native plasmids in *Synechocystis* sp. PCC 6803. Biotechnol Bioeng 2018; 115(9): 2305-14.

[http://dx.doi.org/10.1002/bit.26747] [PMID: 29896914]

[73] Ungerer J, Pakrasi HB. Cpf1 is a versatile tool for CRISPR genome editing across diverse species of Cyanobacteria. Sci Rep 2016; 6(1): 39681.
[http://dx.doi.org/10.1038/srep39681] [PMID: 28000776]

[74] Wendt KE, Ungerer J, Cobb RE, Zhao H, Pakrasi HB. CRISPR/Cas9 mediated targeted mutagenesis of the fast growing cyanobacterium *Synechococcus elongatus* UTEX 2973. Microb Cell Fact 2016; 15(1): 115.
[http://dx.doi.org/10.1186/s12934-016-0514-7] [PMID: 27339038]

[75] Li H, Shen CR, Huang CH, Sung LY, Wu MY, Hu YC. CRISPR-Cas9 for the genome engineering of *Cyanobacteria* and succinate production. Metab Eng 2016; 38: 293-302.
[http://dx.doi.org/10.1016/j.ymben.2016.09.006] [PMID: 27693320]

[76] Ungerer J, Wendt KE, Hendry JI, Maranas CD, Pakrasi HB. Comparative genomics reveals the molecular determinants of rapid growth of the cyanobacterium *Synechococcus elongatus* UTEX 2973. Proc Natl Acad SciA 2018; 115(50): E11761-70.
[http://dx.doi.org/10.1073/pnas.1814912115] [PMID: 30409802]

[77] Niu TC, Lin GM, Xie LR, *et al.* Expanding the potential of CRISPR-Cpf1 based genome editing technology in the cyanobacterium *Anabaena* PCC 7120. ACS Synth Biol 2019; 8(1): 170-80.
[http://dx.doi.org/10.1021/acssynbio.8b00437] [PMID: 30525474]

[78] Naduthodi MIS, Barbosa MJ, Van der Oost J. Progress of CRISPR–Cas based genome editing in photosynthetic microbes. Biotechnol J 2018; 13(9): 1700591.
[http://dx.doi.org/10.1002/biot.201700591] [PMID: 29396999]

[79] Khumsupan P, Donovan S, McCormick AJ. CRISPR/Cas in Arabidopsis: Overcoming challenges to accelerate improvements in crop photosynthetic efficiencies. Physiol Plant 2019; 166(1): 428-37.
[http://dx.doi.org/10.1111/ppl.12937] [PMID: 30706492]

[80] Zhang YT, Jiang JY, Shi TQ, *et al.* Application of the CRISPR/Cas system for genome editing in microalgae. Appl Microbiol Biotechnol 2019; 103(8): 3239-48.
[http://dx.doi.org/10.1007/s00253-019-09726-x] [PMID: 30877356]

[81] Makarova KS, Wolf YI, Alkhnbashi OS, *et al.* An updated evolutionary classification of CRISPR–Cas systems. Nat Rev Microbiol 2015; 13(11): 722-36.
[http://dx.doi.org/10.1038/nrmicro3569] [PMID: 26411297]

[82] Koonin EV, Makarova KS. Origins and evolution of CRISPR-Cas systems. Philosophical Transactions of the Royal Society B 2019; 374(1772): 20180087.

[83] Xiao Y, Wang S, Rommelfanger S, *et al.* Developing a Cas9–based tool to engineer native plasmids in *Synechocystis* sp. PCC 6803. Biotechnol Bioeng 2018; 115(9): 2305-14.
[http://dx.doi.org/10.1002/bit.26747] [PMID: 29896914]

[84] Niu TC, Lin GM, Xie LR, *et al.* Expanding the potential of CRISPR-Cpf1-based genome editing technology in the cyanobacterium *Anabaena* PCC 7120. ACS Synth Biol 2019; 8(1): 170-80.
[http://dx.doi.org/10.1021/acssynbio.8b00437] [PMID: 30525474]

[85] Dong C, Fontana J, Patel A, Carothers JM, Zalatan JG. Synthetic CRISPR-Cas gene activators for transcriptional reprogramming in bacteria. Nat Commun 2018; 9(1): 1-11.
[PMID: 29317637]

[86] Ungerer J, Wendt KE, Hendry JI, Maranas CD, Pakrasi HB. Comparative genomics reveals the molecular determinants of rapid growth of the cyanobacterium *Synechococcus elongatus* UTEX 2973. Proc Natl Acad Sci 2018; 115(50): E11761-70.
[http://dx.doi.org/10.1073/pnas.1814912115] [PMID: 30409802]

[87] Huang HH, Camsund D, Lindblad P, Heidorn T. Design and characterization of molecular tools for a synthetic biology approach towards developing Cyanobacterial biotechnology. Nucleic Acids Res 2010; 38(8): 2577-93.

[http://dx.doi.org/10.1093/nar/gkq164] [PMID: 20236988]

[88] Ramey CJ, Barón-Sola Á, Aucoin HR, Boyle NR. Genome engineering in *Cyanobacteria*: Where we are and where we need to go. ACS Synth Biol 2015; 4(11): 1186-96.
[http://dx.doi.org/10.1021/acssynbio.5b00043] [PMID: 25985322]

[89] Taton A, Unglaub F, Wright NE, *et al.* Broad-host-range vector system for synthetic biology and biotechnology in *Cyanobacteria*. Nucleic Acids Res 2014; 42(17): e136.
[http://dx.doi.org/10.1093/nar/gku673] [PMID: 25074377]

[90] Song X, Wang Y, Diao J, Li S, Chen L, Zhang W. Direct photosynthetic production of plastic building block chemicals from CO_2. Adv Exp Med Biol 2018; 1080: 215-38.
[http://dx.doi.org/10.1007/978-981-13-0854-3_9] [PMID: 30091097]

[91] Li S, Sun T, Xu C, Chen L, Zhang W. Development and optimization of genetic toolboxes for a fast-growing cyanobacterium *Synechococcus elongatus* UTEX 2973. Metab Eng 2018; 48: 163-74.
[http://dx.doi.org/10.1016/j.ymben.2018.06.002] [PMID: 29883802]

[92] Zhou J, Meng H, Zhang W, Li Y. Production of industrial chemicals from co_2 by engineering *Cyanobacteria*. Adv Exp Med Biol 2018; 1080: 97-116.
[http://dx.doi.org/10.1007/978-981-13-0854-3_5] [PMID: 30091093]

[93] Zhou J, Zhang F, Meng H, Zhang Y, Li Y. Introducing extra NADPH consumption ability significantly increases the photosynthetic efficiency and biomass production of *Cyanobacteria*. Metab Eng 2016; 38: 217-27.
[http://dx.doi.org/10.1016/j.ymben.2016.08.002] [PMID: 27497972]

[94] Zhou J, Zhu T, Cai Z, Li Y. From cyanochemicals to cyanofactories: A review and perspective. Microb Cell Fact 2016; 15(1): 2.
[http://dx.doi.org/10.1186/s12934-015-0405-3] [PMID: 26743222]

[95] Ni J, Tao F, Xu P, Yang C. Engineering *Cyanobacteria* for photosynthetic production of C3 platform chemicals and terpenoids from CO_2. Synthetic biology of Cyanobacteria 2018; 1080: 239-59.

[96] Griese M, Lange C, Soppa J. Ploidy in *Cyanobacteria*. FEMS Microbiol Lett 2011; 323(2): 124-31.
[http://dx.doi.org/10.1111/j.1574-6968.2011.02368.x] [PMID: 22092711]

[97] Ungerer J, Pakrasi HB. Cpf1 is a versatile tool for CRISPR genome editing across diverse species of *Cyanobacteria*. Sci Rep 2016; 6(1): 39681.
[http://dx.doi.org/10.1038/srep39681] [PMID: 28000776]

[98] Wendt KE, Ungerer J, Cobb RE, Zhao H, Pakrasi HB. CRISPR/Cas9 mediated targeted mutagenesis of the fast growing cyanobacterium *Synechococcus elongatus* UTEX 2973. Microb Cell Fact 2016; 15(1): 115.
[http://dx.doi.org/10.1186/s12934-016-0514-7] [PMID: 27339038]

[99] Xiao Y, Wang S, Rommelfanger S, *et al.* Developing a Cas9–based tool to engineer native plasmids in *Synechocystis* sp. PCC 6803. Biotechnol Bioeng 2018; 115(9): 2305-14.
[http://dx.doi.org/10.1002/bit.26747] [PMID: 29896914]

[100] Li H, Shen CR, Huang CH, Sung LY, Wu MY, Hu YC. CRISPR-Cas9 for the genome engineering of *Cyanobacteria* and succinate production. Metab Eng 2016; 38: 293-302.
[http://dx.doi.org/10.1016/j.ymben.2016.09.006] [PMID: 27693320]

[101] Wendt KE, Ungerer J, Cobb RE, Zhao H, Pakrasi HB. CRISPR/Cas9 mediated targeted mutagenesis of the fast growing cyanobacterium *Synechococcus elongatus* UTEX 2973. Microb Cell Fact 2016; 15(1): 115.
[http://dx.doi.org/10.1186/s12934-016-0514-7] [PMID: 27339038]

[102] Ungerer J, Pakrasi HB. Cpf1 is a versatile tool for CRISPR genome editing across diverse species of *Cyanobacteria*. Sci Rep 2016; 6(1): 39681.
[http://dx.doi.org/10.1038/srep39681] [PMID: 28000776]

[103] Huang CH, Shen CR, Li H, Sung LY, Wu MY, Hu YC. CRISPR interference (CRISPRi) for gene regulation and succinate production in cyanobacterium S. *elongatus* PCC 7942. Microb Cell Fact 2016; 15(1): 196.
[http://dx.doi.org/10.1186/s12934-016-0595-3] [PMID: 27846887]

[104] Kaczmarzyk D, Cengic I, Yao L, Hudson EP. Diversion of the long-chain acyl-ACP pool in synechocystis to fatty alcohols through CRISPRi repression of the essential phosphate acyltransferase PlsX. Metab Eng 2018; 45: 59-66.
[http://dx.doi.org/10.1016/j.ymben.2017.11.014] [PMID: 29199103]

CHAPTER 10

Genome Editing in *Streptomyces*

Johns Saji[1], **Jibin James**[1], **Ramesh Kumar Saini**[2] **and Shibin Mohanan**[1,*]

[1] *Department of Botany, Nirmala College, Muvattupuzha, Ernakulam, Kerala, India*

[2] *Department of Crop Science, Konkuk University, Seoul, Korea*

Abstract: *Streptomyces* are Gram-positive, filamentous bacteria belonging to the group actinomycetes. This bacterium is important to the modern industrial world because of the presence of 20-50 biosynthetic gene clusters (BGCs). BGCs contain the genes for the production of industrially important natural products (NP), which includes antibiotics, anti-tumor drugs, anti-depressants, *etc.*, naturally originated from this microorganism. Strain improvement is required to enhance the production of these NP in *Streptomyces*. Different methods have been used to enhance NP production and strain improvement. In this chapter, we will be discussing strain improvement of *Streptomyces* species by different genome editing tools. The information, which is put together, includes the basic techniques used for genome editing to the most advanced CRISPR/Cas system associated genome editing in *Streptomyces* (PCR targeting system, Cre-loxP recombination system, I *SceI* meganuclease promoted recombination system and CRISPR/Cas system). The authors have discussed about multiplex automated genome editing (MAGE) tool associated with CRISPR/Cas system.

Keywords: Bacterial gene clusters, CRISPR/Cas, Genetic manipulations, Metabolic engineering, Recombination system, *Streptomyces*.

INTRODUCTION

The genus *Streptomyces* is a Gram-positive bacterium that resembles filamentous fungi and grows in various environmental conditions. Streptomyces are differentiated from other actinomycetes by its filamentous growth and formation of spores in chains. Environmental stress, like nutrient limitation, leading to shifting of *Streptomyces* from the mycelial vegetative phase to the reproductive sporulation phase [1]. *Streptomyces* have linear chromosomes, approximately 8 to 10Mb depending on the species, with high GC content and several linear and circular plasmids. The presence of biosynthetic gene clusters that encodes for enzymes contributes towards the secondary metabolite production having varying

---

* **Corresponding author Shibin Mohanan:** Department of Botany, Nirmala College, Muvattupuzha, Ernakulam, Kerala, India; E-mail: shibin@nirmalacollege.ac.in

Prakash M. Halami & Aravind Sundararaman (Eds.)
All rights reserved-© 2024 Bentham Science Publishers

chemotypes like polyketides, lactams, non-ribosomal peptides, terpenes, *etc.* [2]. These secondary metabolites produced have a wide range of applications like antibiotics, *e.g.,* pristinamycin [3] and daptomycin [4], immune-suppressants (*e.g.,* rapamycin) [5] and FK506 [6], insecticides, *e.g.,* avermectin [7] and milbemycin [8] and anti-tumour drugs daunorubicin [9] and bleomycin [10], which are widely used in agriculture and veterinary/human medicine.

The prolific antibiotic production capability and their significant role in clinical drug production have been exploited. The discovery of natural product (NP) drugs from these now highly exploited bacteria was seriously impaired by conventional screening techniques of synthetic libraries and the low efficiency of conventional top-down screening strategies [2]. The advancement in the next-generation genome sequencing technology and the use of bioinformatics resources to study microbial genomes has lead to a huge leap in unravelling biosynthetic gene clusters for natural products [11 - 13]. Thus has the ability to tap into the possibility of a wide variety of natural products and may lead to drug discovery from the majority of the uncultured microorganisms [14, 15].

Being a significant and most gifted microorganism, *Streptomyces* possesses 20 to 50 biosynthetic gene clusters (BGC) in a single genome [16, 17]. *Streptomyces,* when compared with other common model organisms, like *S. cerevisiae* and *E. coli,* shows poor genetic manipulations and are mostly recalcitrant to genome editing. The natural product (NP) biosynthetic gene clusters are unexplored rich reservoirs for natural compounds, and a majority of these BGCs are either not expressed or poorly expressed hence referred to as silent BGCs [18]. Of late numerous strategies have been developed to activate these BGCs to trigger NP overproduction. These strategies can be grouped into two major groups: (i) induction of BGCs in the native host using gene manipulations (ii) cloning of bacterial gene clusters and subsequent transfer to a surrogate *Streptomyces* host for heterologous expression. In order to achieve the activation of silent BGCs either in native or heterologous *Streptomyces*, highly efficient genome editing techniques are critical as the conventional gene manipulation strategies like DNA deletions, disruption and replacement, use of suicide plasmids with temperature-sensitive replication origin, required selection and screening of single and double cross over recombination events have low efficiency for low genetic engineering at the same time they are time-consuming method [18].

The effect of the low efficiency of conventional gene manipulation was further further compounded by the fact that double cross-over mutations are uncommon in *Streptomyces*, as there is a low level of DNA homologous recombination. Recently, various genome editing technologies (Fig. **1**) adapted from Zhao *et al.* [20], have been developed to overcome these limitations, especially the clustered regularly interspaced short palindromic repeat (CRISPR)/CRISPR-associated protein (Cas)-based techniques, which have significantly enhanced *Streptomycetes* genetic manipulation and accelerated NP development, strain enhancement, and functional genome works [21, 22].

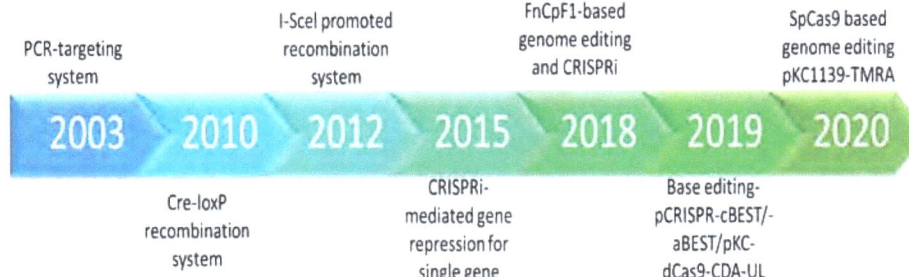

Fig. (1). The evolution of gene-editing technology in *Streptomyces*.

The present chapter aims to look at how genome editing in *streptomyces* has evolved over the years and explore how different works. A brief out line of important steps in different genome editing techniques are discussed along with the pros and cons of using each system for genome editing in *Streptomyces*. The chapter concludes with a brief overview of the possible future prospects of genome editing.

DIFFERENT GENOME EDITING TECHNIQUES

PCR-Targeting System

The PCR-targeting mechanism focuses on high-efficiency recombination between the target region within the *E. coli* genome and a PCR-amplified antibiotic selectable marker flanked on both ends by 40–50 bp homologous extensions [23]. This was first developed for gene knockout in *Escherichia coli* [23]. There are usually three steps in the PCR targeting system 1: The gene within the cosmid is replaced with a disruption cassette bearing a selectable antibiotic marker flanked by FRT or *loxP* site, 2: The mutant cosmid is then transferred into the *S. coelicolor* and screened for mutant strains with double crossover recombination events, 3: The antibiotic-resistant disruption cassette flanked by FRT or *lox P* sites is finally removed by inducing the expression of *tyrosine recombinase FLP* (FLP-FRT) or *Cre* (*Cre-loxP*) to generate unmarked, non-polar mutation (Fig. **2**) [24].

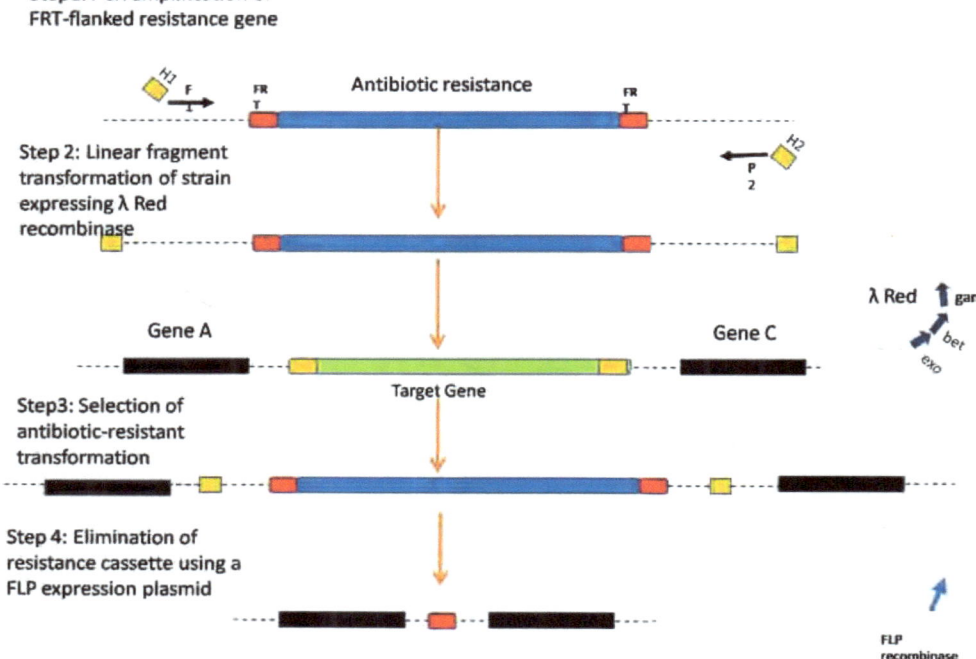

Fig. (2). The λ Red mediated gene replacement with selection marker elimination by site-specific recombinase.

The recombination events in this system is mediated by the λ recombination system (λ Red), which includes the Red α, β, and γ proteins from the λ phage [24]. The λ RED (*gam, bet, exo*) promotes an enhanced rate of recombination in bacteria when using linear DNA, which is otherwise not readily transformable with liner DNA due to the presence of intracellular *recBCD* exonuclease that degrades linear DNA. Using this method, Datsenko & Warner [23] made 40 disruptions on the *E.coli* chromosome by replacing the wild type sequences with PCR generated selectable markers. These selectable markers were generated using primers with 36nt homology extensions.

In *Streptomyces coelicolor* the PCR targeting-based gene disruption method was used to replace large gene clusters for secondary metabolite production. λ-Red was used to promote the recombination in *E. coli* between an antibiotic resistance cassette selectable in *E. coli* and *S. coelicolor DNA* on a cosmid [19] flanked by selectable markers that have been generated by PCR using primers with 30nt homology extensions [19]. As conjugation is more efficient than a transformation in many actinomycetes [25], the inclusion of *oriT* (RK2) in the disruption cassette enables the use of conjugation to introduce the PCR-targeted cosmid DNA into the *S.coelicolor*. This also eliminated the need for FLP- recombinase mediated site specific recombination. The potent methyl-specific restriction system of *S. coelicolor* is circumvented by passing DNA through a methylation-deficient *E. coli* host such as ET12567 [26]. Vectors containing *oriT* [RK2; 27] are mobilizable in trans in *E. coli* by the self-transmissible pUB307 [28, 29] or the non-transmissible pUZ8002, which lacks a *cis*-acting function for its transfer [29].

Further modifications were done to the PCR-based gene targeting system by using λ RED mediated recombination for *Streptomyces*, these gene disruption constructions made it possible to be selected in both *E. coli* and *Streptomyces*. After a single disruption with an oriT-containing cassette, further disruptions can be performed on the same cosmid using oriT-free cassettes containing alternative selective markers. The λ RED recombination plasmid pKD20 was further modified by replacing the ampicillin resistance gene *bla* with the chloramphenicol resistance gene *cat*, generating pIJ790 [19]. This modification permitted the selection in the presence of *Supercos1*-derived cosmids (ampicillin and kanamycin resistance) [19]. This was based on the observation by Murphy *et.al.* [30] that induction of the λ Red genes (*gam*, *bet*, and *exo*) can significantly increase homologous recombination in *E.coli*. The frequency of recombination was much higher when the length of the homologous DNA flanking the transformation marker was increased. The intergeneric transfer of the cosmid DNA from *E. coli* to *Streptomyces* through conjugation was achieved by the inclusion of *oriT* from an *IncP*-group plasmid in the disruption cassette. When compared to the transformation protocols described for *Streptomyces* [19], conjugation is found to be much more efficient and successful when compared to other actinomycetes [25]. This also helped in avoiding the process of protoplast regeneration used during transformation procedures, which is often difficult to develop for other strains and which can cause genetic changes. Gust *et al.* [19] adapted this method for non-polar and in-frame deletion of genes or gene clusters in *S. coelicolor*. This method was also successfully used to remove three endogenous Type *III PKS* genes in the engineered *S. coelicolor* strain M1152, generating an excellent expression host for the discovery and identification of actinobacterial type III polyketide metabolites [31].

Though the method is very effective in deleting genes or gene clusters in some *Streptomyces* strains. The PCR-targeting system has the following three drawbacks, which has restricted its widespread use: (i) a *scar* (*FRT* or *loxP* site) is left on the mutant strain's genome; (ii) a *Streptomyces* cosmid library must be built in advance and which limited the size of DNA fragments (<50kb) (iii) the process consists of four steps that are complicated and time-consuming.

Cre-loxP Recombination System

For genetic engineering in *Streptomyces*, the *Cre-loxP* recombination mechanism has been used in conjunction with the PCR-targeting system to remove antibiotic resistant markers. This was also used independently to knock out huge pieces of DNA in *Streptomyces* [32]. The two major stages involved in this system are: (i) Insertion of two loxP sites into the genome in the same direction to flank the DNA fragment that is to be removed by two step-wise single-crossover events; (ii) *Cre* recombinase expression is triggered to remove the DNA fragment between the two *loxP* sites. This technique is very good for knocking out large DNA fragments, and it has been used to knock out a DNA fragment greater than 1.4 Mb from the *Streptomyces avermitilis* genome [33]. This system was used for deletion mutants with mutations >80% of the wild type chromosome, which resulted in the non-production of endogenous secondary metabolites. The mutant systems were used for heterologous expression of three different biosynthetic gene clusters encoding for streptomycin from *S. griseus*, cephamycin C from *S. clavuligerus* and pladienolide from *S. platensis*. This system, like the PCR-targeting system, leaves a scar (a *loxP* sequence) on the genome, and the genetic modification process is time-consuming.

The *Cre/LoxP* system was not widely used, probably because the *cre* gene was introduced into *Streptomycetes* on the native phage. The expression of two synthetic genes encoding the *Cre* and *Flp* recombinases to delete resistance markers in members of the *Streptomyces* and *Saccharothrix* was also demonstrated [33] (Fig. **3**). Another reason for the limited use of the *Cre-lox* system was the high affinity of *Cre* recombinase to doubly mutated *lox LE/RE* site, as a result of which heterotypic *lox* sites in *actinomycetes* cannot be used for the construction of multiple mutations [34].

Another recombinase called *Dre*, which recognizes a different target site called *rox*, provides an additional tool for marker removal. *Dre* recombinase was first described by Sauer and McDermott [35] in the P1-like transducing bacteriophage D6 isolated from *Salmonella enterica* sero var Oranienburg. The genes encoding *Dre* and *Cre* recombinases share only 39% sequence similarity. Using transfection of mammalian CHO cells, Sauer and McDermott [34] were able to demonstrate that *Dre* recombinase catalyzes site-specific DNA recombination by recognizing *rox* sites, whereas *Cre* recombinase is not able to recognize *rox* sites, which are distinct from *loxP* sites. When compared to the *Cre* recombinase, the *Dre/rox* system was found to be more efficient in resistance marker removal [34].

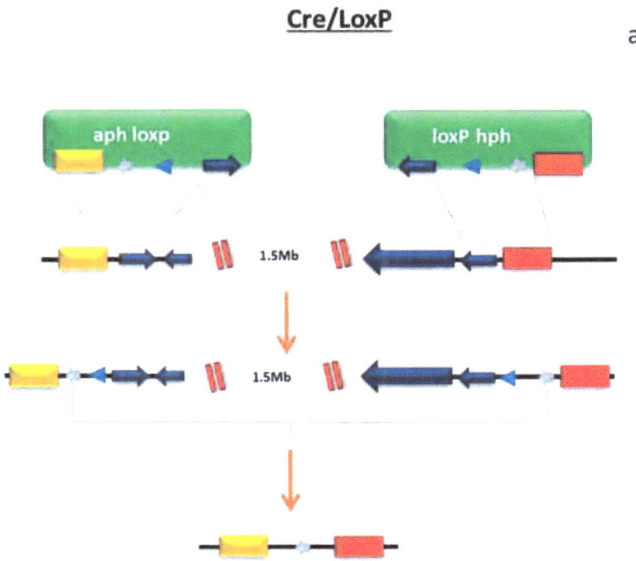

Fig. (3). The strategy for the construction of large-deletion mutants of *S. avermitilis* by the *Cre/LoxP* system.

I-*SceI* Meganuclease-Promoted Recombination System

Traditional methods for in-frame deletion of genes or BGCs that require screening for single- and double-crossover events usually use temperature-sensitive pKC1139 plasmid or the segregationally unstable pJTU1278 plasmid [36, 37]. Antibiotic resistance can be quickly tested for single-crossover mutants. Double-crossover events, on the other hand, are often difficult to acquire, particularly in strains with low intrinsic frequencies of homologous recombination.

The Yeast I-*SceI* meganuclease recognizes a unique 18-bp sequence and induces DNA double-strand breaks (DSBs), which facilitate double-crossover recombination events and are widely used in plant cells, mammalian cells and bacteria [38 - 40].

Lu *etal.* [41] created an *I-SceI*-assisted genome editing technology in *S. coelicolor* by synthesizing the codon-optimized *I-SceI* gene (Fig. **4**). This technology was effective in deleting two BGCs that biosynthesize actinorhodin and undecylprodigiosin [41, 42]. I-SceI cleavage significantly increases the performance of double-crossover events as compared to the conventional gene deletion process.

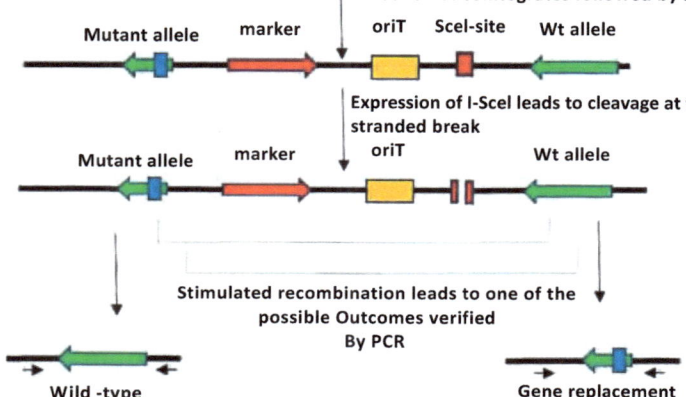

Fig. (4). *I-SceI*-based targeted gene inactivation in *Streptomyces*.

Genome Editing with the CRISPR/Cas Systems

In numerous bacteria and archaea, the CRISPR/Cas system functions as an adaptive immune system. The cleavage of the specific exogenous DNA and the recognition of the Cas protein is mediated by the RNA (ribonucleic acid) harboring a "spacer" sequence of the previously exposed bacteriophages [43 - 45]. The higher accuracy and specificity that is exhibited by the CRISPR/Cas system makes it an excellent choice for precision genome editing [46].

The CRISPR/Cas systems can be categorized into two classes (Class I and Class II) based on the structure and function of Cas proteins. Class I utilizes multiple Cas proteins to cleave foreign DNA, while Class II uses a single Cas protein. The classes are divided into types (type I–VI) [47]. Class I includes type I, III, IV [48]. Type II, V, and VI systems recognize and cleave DNA, type VI can edit RNA, and type III edits both DNA and RNA. The effect of type IV system DNA and RNA is still unknown [22, 49].

The CRISPR/Cas9 system specifically cleaves double-stranded DNA (dsDNA) *in-vitro*, which leads to double-strand breaks (DSBs) [46]. All these systems are unique and have their own characteristics, such as specific protospacer adjacent motif (PAM) regions and Cas proteins of varying sizes, each with a specific cleavage site, as summarized in Fig. (5).

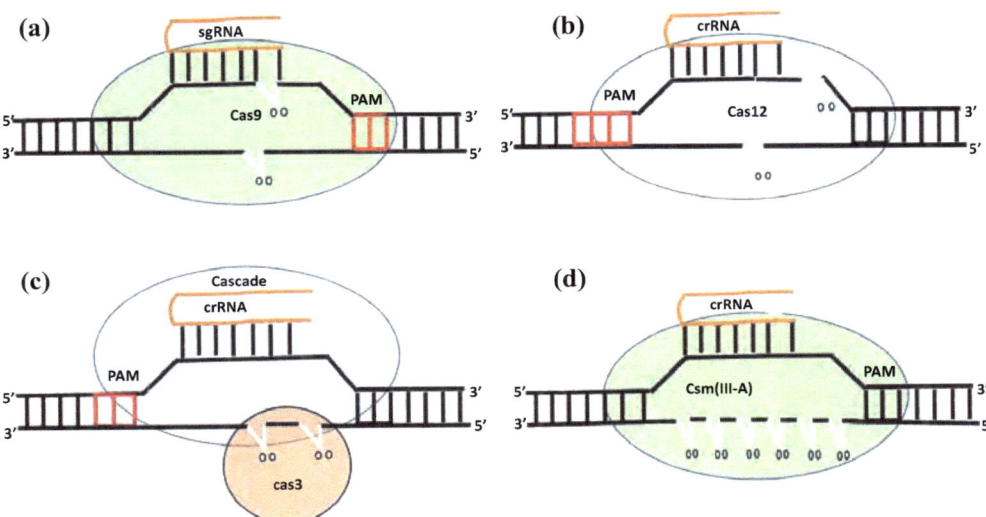

Fig. (5). This figure shows the mechanisms of different types of CRISPR systems: **a)** Type II *Cas9* uses sgRNA to guide *Cas9* protein to cleave both complementary and non-complementary strands, forming a blunt-ended nick in the presence of PAM sequence (NGG). **b)** Type V Cas12a uses crRNA to guide Cas12a protein to cleave both strands, forming a sticky-ended nick in the presence of the PAM sequence (NTTT). **c)** Type I Cas uses crRNA to guide Cas3 protein to cleave the non-complementary strand and create a large gap in the presence of the PAM sequence. **d)** Type III Cas uses sgRNA to guide Csm protein to cleave the non-complementary strand and create short nucleic acid fragments in the absence of a PAM sequence.

The type II CRISPR/Cas9 genome editing system comprises *Cas9* protein, CRISPR RNA (crRNA) and trans-activating crRNA (tracrRNA). The most commonly used *Cas9* protein contains 1368 amino acids which encompasses a REC (recognition lobe) and a NUC (nuclease lobe) [22]. The NUC domain contains a highly conserved RuvC nuclease domain and an HNH nuclease domain. The RuvC domain cleaves the same single strand (non-complementary strand) as the protospacer sequence, while the HNH nuclease domain cleaves a single strand complementary to the crRNA sequence. When acting simultaneously, both of them act at a specific position in the target sequence to produce a blunt end [50]. The PAM region of *Cas9* is at the 3' end of the target sequences whose sequence is 5'-NGG-3'.

TracrRNA is a hairpin RNA transcribed from a repeat region. TracrRNA precursor-crRNA (*pre-crRNA*) and *Cas9* protein form a complex in which tracrRNA is responsible for activating RNase III to promote the maturation of the *pre-crRNA* [51]. Mature *crRNA* combines with tracrRNA and *Cas9* to activate cleavage. A single-stranded guide RNA (*sgRNA*), a fusion of crRNA and tracrRNA, can effectively recognize specific sequences and direct the action of *Cas9* protein [45] which greatly simplifies the process of genome editing [22].

The type V CRISPR/Cas12a genome editing system comprises crRNA and Cas12a protein. The Cas12a protein contains a RuvC endonuclease domain, which sequentially cleaves the non-targeting strand and the targeting strand to form DSBs [52]. Compared to the CRISPR/Cas9 system, this system has several remarkable differences, including the signature protein, PAM sequence and cleavage product (Table **1**).

Table 1. Difference between type I, II, III and V CRISPR/Cas systems.

Classification	Type I	Type II	Type III	Type V
Signature protein	Cas3 (Or Cas3')	Cas9 (1368 amino acids)	Csm(III-A) orCmr(III-B)	Cas12a (1200-1300 amino acids)
Effector	Cascade	crRNA and tracrRNA (sgRNA)	Cascade	crRNA
PAM sequence	3-nt	G-rich sequence,5'-NGG-3'	Without PAM	5'-YTN-3'(FnCas12a) 5'-TTTN-3'(AsCas12a,LbCas12a)
Cleavage product	SSBs	DSB (flat end)	SSBs at every 6-nt	DSB (sticky end with 5 nucleotides protruding)

The maximum number of *cas* genes are seen in *Type* I systems, and these are coded by one or more operons. They contain six proteins, including the Cas3 protein, which has both helicase and nuclease activities and is the main enzyme in the interference phase. Multiple Cas proteins are combined with mature crRNA to form a CRISPR-associated complex for antiviral defence (cascade), which binds to invading foreign DNA and promotes the pairing of crRNA and the complementary strand of exogenous DNA to form an R loop, which is recognized by Cas3 to cleave both the complementary and non-complementary strands [22].

Type III systems contain the Cas10 protein with RNase activity and Cascade, and the function of Cascade resembles type I systems. Cas10 protein plays an important role in the maturation of *crRNA* and cleavage of invading foreign DNA [22]. Type III systems are categorized into four subtypes named A–D. The interference target of type III-A is mRNA, while the interference target of type III-B is the same as that of type I and II CRISPR/Cas systems, which is DNA. However, the interference targets of types III-C and D are unclear [22]. The ribonucleoprotein complexes of type II and V systems are relatively simple compared with those of types I and III. Type II systems only require crRNA, tracrRNA, and *Cas9* protein. The even simpler type V systems only require crRNA and Cas12a protein [22].

Compared to the genome editing methods mediated by tyrosine recombinase *Cre* and meganuclease I-*SceI*, which requires the prior introduction of unique enzyme recognition sites into the genomes, Cas proteins are guided to the specific site on the genome using a transcribed synthetic reference RNA (sgRNA, is a chimera of crRNA and tracrRNA), or just crRNA [53]. The *Streptococcus pyogenes* derived *Cas9*, which belongs to Class 2 type II, is the most commonly used Cas endonuclease in *Streptomyces* [2, 54]. *Cpf1* (also known as Cas12a), derived from *Francisella novicida* (Class 2 type V), has recently been used for *Streptomyces* engineering [2, 53, 54].

The CRISPR/Cas genome editing technology has greatly aided *Streptomyces'* genetic engineering. Several CRISPR/Cas-derived technologies, such as the CRISPR interference (CRISPRi)-mediated gene repression tool based on *dCas9* (a nuclease-deficient *Cas9* with two mutations of D10A and H840A) [55] or ddCpf1 (a nuclease-deficient Cpf1 with the mutation of E1006A) [54], and the base editors (BEs) *dCas9* or *Cas9n* (a nickase variant of *Cas9* with the mutation of D10A) have recently been produced for selective base mutagenesis [2, 56, 57].

CRISPR/Cas9 and HRD-mediated Genome Editing

The traditional methods of genome editing in *Streptomyces* were done through homologous recombination with a self-replicative or suicide or temperature-sensitive plasmid using a λ-red recombination system, which generally required about a week for an insertion [22]. The development of CRISPR/Cas9 systems shortened the gene-editing cycle to 3 days and the method was much more efficient. Cobb *et al.* [58] first demonstrated the use CRISPR/Cas9 system in *Streptomyces* for genome editing and developed two sets of CRISPR/Cas9 genome editing systems, *pCRISPomyces-1* and *pCRISPomyces-2*, and used homology-directed repair (HDR) (Fig. **6**) to achieve precise deletion of various DNA sizes (ranging from 20 bp–31.4 kb) (including individual genes, double genes simultaneously, and single antibiotic BGCs) with an efficiency of 21–100%. The *pCRISPomyces-2* system has a strong promoter of codon-modified

Cas9 nuclease, a sgRNA expression cassette and a 2 kb homology repair template (HRT). This system specifically generates a double-stranded break (DSB) at the targeted site by using the *Cas9* nuclease, then guided by the sgRNA harboring a custom-designed spacer, and the resulting chromosome break is repaired by the homology-dependent repair (HDR) system with the help of HRT and chromosomal deletions ranging from 20bp to 31kb with 70 to 100% efficiency [2].

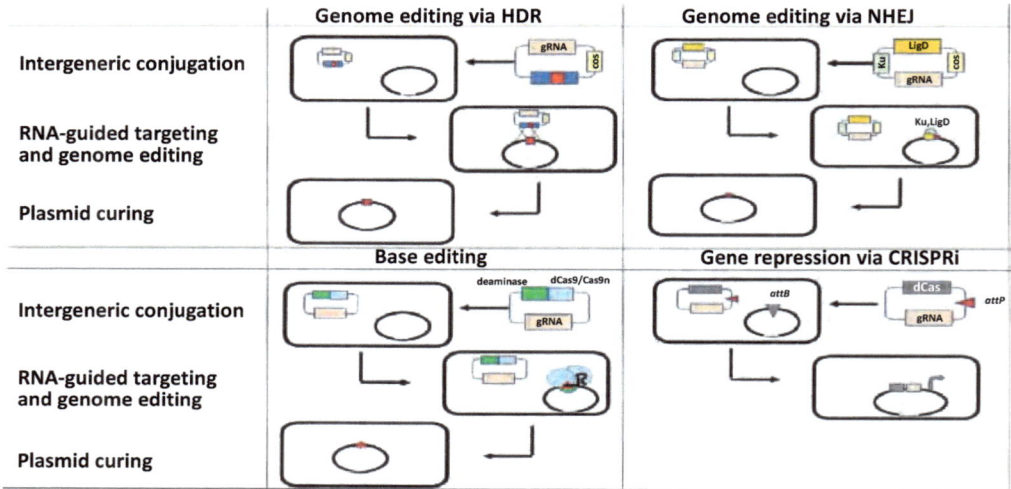

Fig. (6). A brief illustration of the CRISPR/Cas system-based technologies for genome editing in *Streptomyces*.

- **Coloumn 1**: Homology-directed repair for genome editing (HDR). By intergeneric conjugation, an editing plasmid containing homologous repair prototype and the Cas endonuclease expression cassettes and a short reference RNA (gRNA) is inserted into *Streptomyces*. With the aid of gRNA, the Cas nuclease cleaves the target site on the chromosome, resulting in a double-strand separation (DSB). In the presence of a homologous repair template, HDR repairs the DSB. Then specific mutations are added, such as deletions, insertions, and point mutations. The edited strain could be subjected to further genome editing after plasmid curing.
- **Column 2**: Non-homologous end joining genome editing (NHEJ). Intergeneric conjugation is used to insert an editing plasmid comprising the Cas endonuclease expression cassettes, a short gRNA, and the ku/ligD genes into *Streptomyces*. The DSB is repaired by the actions of LigD and Ku after RNA-guided DNA cleavage by the Cas nuclease. The editing plasmid is then cured to allow for the next round of genome editing.

- **Column 3**: Base editing. Base substitutions are induced by fusing a deaminase gene with the *dCas9* or *Cas9n* gene. The fusion protein could achieve RNA-guided base editing within a particular target window after introducing the editing plasmid into *Streptomyces* by intergeneric conjugation. The edited strain could be subjected to further genome editing after plasmid curing.
- **Column 4**: CIRPSRi-mediated gene repression. *Streptomyces* is conjugated with an integrative plasmid containing the expression cassettes of the nuclease-deficient Cas (dCas) gene and gRNA. The plasmid is inserted into the genome after site-specific recombination. By interfering with transcriptional initiation or elongation, the dCas/gRNA complex could repress the transcription of target genes.

CRISPR/Cas9 technology was used to activate silent BGCs of various groups in five native *Streptomyces* hosts, resulting in the discovery of novel metabolites, including a novel pentangular type II polyketide in *Streptomyces viridochromogenes*. This was followed by the successful use of CRISPR/Cas9 in *Streptomyces* [59].

Lu group developed a CRISPR/Cas system, *pKCCas9dO* encoding codon-optimized *Cas9* with two homology-directed repair templates and a target-specific guide RNA for gene editing in *S. coelicolor M145* [22, 60]. This system was successfully used for genome editing at different levels, including deleting individual genes (*actII-orf, red and glnR*) and single antibiotic BGCs with sizes of 21.3, 31.6 and 82.8Kb with high efficiency ranging from 60-100% [61]. This system was also used to incorporate point mutations in the *rpsL* gene with a 64% success rate [22]. Tong *et al*. [66] developed the CRISPR/Cas9 editing method (*pCRISPR-Cas9*) for the precise deletion of human and multiple genes by HDR in *S. coelicolor* at the same time. A combined system of CRISPR/Cas9-CodA(sm), using *CodA(sm)*, the D314A mutant of cytosine deaminase, was developed to convert 5-fluorocytosine to toxic 5-fluorouracil [21, 22]. This approach of counter-selecting progenies that had lost the recombinant plasmid dramatically accelerated the selection process [62]. The self-replicating behaviour with a high copy number of chromosomes (50 per chromosome) of this delivery vector could produce a huge number of strand plasmid DNA and several template DNAs, which results in the high efficiency of double cross-over recombination and target mutant frequencies [62]. The knock-in strategy was used with CRISPR/Cas9 for efficient and precise insertion of the constitutive promoters upstream of pathway-specific activators or main biosynthetic operons and thus triggering the production of NPs in multiple *Streptomyces* species of different classes [2, 22, 60, 63].

Two CRISPR/Cas9 genome editing tools derived from *pWHU2653* have been created by replacing the *ermEp** promoter of *Cas9* with the thiostrepton-inducible promoter *tipAp*, Mo *et al.* created this by editing method *pMWCas9* based on *pWHU2653*, which greatly improved DNA transformation speed [48]. Importantly, *pMWCas9* was also used to successfully remove highly repetitive DNA sequences in *Saccharopolyspora erythraea*, including the *eryAIII* gene from erythromycin polyketide synthase (PKS). The *pCRISPR-Cas9* editing method, on the other hand, was unable to remove the target gene, which may be due to unanticipated DNA recombination triggered by the *pSG5* replicon [64]. By replacing the counter-selection marker *CodA*(sm) with two reporter schemes, *GusA* and *IdgS*, Wang *et al.* [48] created an improved dual-functional chromogenic screening CRISPR/Cas9 tool (*pQS-gusA* and *pQS-idgS*) based on *pWHU2653* [49]. Following the chromogenic screening, these two reporter systems increased the performance of both genome editing and plasmid curing by up to 100% and streamlined the plasmid curing method much further.

CRISPR/Cas9 and NHEJ Mediated Genome Editing

Non-homologous end-joining (NHEJ) is a double-strand break (DSB) repair pathway which does not require any homologous template and can ligate two DNA ends together. The basic NHEJ machinery in bacteria involves two partners: the Ku protein, a DNA end binding protein for DSB recognition and the multifunctional LigD protein composed of a ligase, a nuclease and a polymerase domain, for end processing and ligation of the broken ends. Hoff *et al.* [65] reported the existence of *lig*D in *Streptomyces*. This method was used for targeting two genes, *actIORF1* (SCO5087) and *actVB* (SCO5092), from the actinorhodin biosynthetic gene cluster in *S. coelicolor* A3 [66].

The CRISPR-Cas9 system successfully inactivated the targeted genes even in the absence of HDR templates by the NHEJ pathway. The repair was error prone and created a library of deletions of variable sizes around the targeted sequences. Tong *et al.* [66], showed that the random deletions around the target site were due to the absence of a *LigD* in *S. coelicolor*. It was also observed that reconstitution of this defective NHEJ system by complementing *Streptomyces carneus* derived *ScaligD* increased editing efficiency up to 77% and qualified the mutations to 1–3 bp deletion/insertion/substitution in most of the cases. The presence of HRT helped in achieving 100% efficiency in high-precision genome editing.

Later a modified editing method was created from the *pCRISPR-Cas9* plasmid by combining the *S. carneus*-derived ligD (required for NHEJ) and a homology template (required for HDR) into a single plasmid, allowing any DNA repair pathway to occur [67]. This method achieved gene deletion in the BGCs responsible for the biosynthesis of streptothricin or streptomycin in 11 *actinomycete* strains with up to 100% efficiency by designing one *sgRNA* harboring a conserved protospacer sequence targeting a BGC gene of interest. The inactivation of typical BGCs by CRISPR/Cas9 allows for the discovery of new NPs in a larger proportion of *actinomycetes* strain collections (many of which are *Streptomyces*).

The high toxicity of *Cas9* to the host resulting from off-target DNA cleavage and non-target DNA binding in the absence of sgRNAs, which hinders its application in *Streptomyces* with low DNA transformation ability, is a major challenge for CRISPR/Cas9-mediated genome editing. Wang *et al.* [49] developed an updated CRISPR/Cas9 framework based on *pWHU2653* (dubbed *pWHU2653*-TRMA) to solve this problem by modulating *Cas9* operation at multiple levels: (i) instead of the heavy constitutive promoter, the inducible promoter tipAp is used to control *Cas9* expression (at the transcriptional level); (ii) Theophylline-inducible riboswitch and Mag-based blue light-inducible reconstitution system were introduced to control *Cas9* function at the translational and protein levels, respectively [2, 68]. Since DSB repair is an ATP-dependent mechanism, the gene encoding the ATP synthase-subunit AtpD was added for overexpression to improve editing performance. When compared to *pWHU2653*, triple *Cas9* controls reduced toxicity and increased DNA transition performance by over 250-fold in *S. coelicolor* under non-induction conditions. A related *pKC1139-TRMA* based on *pKC1139* was also created.

The counter-selection marker *CodA(sm)* is not optimal for *Streptomyces* with high resistance. After simultaneous induction with thiostrepton, theophylline, and blue light for *Cas9* operation reconstitution, the *pKC1139-TRMA* was used to achieve individual deletion of actII-orf4 and redD in *S. coelicolor* with efficiencies ranging from 35 to 80%. The uncoupling of DNA transformation and Cas9-mediated DNA cleavage was possible with the *pWHU2653-TRMA* and *pKC1139-TRMA* editing systems, which greatly improved the genetic engineering of *Streptomyces* species with low DNA transformation ability.

The presence of numerous multicopy genes and mobile genetic elements with identical or very similar DNA sequences found in the genomes of *Streptomyces* strains makes editing extremely difficult [22]. Najah *et al.* [69] created a generic two-step CRISPR/Cas9 editing technique, which is identical to *I-SceI* meganuclease-assisted genome editing technology [41], to solve this problem. By single-crossover recombination, a non-replicative plasmid (bait DNA) carrying the homologous arms flanking the target gene was first inserted into the genome. The bait DNA (such as the antibiotic-resistance gene) was then cleaved by another plasmid expressing *Cas9* and *sgRNA*, resulting in double-crossover recombination events (Fig. **6**). Using this technology, the native copies of two xenogeneic silencers, *lsr2* paralogs, were removed from *Streptomyces ambofaciens*. This method can be used to modify one copy of multicopy genes directly, as well as to investigate gene essentiality in *Streptomyces*.

Heterologous Expression and BGCS Cloning using CRISPR/Cas9

Various BGC awakening approaches have been used for the discovery of NPs in *Streptomyces*, like global regulator/pathway-specific, ribosome engineering and promotor refactoring. However, most of these methods involve genetic modification of native strains, so application in genetically stubborn strains or environmental DNA BGCs are limited [11, 12, 48]. Strategies for heterologous BGC expression in a genetically manipulable host helps to overcome this barrier perfectly, but modifying and cloning of large-sized BGCs remains difficult (sometimes over 100 kb). Traditional methods typically use randomly digested genomic libraries to clone large-sized DNAs; however, the screening procedure is often laborious and difficult to bundle intact BGCs over 100 kb in a single vector. Restriction enzyme (REs) was used to excise target BGCs from the genome. Direct cloning of these RE generated NP BGCs genome fragments was achieved by Linear-linear homologous recombination mediated by *RecE/T* [13, 34].

However, for wider use, these REs-dependent methods are severely limited because ideal RE cutting sites do not typically appear near BGC terminals. This restriction is perfectly solved by the CRISPR/Cas9 system, which cleaves the DNA directed by a synthetic sgRNA, allowing cloning of the large-sized BGCs. To linearize a large vector (22 kb), Wang *et al.* [70] tentatively implemented CRISPR/Cas9 system as REs *in vitro* and then efficiently assembled using Gibson assembly with a small DNA. Jiang *et al.*, [71] developed Cas9-assisted chromosome segment targeting (CATCH) for the precision acquiring of large-sized DNAs harboring NP BGCs, which enables target cloning of intact BGCs up to 100 kb cleave by CRISPR/Cas9 at unique sites driven by custom-designed sgRNAs and subsequent Gibson assembly target. At the same time, Lee *et al.* [72] merged CRISPR/Cas9 with TAR cloning using homologous yeast recombination to target CRISPR/Cas9 capture, releasing chromosomal segments and significantly accelerating TAR cloning capture performance by up to 32% [72]. Shortly afterwards, coupling of TAR cloning with the CRISPR/Cas9 system further extended the ability to handle even megabase-sized DNA segments. Cas9-facilitated homologous recombination assembly (*CasHRA*) was created by Zhou *et al.* [73], which co-introduces broad circular DNAs into *S. cerevisiae* and releases CRISPR/Cas9 target DNA segments for corresponding integration through homologous recombination. It provides an alternative for assembling large-sized BGCs over 100 kb using DNAs collected from cDNA cosmid libraries or *Streptomyces*. It requires assembly procedures, though, which appears to be time-consuming.

Genome Editing with the Assistance of Cpf1

The requirement of a G-rich protospacer-adjacent motif (PAM) sequence (5j-NGG-3j) for target sequence recognition by *Cas9* is one of the most important limitations of the CRISPR/Cas9 genome editing tool. As the *Streptomyces* genome is having high GC content (>70%), the PAM sequence is frequently distributed across *Streptomyces* genomes (*e.g.*, 260 targets per 1000 bp in *S. coelicolor*) [55]. The PAM series, on the other hand, might not be present in AT-rich DNA regions. The method of plasmid construction for independent transcription of multiple *sgRNAs* (each with its promoter and terminator) and the implementation of homologous DNA templates for DSB repair is complicated and time-consuming to introduce multiplex gene editing. To address these limitations, *Cpf1* from *Francisella novicida* (*FnCpf1*) recognizes T-rich PAM sequences (5j-TTV-3) have been developed for *Streptomyces* genome engineering [55]. *Cpf1* has RNase activity for pre-crRNA processing, and multiple guidance crRNAs can be expressed with just one promoter, which is beneficial for multiplex genome editing [74]. In *S. coelicolor,* HDR was used to achieve high-efficiency (75–95%) precision deletion of single or double genes at the same time. Codon-optimized *ligD* and *ku* genes from *Mycobacterium smegmatis* to use to reassemble the NHEJ pathway. NHEJ-assisted DSB repair may result in random-sized DNA deletions when target genes or gene clusters are inactivated. It was also found that *Cas9* from *S. pyogenes* and Cpf1 from *F. novicida* were not appropriate in the seven *Streptomyces* species studied. The 5-oxomilbemycin A3/A4-producing strain *Streptomyces hygroscopicus* SIPI-KF, which cannot be edited by *Cas9* due to its high toxicity, was successfully gene deleted using FnCpf1. Since *S. pyogenesCas9* cannot edit several *Streptomyces* strains, Yeo *et al.* used alternative CRISPR-Cas systems based on the *pCRISPomyces-2* method and demonstrated that *Cas9* from *Streptococcus thermophilus* CRISPR1 (Sth1Cas9, PAM: NNAGAA and NNGGAA), *Cas9* from *Staphylococcus aureus* (SaCas9, PAM: NNGRRT), and Cpf1 from *F. novicida* (FnCpf1) are functional in multiple *Streptomycetes*, which enables efficient HDR-mediated DNA knock-in and gene deletion [55]. The Cpf1- and alternative Cas9-assisted genome editing technologies can efficiently edit strains that cannot be edited by *Cas9* from *S. pyogenes*, such as *Streptomyces sp.* NRRL S-244. Hence, they complement the current Cas9-based tools. The presence of a diverse CRISPR/Cas toolbox will facilitate NP discovery and overproduction in *Streptomyces* as well as other *actinomycetes.*

Transcriptional Repression using dCas (CRISPRi)

The CRIPSRi method, which is focused on the nuclease-defective Cas nuclease (such as *dCas9* and *ddCpf1*), has been shown to be an effective tool for functional genome study and metabolic engineering in bacteria [57]. *Streptomyces* has recently produced three related CRISPRi instruments. By replacing *Cas9* with *dCas9*, Tong *et al.* created an inducible CRISPRi based on the genome-editing method *pCRISPR-Cas9*, which allowed the effective repression of single genes upon induction [11]. The *dCas9/sgRNA* complex was expressed in this CRISPRi system using the replicative plasmid *pGM1190*, which contains a temperature-sensitive replicon *pSG5*, and the *dCas9* gene is regulated by the thiostrepton-inducible promoter (tipAp). Based on *dCas9* and ddCpf1 [54], two CRISPRi tools in *S. coelicolor* were developed. In these two structures, the *dCas9/sgRNAs* or *ddCpf1/crRNAs* complex was expressed using *pSET152*, an integrative plasmid, and both *dCas9* and *sgRNAs* or *ddCpf1* and *crRNAs* were planned to be controlled by constitutive promoters. They were able to achieve high-efficiency simultaneous repression of up to four genes using these two CRISPRi systems. The latter two systems can have two advantages over the inducible CRISPRi tool based on the replicative plasmid (*e.g.*, *pGM1190*). Firstly, since they are inserted into the genome, their suppression effects are likely to be stable. Secondly, they have a broader use since *pSET152* transformation is more efficient than replicative plasmids. Meanwhile, all *Streptomyces* strains (whose genome sequences are available so far) have the C31 *attB* site for *pSET152* incorporation [75]. It's worth noting that using the *dCas9*-based method to repress several targets simultaneously necessitates a time-consuming process to create multiple *sgRNA* speech cassettes with individual promoters and terminators. Because of ddCpf1's ability to process *pre-crRNA*, only a single personalized CRISPR array with one promoter is needed, which saves time and effort. As a result, for multiplexed gene repression, the *ddCpf1*-based CRISPRi system outperforms *dCas9*-based systems.

Editors Based on Cas9 Variants (dCas9 or Cas9n)

CRISPR-guided Base Editors (BEs) have been successfully used for genome editing in mammalian cells [76], animals [77], plants [20], and bacteria [78] as they permit effective targeting of single nucleotide resolution DNA mutagenesis. BEs work by fusing a Cas9 version, such as *dCas9* (D10A and H840A) or *Cas9n* (D10A), with a base deaminase to deaminate the target bases' exocyclic amine, resulting in base substitutions [78]. BEs do not generate DNA DSBs and do not focus on cellular HDR or NHEJ DNA repair pathways, unlike the CRISPR/Cas-based genome editing tools described above. As a consequence, by-products associated with DSBs, such as minor insertions or deletions (indels), are minimized [59]. There are two types of DNA BEs currently in use: cytosine base editors (CBEs) and adenine base editors (ABEs) [80]. ABE can convert adenosine (A) to guanosine (G), while CBE can convert cytidine (C) to thymidine (T). In *Streptomyces*, both forms of BEs have been established for genome editing. Tong *et al.* created the *CRISPR-cBEST* (belonging to CBE) and CRISPR-aBEST (belonging to ABE) base editing structures by fusing the rat APOBEC1 (rAPOBEC1) cytidine deaminase and the adenosine deaminase *ecTadA* to the N-terminus of the codon-optimized *Cas9n* [2, 56, 81]. A codon-optimized uracil glycosylase inhibitor (UGI) from *Bacillus phage AR9* is connected to the C-terminus of *Cas9n* in *CRISPR-cBEST* to inhibit the action of the uracil-DNA glycosylase (UDG) and improve editing performance. *CRISPR-cBEST* transformed cytidine to thymidine with 100% frequency in *S. coelicolor* within a 7-base target window (11 to 17 bp upstream of the PAM sequence). Within a 6-base target window, 12 to 17 bp upstream of PAM, CRISPR-aBEST can transform adenosine to guanosine. *CRISPR-cBEST* editing is prioritized as TC > CC > AC >GC, while CRISPR-aBEST editing is prioritized as TA > GA > AA > CA. *CRISPR-cBEST* outperformed CRISPR-aBEST in terms of editing performance and off-target effects. *CRISPR-cBEST* was successfully used to insert STOP codons into the engineered DNA positions in *Streptomyces griseus* with precision and also used for simultaneously targeted mutagenesis of two similar gene copies of the gene kirN in *Streptomyces collinus Tü365*. Finally, simultaneous editing of three separate sites at frequencies up to 100 percent by incorporating the Csy4-based RNA processing method was also achieved between sgRNA for the expression of multiple sgRNAs with one promoter and terminator. CBE system *dCas9*-CDA-ULstr derived from the CBE *dCas9*-CDA-UL formed in *E. coli* was also produced [81], which includes *dCas9*, PmCDA1 (a Petromyzon marinus activation-induced cytidine deaminase (AID) ortholog), UGI, and the degradation tag (LVA) [58]. Single-, double-, and triple-point mutations (cytidine to thymidine) at target sites in *S. coelicolor* were achieved with high efficiencies of up to 100%, 60%, and 20%, respectively, using *dCas9*-CDA-ULstr [57]. In the industrial strain *Streptomyces rapamycinicus*, this CBE was also useful for highly effective base editing. *dCas9*-CDA-ULstr has a 5-base editing time, 16 to 20 bp upstream of the PAM chain, compared to CRISPR-cBEST, which has a seven-base editing window. Furthermore, for cytidines followed by guanosines, *dCas9*-CDA-ULstr had a higher editing efficiency (70–100%) than *rAPOBEC1*-derived *CRISPR-cBEST* (0–60%), which is a benefit for base editing in *Streptomyces* with high GC contents in their genomes [57]. BEs have been developed to provide alternative and effective genome editing methods in *Streptomyces*. Four amino acid codons, in particular, *Arg* (CGA), *Gln* (CAA and CAG), and Trpcodons (TGG, goal C in the non-coding strand), can be effectively mutated to STOP codons (TGA, TAA, and TAG) using CBEs, resulting in gene inactivation [80]. CBE-based STOP codon incorporation saves time and is less labour-intensive when compared to HDR-mediated gene deletion since the repair models do not need to be cloned [82]. As a result, CBEs can make functional genome research and metabolic engineering-based strain improvement in *Streptomyces*, especially those with weak HR ability, significantly easier.

Genome Editing using Multiplex Automated Genome Editing (MAGE) Tool

The whole-genome analysis of *Streptomyces* reveals that it possesses a large number of silent biosynthetic gene clusters (BGCs) that would help to identify unique metabolite pathways [83, 84]. And using these pathways, several useful biosynthetic products like antibiotics, anti-cancer agents, immune depressants, herbicides, *etc.*, were produced [79, 85 - 87]. So, there was a need for a well advanced system of genome editing for the better production of these biomolecules.

Before the discovery of the multiplex automated genome editing (MAGE) tool, scientists could only edit one site of an organism at a time. This disadvantage became the reason for the evolution of MAGE as it enables scientists to quickly edit an organism's DNA to produce multiple changes across the genome. The changes produced by MAGE, whether it is addition or deletion are grouped into three types 1) many target sites and a single type of mutation, 2) single target site and many genetic mutations and, 3) many target sites and many genetic mutations.

Multiplex Genome Editing of Streptomyces Species using Engineered CRISPR/Cas System

The engineered CRISPR system for multiplex genome editing shows better efficiency with the usage of sgRNA (synthetic guide RNA synthesized by the fusion of crRNA and tracrRNA). This increases the efficiency of genome deletion ranging from 20bp to 30kb. The CRISPR-Cas system is a powerful tool for gene deletion [61, 58], single/double site mutation [88], reversible gene expression control [11], and activation of silent BCGs in several *Streptomyces* strains [60].

Multiplex Genome Editing using Engineered CRISPR/Cas9 System.

Engineered CRISPR/Cas system for rapid multiplex genome editing can be used for chromosomal deletion ranging from the size 20bp to 30kb with very high efficiency (~70- 100%). The designed *pCRISPomyces* plasmids acted as a suicide plasmid to carry out a double crossover integration and these engineered plasmids are of two types, the type 1 plasmid is less efficient than the type 2 [58].

The CRISPR/Cas system from *S. pyogenes* can effectively be transferred into a variety of hosts like *E. coli*, yeast and human cell lines. The engineered pCRISPomyces plasmids were transferred to the host cell by electroporation. The transformed cells were identified using selective markers. Then, these transformed cells were used to edit the *Streptomyces* species by conjugation [58].

Multiplex Genome Editing using Engineered CRISPR/Cas12a System

The clustered regularly interspaced short palindromic repeats/CRISPR-associated protein 9 (CRISPR/Cas 9) system is a great genome editing tool for *Streptomyces* strains. But it has failed to work in certain newly discovered strains and some important industrial strains [89]. It was also observed that the protospacer adjacent motif (PAM) recognition scope of this system sometimes limits its applications for generating precise site mutations and insertions [89]. This led to the development of three efficient *CRISPR-FnCas12a* systems for multiplex genome editing in several *Streptomyces* strains with each system exhibiting unique advantages for different applications [89]. The *CRISPR-FnCas12a1* system was efficiently applied in the industrial strain *Streptomyces hygroscopicus*, in which *SpCas9* does not work well. The CRISPR-FnCas12a2 system was used for deleting large fragments ranging from 21.4 kb to 128 kb and the *CRISPRFnCas12a3* system (Fig. 7) employing the engineered FnCas12a mutant EP16, which recognizes a broad spectrum of PAM sequences, was used to precisely perform site mutations and insertions. The *CRISPR-FnCas12a3* system addressed the limitation of TTN PAM recognition in *Streptomyces* strains with high GC contents [60, 89].

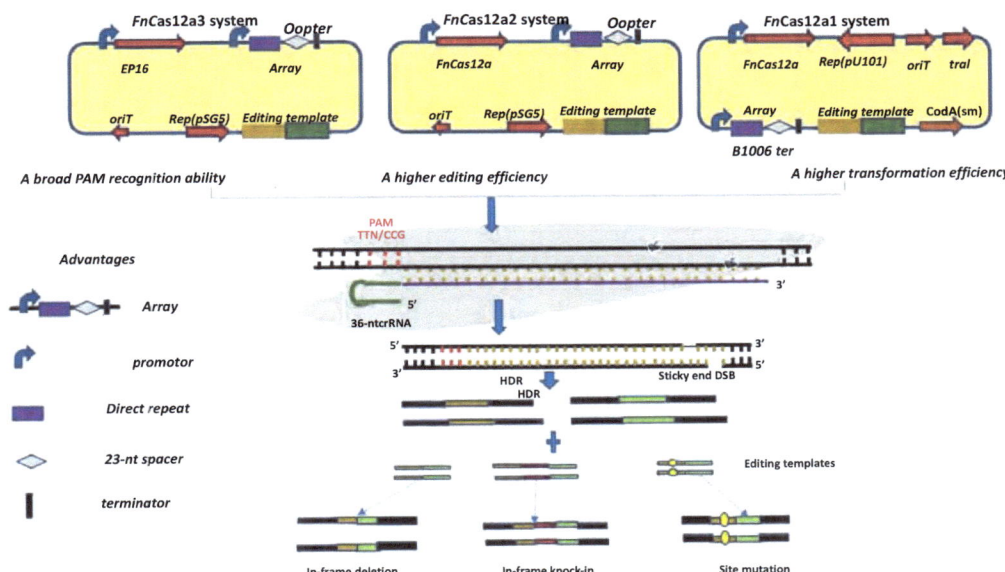

Fig. (7). Multiplex genome editing using engineered CRISPR/Cas12a system.

CRISPR-Cpf1 Assisted Multiplex Genome Editing and Transcriptional Repression in Streptomyces

The *Streptococcus pyogenes* (Sp) *CRISPR-Cas9*-assisted genome-editing tool developed for rapid genetic engineering in *Streptomyces* had several limitations, including the toxicity of *SpCas9* expression in some important industrial *Streptomyces* strains and the need for complex expression constructs when targeting multiple genomic loci [90]. To address these problems, a high-efficiency CRISPR-Cpf1 system (from *Francisella novicida*) for multiplex genome editing and transcriptional repression in *Streptomyces* was developed using an all-in-one editing plasmid with homology-directed repair (HDR), the edited CRISPR-Cpf1 system precisely deletes single or double genes at efficiencies of 75-95% in *Streptomyces coelicolor* [90]. When no templates for HDR are present, random sized DNA deletions are achieved by *FnCpf1-induced* double-strand breaks (DSBs) repair by a reconstituted non-homologous end joining (NHEJ) pathway. This was followed by the development of a DNase-deactivated *Cpf1* (*ddCpf1*)-based integrative CRISPRi system for robust, multiplex gene repression using a single customized crRNA array [91]. FnCpf1 efficiently promotes HDR-mediated gene deletion in the 5-oxomilbemycin-producing strain *Streptomyces hygroscopicus* SIPI-KF, in which *SpCas9* does not work well. Thus making FnCpf1 is a powerful and indispensable addition to *Streptomyces* CRISPR toolbox [91].

CONCLUSION

The genome mining strategy, which includes whole genome sequencing, heterologous production of silent BGCs, and characterization of target metabolites, is becoming increasingly important in reinvigorating drug development processes in the post-genomic age. Rapidly unlocking these cryptic BGCs using simple and high-throughput techniques is currently a primary aim for genomics-driven NPs discovery. Great advancements in technologies based on DNA recombination systems and synthetic biology have offered renewed motivation for the study of microbial NPs for much of the last two decades. Due to exceptional qualities such as greater sequence specificity, artificial directed targeting, and high editing efficiency, the CRISPR/Cas9 system is found to be a powerful technology for genome editing or BGC refactoring. Though the genome editing applications for *Streptomyces* are presently limited, especially for strains that haven't been thoroughly researched. As a result, more efficient and convenient CRISP/Cas tools are essential. For instance, diversified *CRISPR/Cas* systems likeCpf1 [92], the newly identified class 2 type V *CRISPR/Cas* protein, xCRISPR/Cas9, Cas12, Cas13 systems with broad PAM compatibility could be introduced for diverse applications in *Streptomyces*, to advance the researches on

NP drugs and open a new era for NP drugs discovery [93]. For many bacteria, however, continuous expression of any foreign protein with DNA-binding/editing activity appears to be particularly toxic. In prokaryotes, CRISPR's natural function as an adaptive immune system is closely restricted. Further studies into the ability to utilize indigenous CRISPR loci (spacers and Cas proteins) for genome editing would be the most efficient approach to prevent foreign CRISPR systems in bacteria. This method would need a case-by-case analysis and adjustment for each native CRISPR effector.

In summary, the prospects for continued advancements in methods to genetically modify *Streptomyces* for enhanced secondary metabolite production is beneficial. This bodes well for the future use of "biosynthetic dark matter" in the development of human drugs, animal health, and crop protection. The open exchange of ideas and genetic tools between academic and industrial researchers has helped in the development of the current state-of-the-art systems. Future successes are dependent on the continuation of the international cooperation that has characterized the *Streptomyces* community.

ACKNOWLEDGEMENTS

The authors are indebted to all the faculty members of the Department of Botany, Nirmala College for their help and support. The authors thank Dr. Prakash M. Halami, Senior Principal Scientist & Professor-AcSIR, Department of Microbiology Fermentation technology, CSIR- Central Food Technological Research Institute, Mysore for the invitation to write the book chapter and all the help in the drafting of the chapter and for his critical comments. The help and support of the Library staff of Nirmala College and Konkuk University are acknowledged.

REFERENCES

[1] Van Kulen G, Dyson PJ. Chapter Six: Production of Specialized Metabolites by *Streptomyces coelicolor* A3(2). In: Sima S, Geoffery Micheal G, Eds. Advances in Applied Microbiologyi. Elsevier 2014; 89: pp. 217-66.

[2] Zhao Y, Li G, Chen Y, Lu Y. Challenges and advances in genome editing technologies in *Streptomyces*. Biomolecules 2020; 10(5): 734.
[http://dx.doi.org/10.3390/biom10050734] [PMID: 32397082]

[3] Mast Y, Weber T, Gölz M, *et al.* Characterization of the 'pristinamycin supercluster' of *Streptomyces pristinaespiralis*. Microb Biotechnol 2011; 4(2): 192-206.
[http://dx.doi.org/10.1111/j.1751-7915.2010.00213.x] [PMID: 21342465]

[4] Baltz RH. Daptomycin: Mechanisms of action and resistance, and biosynthetic engineering. Curr Opin Chem Biol 2009; 13(2): 144-51.
[http://dx.doi.org/10.1016/j.cbpa.2009.02.031] [PMID: 19303806]

[5] Yoo YJ, Kim H, Park SR, Yoon YJ. An overview of rapamycin: From discovery to future perspectives. J Ind Microbiol Biotechnol 2017; 44(4-5): 537-53.

[http://dx.doi.org/10.1007/s10295-016-1834-7] [PMID: 27613310]

[6] Barreiro C, Martínez-Castro M. Trends in the biosynthesis and production of the immunosuppressant tacrolimus (FK506). Appl Microbiol Biotechnol 2014; 98(2): 497-507.
[http://dx.doi.org/10.1007/s00253-013-5362-3] [PMID: 24272367]

[7] Yoon YJ, Kim ES, Hwang YS, Choi CY. Avermectin: Biochemical and molecular basis of its biosynthesis and regulation. Appl Microbiol Biotechnol 2004; 63(6): 626-34.
[http://dx.doi.org/10.1007/s00253-003-1491-4] [PMID: 14689246]

[8] Li L, Zheng G, Chen J, Ge M, Jiang W, Lu Y. Multiplexed site-specific genome engineering for overproducing bioactive secondary metabolites in actinomycetes. Metab Eng 2017; 40: 80-92.
[http://dx.doi.org/10.1016/j.ymben.2017.01.004] [PMID: 28088540]

[9] Vasanthakumar A, Kattusamy K, Prasad R. Regulation of daunorubicin biosynthesis in *Streptomyces peucetius* - feed forward and feedback transcriptional control. J Basic Microbiol 2013; 53(8): 636-44.
[http://dx.doi.org/10.1002/jobm.201200302] [PMID: 23417448]

[10] Shen B, Du L, Sanchez C, Edwards DJ, Chen M, Murrell JM. The biosynthetic gene cluster for the anticancer drug bleomycin from *Streptomyces verticillus* ATCC15003 as a model for hybrid peptide-polyketide natural product biosynthesis. J Ind Microbiol Biotechnol 2001; 27(6): 378-85.
[http://dx.doi.org/10.1038/sj.jim.7000194] [PMID: 11774003]

[11] Tong Y, Charusanti P, Zhang L, Weber T, Lee SY. CRISPR-Cas9 based engineering of actinomycetal genomes. ACS Synth Biol 2015; 4(9): 1020-9.
[http://dx.doi.org/10.1021/acssynbio.5b00038] [PMID: 25806970]

[12] Rutledge PJ, Challis GL. Discovery of microbial natural products by activation of silent biosynthetic gene clusters. Nat Rev Microbiol 2015; 13(8): 509-23.
[http://dx.doi.org/10.1038/nrmicro3496] [PMID: 26119570]

[13] Nah HJ, Pyeon HR, Kang SH, Choi SS, Kim ES. Cloning and heterologous expression of a large-sized natural product biosynthetic gene cluster in *Streptomyces* species. Front Microbiol 2017; 8: 394-4.
[http://dx.doi.org/10.3389/fmicb.2017.00394] [PMID: 28360891]

[14] Banik JJ, Brady SF. Recent application of metagenomic approaches toward the discovery of antimicrobials and other bioactive small molecules. Curr Opin Microbiol 2010; 13(5): 603-9.
[http://dx.doi.org/10.1016/j.mib.2010.08.012] [PMID: 20884282]

[15] Katz M, Hover BM, Brady SF. Culture-independent discovery of natural products from soil metagenomes. J Ind Microbiol Biotechnol 2016; 43(2-3): 129-41.
[http://dx.doi.org/10.1007/s10295-015-1706-6] [PMID: 26586404]

[16] Challis GL. Exploitation of the *Streptomyces coelicolor* A3(2) genome sequence for discovery of new natural products and biosynthetic pathways. J Ind Microbiol Biotechnol 2014; 41(2): 219-32.
[http://dx.doi.org/10.1007/s10295-013-1383-2] [PMID: 24322202]

[17] Baltz RH. Gifted microbes for genome mining and natural product discovery. J Ind Microbiol Biotechnol 2017; 44(4-5): 573-88.
[http://dx.doi.org/10.1007/s10295-016-1815-x] [PMID: 27520548]

[18] Onaka H. Novel antibiotic screening methods to awaken silent or cryptic secondary metabolic pathways in actinomycetes. J Antibiot 2017; 70(8): 865-70.
[http://dx.doi.org/10.1038/ja.2017.51] [PMID: 28442735]

[19] Gust B, Challis GL, Fowler K, Kieser T, Chater KF. PCR-targeted *Streptomyces* gene replacement identifies a protein domain needed for biosynthesis of the sesquiterpene soil odor geosmin. Proc Natl Acad Sci 2003; 100(4): 1541-6.
[http://dx.doi.org/10.1073/pnas.0337542100] [PMID: 12563033]

[20] Mishra R, Joshi RK, Zhao K. Base editing in crops: Current advances, limitations and future implications. Plant Biotechnol J 2020; 18(1): 20-31.
[http://dx.doi.org/10.1111/pbi.13225] [PMID: 31365173]

[21] Tao W, Yang A, Deng Z, Sun Y. CRISPR/Cas9-based editing of *streptomyces* for discovery, characterization, and production of natural products. Front Microbiol 2018; 9: 1660-0.
[http://dx.doi.org/10.3389/fmicb.2018.01660] [PMID: 30087666]

[22] Liu Z, Dong H, Cui Y, Cong L, Zhang D. Application of different types of CRISPR/Cas-based systems in bacteria. Microb Cell Fact 2020; 19(1): 172.
[http://dx.doi.org/10.1186/s12934-020-01431-z] [PMID: 32883277]

[23] Datsenko KA, Wanner BL. One-step inactivation of chromosomal genes in *Escherichia coli* K-12 using PCR products. Proc Natl Acad Sci 2000; 97(12): 6640-5.
[http://dx.doi.org/10.1073/pnas.120163297] [PMID: 10829079]

[24] Murphy KC. Phage Recombinases and Their Applications Advances in Virus Research. Elsevier 2012; pp. 367-414.

[25] Matsushima P, Broughton MC, Turner JR, Baltz RH. Conjugal transfer of cosmid DNA from *Escherichia coli* to *Saccharopolyspora spinosa*: Effects of chromosomal insertions on macrolide A83543 production. Gene 1994; 146(1): 39-45.
[http://dx.doi.org/10.1016/0378-1119(94)90831-1] [PMID: 8063103]

[26] MacNeil DJ, Gewain KM, Ruby CL, Dezeny G, Gibbons PH, MacNeil T. Analysis of *Streptomyces avermitilis* genes required for avermectin biosynthesis utilizing a novel integration vector. Gene 1992; 111(1): 61-8.
[http://dx.doi.org/10.1016/0378-1119(92)90603-M] [PMID: 1547955]

[27] Pansegrau W, Lanka E, Barth PT, *et al.* Complete nucleotide sequence of Birmingham IncP α plasmids. Compilation and comparative analysis. J Mol Biol 1994; 239(5): 623-63.
[http://dx.doi.org/10.1006/jmbi.1994.1404] [PMID: 8014987]

[28] Robinson MK, Bennett PM, Grinsted J, Richmond MH. The stable carriage of two TnA units on a single replicon. Mol Gen Genet 1978; 160(3): 339-46.
[http://dx.doi.org/10.1007/BF00332978] [PMID: 672895]

[29] Flett F, Mersinias V, Smith CP. High efficiency intergeneric conjugal transfer of plasmid DNA from *Escherichia coli* to methyl DNA-restricting streptomycetes. FEMS Microbiol Lett 1997; 155(2): 223-9.
[http://dx.doi.org/10.1111/j.1574-6968.1997.tb13882.x] [PMID: 9351205]

[30] Murphy KC, Campellone KG, Poteete AR. PCR-mediated gene replacement in *Escherichia coli*. Gene 2000; 246(1-2): 321-30.
[http://dx.doi.org/10.1016/S0378-1119(00)00071-8] [PMID: 10767554]

[31] Thanapipatsiri A, Claesen J, Gomez-Escribano JP, Bibb M, Thamchaipenet A. A *Streptomyces coelicolor* host for the heterologous expression of Type III polyketide synthase genes. Microb Cell Fact 2015; 14(1): 145-5.
[http://dx.doi.org/10.1186/s12934-015-0335-0] [PMID: 26376792]

[32] Komatsu M, Uchiyama T, Ōmura S, Cane DE, Ikeda H. Genome-minimized *Streptomyces* host for the heterologous expression of secondary metabolism. Proc Natl Acad Sci 2010; 107(6): 2646-51.
[http://dx.doi.org/10.1073/pnas.0914833107] [PMID: 20133795]

[33] Fedoryshyn M, Petzke L, Welle E, Bechthold A, Luzhetskyy A. Marker removal from actinomycetes genome using Flp recombinase. Gene 2008; 419(1-2): 43-7.
[http://dx.doi.org/10.1016/j.gene.2008.04.011] [PMID: 18550297]

[34] Fu J, Bian X, Hu S, *et al.* Full-length RecE enhances linear-linear homologous recombination and facilitates direct cloning for bioprospecting. Nat Biotechnol 2012; 30(5): 440-6.
[http://dx.doi.org/10.1038/nbt.2183] [PMID: 22544021]

[35] Sauer B, McDermott J. DNA recombination with a heterospecific Cre homolog identified from comparison of the pac-c1 regions of P1-related phages. Nucleic Acids Res 2004; 32(20): 6086-95.
[http://dx.doi.org/10.1093/nar/gkh941] [PMID: 15550568]

[36] Bierman M, Logan R, O'Brien K, Seno ET, Nagaraja Rao R, Schoner BE. Plasmid cloning vectors for the conjugal transfer of DNA from *Escherichia coli* to *Streptomyces* spp. Gene 1992; 116(1): 43-9.
[http://dx.doi.org/10.1016/0378-1119(92)90627-2] [PMID: 1628843]

[37] He Y, Wang Z, Bai L, Liang J, Zhou X, Deng Z. Two pHZ1358-derivative vectors for efficient gene knockout in *streptomyces*. J Microbiol Biotechnol 2010; 20(4): 678-82.
[http://dx.doi.org/10.4014/jmb.0910.10031] [PMID: 20467238]

[38] Monteilhet C, Perrin A, Thierry A, Colleaux L, Dujon B. Purification and characterization of the *in vitro* activity of I- *Sce* I, a novel and highly specific endonuclease encoded by a group I intron. Nucleic Acids Res 1990; 18(6): 1407-13.
[http://dx.doi.org/10.1093/nar/18.6.1407] [PMID: 2183191]

[39] Plessis A, Perrin A, Haber JE, Dujon B. Site-specific recombination determined by I-SceI, a mitochondrial group I intron-encoded endonuclease expressed in the yeast nucleus. Genetics 1992; 130(3): 451-60.
[http://dx.doi.org/10.1093/genetics/130.3.451] [PMID: 1551570]

[40] Choulika A, Perrin A, Dujon B, Nicolas JF. Induction of homologous recombination in mammalian chromosomes by using the I-SceI system of *Saccharomyces cerevisiae*. Mol Cell Biol 1995; 15(4): 1968-73.
[http://dx.doi.org/10.1128/MCB.15.4.1968] [PMID: 7891691]

[41] Lu Z, Xie P, Qin Z. Promotion of markerless deletion of the actinorhodin biosynthetic gene cluster in *Streptomyces coelicolor*. Acta Biochim Biophys Sin 2010; 42(10): 717-21.
[http://dx.doi.org/10.1093/abbs/gmq080] [PMID: 20810535]

[42] Fernández-Martínez LT, Bibb MJ. Use of the Meganuclease I-SceI of *Saccharomy cescerevisiae* to select for gene deletions in actinomycetes. Sci Rep 2014; 4(1): 7100-0.
[http://dx.doi.org/10.1038/srep07100] [PMID: 25403842]

[43] Barrangou R, Fremaux C, Deveau H, *et al.* CRISPR provides acquired resistance against viruses in prokaryotes. Science 2007; 315(5819): 1709-12.
[http://dx.doi.org/10.1126/science.1138140] [PMID: 17379808]

[44] Grissa I, Vergnaud G, Pourcel C. The CRISPRdb database and tools to display CRISPRs and to generate dictionaries of spacers and repeats. BMC Bioinformatics 2007; 8(1): 172-2.
[http://dx.doi.org/10.1186/1471-2105-8-172] [PMID: 17521438]

[45] Horvath P, Barrangou R. CRISPR/Cas, the immune system of bacteria and archaea. Science 2010; 327(5962): 167-70.
[http://dx.doi.org/10.1126/science.1179555] [PMID: 20056882]

[46] Jinek M, Chylinski K, Fonfara I, Hauer M, Doudna JA, Charpentier E. A programmable dual-RN--guided DNA endonuclease in adaptive bacterial immunity. Science 2012; 337(6096): 816-21.
[http://dx.doi.org/10.1126/science.1225829] [PMID: 22745249]

[47] Makarova KS, Wolf YI, Alkhnbashi OS, *et al.* An updated evolutionary classification of CRISPR–Cas systems. Nat Rev Microbiol 2015; 13(11): 722-36.
[http://dx.doi.org/10.1038/nrmicro3569] [PMID: 26411297]

[48] Mohanraju P, Makarova KS, Zetsche B, Zhang F, Koonin EV, van der Oost J. Diverse evolutionary roots and mechanistic variations of the CRISPR-Cas systems. Science 2016; 353(6299): aad5147.
[http://dx.doi.org/10.1126/science.aad5147] [PMID: 27493190]

[49] Mo J, Wang S, Zhang W, *et al.* Efficient editing DNA regions with high sequence identity in actinomycetal genomes by a CRISPR-Cas9 system. Synth Syst Biotechnol 2019; 4(2): 86-91.
[http://dx.doi.org/10.1016/j.synbio.2019.02.004] [PMID: 30891508]

[50] Jiang F, Doudna JA. CRISPR–Cas9 structures and mechanisms. Annu Rev Biophys 2017; 46(1): 505-29.
[http://dx.doi.org/10.1146/annurev-biophys-062215-010822] [PMID: 28375731]

[51] Deltcheva E, Chylinski K, Sharma CM, *et al.* CRISPR RNA maturation by trans-encoded small RNA and host factor RNase III. Nature 2011; 471(7340): 602-7.
[http://dx.doi.org/10.1038/nature09886] [PMID: 21455174]

[52] Gao P, Yang H, Rajashankar KR, Huang Z, Patel DJ. Type V CRISPR-Cas Cpf1 endonuclease employs a unique mechanism for crRNA-mediated target DNA recognition. Cell Res 2016; 26(8): 901-13.
[http://dx.doi.org/10.1038/cr.2016.88] [PMID: 27444870]

[53] Alberti F, Corre C. Editing streptomycete genomes in the CRISPR/Cas9 age. Nat Prod Rep 2019; 36(9): 1237-48.
[http://dx.doi.org/10.1039/C8NP00081F] [PMID: 30680376]

[54] Zhao Y, Li L, Zheng G, *et al.* CRISPR/*dCas9*-mediated multiplex gene repression in *Streptomyces*. Biotechnol J 2018; 13(9): 1800121.
[http://dx.doi.org/10.1002/biot.201800121] [PMID: 29862648]

[55] Yeo WL, Heng E, Tan LL, *et al.* Characterization of Cas proteins for CRISPR-Cas editing in streptomycetes. Biotechnol Bioeng 2019; 116(9): 2330-8.
[http://dx.doi.org/10.1002/bit.27021] [PMID: 31090220]

[56] Tong Y, Whitford CM, Robertsen HL, *et al.* Highly efficient DSB-free base editing for *streptomycetes* with CRISPR-BEST. Proc Natl Acad Sci 2019; 116(41): 20366-75.
[http://dx.doi.org/10.1073/pnas.1913493116] [PMID: 31548381]

[57] Zhao Y, Tian J, Zheng G, *et al.* using *dCas9*-CDA-ULstr using a *dCas9*-cytidine deaminase fusion in *Streptomyces*. Sci China Life Sci 2019; 63(7): 1053-62.
[http://dx.doi.org/10.1007/s11427-019-1559-y] [PMID: 31872379]

[58] Cobb RE, Wang Y, Zhao H. High-efficiency multiplex genome editing of *Streptomyces* species using an engineered CRISPR/Cas system. ACS Synth Biol 2015; 4(6): 723-8.
[http://dx.doi.org/10.1021/sb500351f] [PMID: 25458909]

[59] Komor AC, Badran AH, Liu DR. Editing the genome without double-stranded DNA breaks. ACS Chem Biol 2018; 13(2): 383-8.
[http://dx.doi.org/10.1021/acschembio.7b00710] [PMID: 28957631]

[60] Zhang MM, Wong FT, Wang Y, *et al.* CRISPR–Cas9 strategy for activation of silent *Streptomyces* biosynthetic gene clusters. Nat Chem Biol 2017; 13(6): 607-9.
[http://dx.doi.org/10.1038/nchembio.2341] [PMID: 28398287]

[61] Huang H, Zheng G, Jiang W, Hu H, Lu Y. One-step high-efficiency CRISPR/Cas9-mediated genome editing in *Streptomyces*. Acta Biochim Biophys Sin 2015; 47(4): 231-43.
[http://dx.doi.org/10.1093/abbs/gmv007] [PMID: 25739462]

[62] Zeng H, Wen S, Xu W, *et al.* Highly efficient editing of the actinorhodin polyketide chain length factor gene in *Streptomyces coelicolor* M145 using CRISPR/Cas9-CodA(sm) combined system. Appl Microbiol Biotechnol 2015; 99(24): 10575-85.
[http://dx.doi.org/10.1007/s00253-015-6931-4] [PMID: 26318449]

[63] Luo Y, Enghiad B, Zhao H. New tools for reconstruction and heterologous expression of natural product biosynthetic gene clusters. Nat Prod Rep 2016; 33(2): 174-82.
[http://dx.doi.org/10.1039/C5NP00085H] [PMID: 26647833]

[64] Wlodek A, Kendrew SG, Coates NJ, *et al.* Diversity oriented biosynthesis *via* accelerated evolution of modular gene clusters. Nat Commun 2017; 8(1): 1206-6.
[http://dx.doi.org/10.1038/s41467-017-01344-3] [PMID: 29089518]

[65] Hoff G, Bertrand C, Zhang L, *et al.* Multiple and variable NHEJ-like genes are involved in resistance to dna damage in *Streptomyces ambofaciens*. Front Microbiol 2016; 7: 1901-1.
[http://dx.doi.org/10.3389/fmicb.2016.01901] [PMID: 27965636]

[66] Weber T, Charusanti P, Musiol-Kroll EM, *et al.* Metabolic engineering of antibiotic factories: New tools for antibiotic production in actinomycetes. Trends Biotechnol 2015; 33(1): 15-26.
[http://dx.doi.org/10.1016/j.tibtech.2014.10.009] [PMID: 25497361]

[67] Culp EJ, Yim G, Waglechner N, Wang W, Pawlowski AC, Wright GD. Hidden antibiotics in actinomycetes can be identified by inactivation of gene clusters for common antibiotics. Nat Biotechnol 2019; 37(10): 1149-54.
[http://dx.doi.org/10.1038/s41587-019-0241-9] [PMID: 31501558]

[68] Wang K, Zhao QW, Liu YF, *et al.* Multi-layer controls of *Cas9* activity coupled with ATP synthase over-expression for efficient genome editing in *Streptomyces*. Front Bioeng Biotechnol 2019; 7: 304-4.
[http://dx.doi.org/10.3389/fbioe.2019.00304] [PMID: 31737622]

[69] Najah S, Saulnier C, Pernodet JL, Bury-Moné S. Design of a generic CRISPR-Cas9 approach using the same sgRNA to perform gene editing at distinct loci. BMC Biotechnol 2019; 19(1): 18-8.
[http://dx.doi.org/10.1186/s12896-019-0509-7] [PMID: 30894153]

[70] Wang JW, Wang A, Li K, *et al.* CRISPR/Cas9 nuclease cleavage combined with Gibson assembly for seamless cloning. Biotechniques 2015; 58(4): 161-70.
[http://dx.doi.org/10.2144/000114261] [PMID: 25861928]

[71] Jiang W, Zhao X, Gabrieli T, Lou C, Ebenstein Y, Zhu TF. Cas9-assisted targeting of chromosome segments catch enables one-step targeted cloning of large gene clusters. Nat Commun 2015; 6(1): 8101-1.
[http://dx.doi.org/10.1038/ncomms9101] [PMID: 26323354]

[72] Lee NCO, Larionov V, Kouprina N. Highly efficient CRISPR/Cas9-mediated TAR cloning of genes and chromosomal loci from complex genomes in yeast. Nucleic Acids Res 2015; 43(8): e55-5.
[http://dx.doi.org/10.1093/nar/gkv112] [PMID: 25690893]

[73] Zhou J, Wu R, Xue X, Qin Z. CasHRA (Cas9-facilitated Homologous Recombination Assembly) method of constructing megabase-sized DNA. Nucleic Acids Res 2016; 44(14): e124.
[http://dx.doi.org/10.1093/nar/gkw475] [PMID: 27220470]

[74] Zetsche B, Heidenreich M, Mohanraju P, *et al.* Erratum: Multiplex gene editing by CRISPR–Cpf1 using a single crRNA array. Nat Biotechnol 2017; 35(2): 178-8.
[http://dx.doi.org/10.1038/nbt0217-178b] [PMID: 28178246]

[75] Baltz RH. *Streptomyces* temperate bacteriophage integration systems for stable genetic engineering of actinomycetes (and other organisms). J Ind Microbiol Biotechnol 2012; 39(5): 661-72.
[http://dx.doi.org/10.1007/s10295-011-1069-6] [PMID: 22160317]

[76] Katayama K, Mitsunobu H, Nishida K. Mammalian synthetic biology by CRISPRs engineering and applications. Curr Opin Chem Biol 2019; 52: 79-84.
[http://dx.doi.org/10.1016/j.cbpa.2019.05.020] [PMID: 31254926]

[77] Rees HA, Liu DR. Base editing: precision chemistry on the genome and transcriptome of living cells. Nat Rev Genet 2018; 19(12): 770-88.
[http://dx.doi.org/10.1038/s41576-018-0059-1] [PMID: 30323312]

[78] Arazoe T, Kondo A, Nishida K. Targeted nucleotide editing technologies for microbial metabolic engineering. Biotechnol J 2018; 13(9): 1700596.
[http://dx.doi.org/10.1002/biot.201700596] [PMID: 29862665]

[79] Kim JS. Precision genome engineering through adenine and cytosine base editing. Nat Plants 2018; 4(3): 148-51.
[http://dx.doi.org/10.1038/s41477-018-0115-z] [PMID: 29483683]

[80] Molla KA, Yang Y. CRISPR/Cas-mediated base editing:Technical considerations and practical applications. Trends Biotechnol 2019; 37(10): 1121-42.
[http://dx.doi.org/10.1016/j.tibtech.2019.03.008] [PMID: 30995964]

[81] Banno S, Nishida K, Arazoe T, Mitsunobu H, Kondo A. Deaminase-mediated multiplex genome editing in *Escherichia coli*. Nat Microbiol 2018; 3(4): 423-9.
[http://dx.doi.org/10.1038/s41564-017-0102-6] [PMID: 29403014]

[82] Billon P, Bryant EE, Joseph SA, *et al.* CRISPR-mediated base editing enables efficient disruption of eukaryotic genes through induction of stop codons. Mol Cell 2017; 67(6): 1068-1079.e4.
[http://dx.doi.org/10.1016/j.molcel.2017.08.008] [PMID: 28890334]

[83] Saha S, Zhang W, Zhang G, *et al.* Activation and characterization of a cryptic gene cluster reveals a cyclization cascade for polycyclic tetramate macrolactams. Chem Sci 2017; 8(2): 1607-12.
[http://dx.doi.org/10.1039/C6SC03875A] [PMID: 28451290]

[84] Yan Y, Liu Q, Zang X, *et al.* Resistance-gene-directed discovery of a natural-product herbicide with a new mode of action. Nature 2018; 559(7714): 415-8.
[http://dx.doi.org/10.1038/s41586-018-0319-4] [PMID: 29995859]

[85] Xu DB, Ye WW, Han Y, Deng ZX, Hong K. Natural products from mangrove actinomycetes. Mar Drugs 2014; 12(5): 2590-613.
[http://dx.doi.org/10.3390/md12052590] [PMID: 24798926]

[86] Lv M, Ji X, Zhao J, *et al.* Characterization of a C3 deoxygenation pathway reveals a key branch point in aminoglycoside biosynthesis. J Am Chem Soc 2016; 138(20): 6427-35.
[http://dx.doi.org/10.1021/jacs.6b02221] [PMID: 27120352]

[87] Tan GY, Deng K, Liu X, *et al.* Heterologous biosynthesis of spinosad: An omics-guided large polyketide synthase gene cluster reconstitution in *Streptomyces*. ACS Synth Biol 2017; 6(6): 995-1005.
[http://dx.doi.org/10.1021/acssynbio.6b00330] [PMID: 28264562]

[88] Jia H, Zhang L, Wang T, Han J, Tang H, Zhang L. Development of a CRISPR/Cas9-mediated gene-editing tool in *Streptomyces rimosus*. Microbiology 2017; 163(8): 1148-55.
[http://dx.doi.org/10.1099/mic.0.000501] [PMID: 28742008]

[89] Jun Z, Dan Z, Jie Z, *et al.* Efficient multiplex genome editing in *Streptomycesvia* engineered CRISPR-Cas12a systems. Front Bioeng Biotechnol 2020; 726(8).

[90] Li L, Wei K, Zheng G, *et al.* CRISPR-Cpf1-assisted multiplex genome editing and transcriptional repression in *streptomyces*. Appl Environ Microbiol 2018; 84(18): e00827-18.
[http://dx.doi.org/10.1128/AEM.00827-18] [PMID: 29980561]

[91] Lei C, Li S-Y, Liu J-K, Zheng X, Zhao GP, Wang J. The CCTL (Cpf1-assisted Cutting and Taq DNA ligase-assisted Ligation) method for efficient editing of large DNA constructs *in vitro*. Nucleic Acids Res 2017; 45(9): e74-4.
[PMID: 28115632]

[92] Zetsche B, Gootenberg JS, Abudayyeh OO, *et al.* Cpf1 is a single RNA-guided endonuclease of a class 2 CRISPR-Cas system. Cell 2015; 163(3): 759-71.
[http://dx.doi.org/10.1016/j.cell.2015.09.038] [PMID: 26422227]

[93] Choi KR, Lee SY. CRISPR technologies for bacterial systems: Current achievements and future directions. Biotechnol Adv 2016; 34(7): 1180-209.
[http://dx.doi.org/10.1016/j.biotechadv.2016.08.002] [PMID: 27566508]

SUBJECT INDEX

A

Acid 2, 6, 11, 14, 15, 21, 112, 115, 116, 124, 131, 188, 189, 190, 192, 199, 202, 233, 235, 236, 237, 238, 253, 256, 257, 265, 285
  abscisic 14, 112, 115
  butyric 237, 238
  citric 189
  crotonic 256
  cyclopiazonic 2
  dimethylarsinous 202
  dipicolinic 15
  docosahexaenoic 21
  eicosapentaenoic 21
  eicosatetraenoic 21
  jasmonic 14, 116
  lactic 253, 257
  linoleic 21
  linolenic 21
  malic 11
  octanoic 237
  oleic 21
  organic 11, 188, 190, 192, 233, 235, 265
  oxalic 11
  phenolic 131
  propionic 236
  ribonucleic 285
  salicylic 6, 14, 124
  stearidonic 21
  succinic 11
  teichoic 199
Activity 55, 98, 157, 158, 173, 229, 256
  antimicrobial 98
  biocatalytic 229
  catalytic 256
  degradative 157
  galactosidase 55
  metabolic 158, 173
Advanced oxidation processes (AOPs) 192
Age-related macular degeneration 74, 97
Agents 69, 83, 297
  anti-cancer 297
  antibacterial 83
  infectious 69
Agricultural production systems 8
Agrobacterium-mediated 49, 50, 52, 53, 54
  rice transformation 50
  transformation method 49, 52, 53, 54
Air quality index (AQI) 187
Antibiotic(s) 68, 70, 75, 78, 79, 80, 82, 85, 89, 95, 98, 198, 204, 226, 265, 284
  drugs 98
  pressure 265
  resistance 68, 70, 75, 82, 85, 95, 198, 204, 226, 284
  resistance genes 89
  resistance mechanisms 80
  -resistant pathogens 79
  traditional 78
Antimicrobial 95, 125
  peptides 125
  therapy 95
Applications 96, 231
  cancer immunotherapy 96
  industrial 231
Aspergillus oryzae 16, 131
Assays, rapid infection 84
ATP-dependent mechanism 292
Atrophic rhinitis 85

B

Bacillus thuringiensis 19, 20, 198, 199
Bacteria 2, 8, 9, 11, 57, 75, 88, 109, 118, 193, 199, 230, 259, 262
  aerobic 193
  antibiotic-resistant 75
  chromium-loaded 199
  engineer methanotrophic 259
  engineering probiotic 88
  nitrogen fixing 8, 11, 230
  photosynthetic 262
  phytopathogenic 57

soil 2, 9, 109, 118
Bacterial 54, 70, 79, 89, 90, 97, 122, 198, 223, 249
  adhesins 79
  genome editing 54
  genomes 70, 90, 223
  hosts 70, 89, 90
  immune systems 97, 122
  metalloregulatory protein 198
  protein 249
Bacterial infections 57, 70, 78, 79, 82, 87, 89, 91, 127, 163
  resistant 78
Bacterial pathogens 53, 54, 57, 60, 61, 78, 79, 87, 88, 90, 95, 126, 128
  drug-resistant 90
Bacterium 59, 75, 83, 167, 231, 232, 278
  methanotrophic 231, 232
Banana streak virus (BSV) 129
Base editing techniques 106
Biodegradation 153, 155, 156, 160, 165, 172, 173, 195, 225, 226
  pathways 160
  process 173
Biodiesel production 3, 22
Bioenergy production 22
Bioethanol production 22
Biofertilizers 1, 5, 8
Biofilm properties 119
Biofuel(s) 87, 224, 236, 248, 262, 263, 269
  liquid 248
  production 87, 224, 236
Biological nitrogen fixation (BNF) 8, 107, 132
Biomarkers, environmental 231
Biomass 160, 199, 235
  agricultural 235
Bioprocesses 223, 224, 235
  anaerobic 223, 224
Bioremediation 152, 155, 158, 159, 160, 172, 174, 193, 194, 206, 207, 224
  effective 207
  food waste 224
  of contaminants 152, 160, 172
  pathways 174
  techniques 155, 158, 159, 193, 194, 206
Bioremediation processes 150, 153, 154, 155, 159, 160, 166, 167, 168, 169, 172, 174
  microbial 169
Bioremediators 153
  essential 153

Biosynthesis 17, 19, 49, 50, 52, 55, 78, 85, 90, 111, 118, 130, 131, 186, 254
  fatty acid 78, 254
  glycine betaine 118
  lignocellulose 131
Biosynthetic 16, 89, 130, 278, 279, 283, 284, 291, 292, 297
  gene clusters (BGCs) 278, 279, 283, 284, 291, 292, 297
  pathways 16, 89, 130
Biotic stress 107, 116, 133
  resistance 107, 116
  tolerance 133
Broad-spectrum antibiotic therapy 79

C

Cancer 95, 96, 184, 185
  lung 96
Cascade 19, 76
  activating signalling 19
  immune reaction 76
Catabolic pathways 202, 240
Cellulase production 131
Chemicals 87, 153, 173, 225
  harmful 173
  industrial 87
  organic 225
  toxic 153
Chemisorption 199
Chemotherapy 56, 96, 97
Chloroplasts 263
Cismuconic acid production 235
Cloning 91, 161
  and sequencing of ribosomal DNA 161
  vectors 91
Contaminants 150, 152, 153, 154, 155, 156, 157, 158, 159, 161, 164, 169, 172, 173, 184, 190, 195, 196, 202, 206, 226
  aerobic 159
  biodegrade 195
  chemical 164
  degrading 206
  halogenated 172
  heavy metal 184, 202
  removal process 173
  toxic 159
Crop(s) 17, 19, 128
  developing fungal-resistance 17
  developing insect resistance 19

disease resistance 128
herbicide-resistant 19
Cucumber mosaic virus (CMV) 129
Cytidine 121, 123, 129, 296
   base editor 121, 129
   deaminase 123, 296
Cytosine base editors (CBEs) 56, 121, 123, 295, 296

D

Damage-associated molecular patterns (DAMPs) 125
Databases 171, 252
   biochemical 252
   metagenomic 171
Defense pathways 107, 114, 125
Degradation 152, 153, 225
   activities 152
   environmental 225
   pathways 153
Detoxification pathway 184
Developing disease resistance 129
Diseases 2, 15, 17, 53, 57, 58, 59, 69, 77, 78, 80, 81, 84, 87, 91, 93, 94, 95, 96, 186, 187, 223, 224, 239
   bacterial 53, 57, 69, 84
   bacterial-mediated 78
   cancer 96
   infectious 87, 94, 239
   life-threatening 95
   pulmonary 187
   skin 187
   urinary 80
   yellow mosaic 186
Disorders 152, 239, 240
   genetic 152
   metabolic 239
DNA 18, 46, 47, 49, 51, 73, 74, 76, 77, 79, 93, 95, 122, 123, 152, 163, 223, 231, 249, 250, 251, 263, 267, 268, 279, 282, 286, 291, 292, 293, 294, 299
   binding proteins 93
   broad circular 293
   cleavage activity 49
   cleave 286
   -cleaving nuclease 163
   cosmid 282
   deletions 279
   double-stranded 286

   efficient HDR-mediated 294
   homologous recombination 279
   infecting pathogen 51
   phage 73, 223
   plasmid 73, 251
   recombination systems 299
   replication 79
   target pathogens 95
   targeted 163, 249
   transfer 18, 250, 263
   transformation 292
DNA-binding proteins 46, 92
   engineered 46
DNA repair 120, 291
   pathway 291
DNA synthesis 89, 90
   sequences 90
   techniques 89
DNase, inactivating 268
Double-stranded breaks (DSBs) 48, 71, 120, 204, 284, 286, 287, 289, 291, 295, 299
Draught 5, 9, 13, 20, 21, 112, 114, 118
   resistance 5
   and salinity stress 13
   tolerance 9, 20, 21, 112, 114, 118
Drought stress 16, 112, 114, 115, 116, 117, 118
   mitigated 117
   responses 115

E

Editing 53, 55, 56, 59, 72, 107, 126, 127, 128, 131, 132, 133, 134, 239, 266, 269
   genomic 72
   metabolic pathway 107
Editing tools 68, 70, 87, 163, 226
   genomic 226
   robust gene 163
Encode sucrose transporter 57
Energy, solar 262, 264
Engineering 92, 258
   cellular systems 92
   methane monooxygenase 258
Environmental 159, 169, 239, 240, 278
   pollutants 159, 169, 240
   rejuvenation process 239
   stress 278
Enzyme methane monooxygenase 231
Epstein-Barr virus (EBV) 95

F

Factors 19, 79, 81, 106, 108, 110, 156, 169, 172, 187, 188, 191
  abiotic stress 19
  heat shock 19
Fatty acid 224, 239, 254, 269
  biosynthesis regulation 254
  metabolism 269
  production 224, 239, 254
Fibronectin-binding protein 84
Floral dip method 47
FTIR spectroscopy 199
Fungal pathogens 16, 126
Fungi 14, 15, 128, 278
  antagonistic 15
  entomopathogenic 14, 15
  filamentous 128, 278
Fungicide 3, 13, 15
  chemical 3

G

Gene(s) 18, 45, 46, 123, 204, 229, 230, 256, 266, 290
  deaminase 290
  endonuclease 45
  glutamine synthetase 230
  hemoglobin 256
  herbicide-tolerance 46
  mercury transport 204
  metabolic 229
  regulation 123
  toxin 266
  transfer methods 18
Gene editing 98, 125, 150, 160, 161, 224
  techniques 150, 224
  technology 98
  tools 125, 150, 160, 161
Genetic engineering 164, 166
  of microorganisms 164
  technology 166
Genetic tools 49, 54, 93, 253, 265, 300
Genetically 164, 165, 167, 185, 189, 203
  engineered microorganisms (GEMs) 164, 165, 189, 203
  modified foods (GMF) 185
  modified microorganisms (GMM) 165, 167, 203
Genome 90, 96, 123
  automated 90
  cancer 96
  prokaryotic 123
Genome editing 23, 51, 54, 61, 125, 133, 150, 159, 167, 174, 189, 279, 280, 288, 294
  methods 288
  targets 51, 125
  techniques 23, 61, 125, 159, 280
  technologies 54, 133, 150, 167, 174, 189, 279, 294
Germination protease 85
Gibberlin oxidase 132
Glutamine synthetase enzymes 19
Glycan biosynthesis 131
Glycosyl hydrolases (GH) 233
Groundwater 156, 164, 190, 191, 192, 193, 195, 197, 203
  bioremediation 203
  contaminated 192, 195, 197
Growth promotion 3, 7, 113
  microbe-mediated plant 113
Gut 223, 239
  dysbiosis 239
  microbiome 223, 239

H

Harmful plant-microbe interactions 110
Hazardous waste 151
Heat shock proteins (HSPs) 188
Heavy metal 188, 198, 201, 204
  mechanisms 198
  resistance 198, 204
  stresses 188, 201
HIV infection 185
HLB 58
  disease 58
  infection 58
Homology repair template (HRT) 289, 291
Host 60, 109
  defense mechanisms 60
  microbiota, lectin compounds influence 109
Human papilloma virus (HPV) 186

I

Immune responses 109, 110, 116
Immunity 53, 60, 72
  mediated 72

Industries 95, 151, 158, 186, 235, 247, 248
  agricultural 186
  biofuel 235
  manufacturing 158
Infection 17, 59, 60, 72, 74, 75, 78, 79, 80, 82, 83, 84, 85, 94, 95, 96
  acute systemic 85
  chronic 84
  inhibiting 17
  massive 83
  neoplastic 96
  urinary tract 79

L

Liposomes 19, 263
Lysis protein 88

M

Membrane 18, 188, 201, 202, 204
  -associated transport protein 204
  -bound heavy metal-transporting ATPases 188
  cytoplasmic 201
  plasma 202
Metabolic pathways 21, 22, 23, 130, 131, 133, 165, 168, 225, 228, 229, 256, 257, 259
Metal-binding proteins 189, 200
Methane 247, 248, 258
  bioconversion 248, 258
  monooxygenase 247
Methanobactin transport proteins 253
Microbes 10, 106, 112, 123, 130
  growth-promoting 106, 123
  plant-associated 10, 106, 112, 130
Microbial communities 93, 109, 158, 169, 205, 223, 225
Microorganisms commensal 81
Mitogen activated protein kinase 111
Mobile genetic elements (MGEs) 73, 98, 205, 292
Monoclonal antibodies 185
Mycobacterium tuberculosis 56, 200

N

Nitrogen 132, 222, 229, 230
  metabolizing bacteria 222, 229, 230
  use efficiency (NUE) 132

Nutrient transporters 132

O

Organic pollutants 157, 159, 195, 229
  toxic 159
Osmotic stress tolerance 21
Oxygenic photosynthesis 2, 262

P

Pathogenic bacteria 61, 69, 70, 74, 75, 79, 249
  drug-resistant 69
Pathogens 44, 77, 80, 84, 116
  interfering 44
  necrotrophic 116
  opportunistic 77, 80, 84
Pathways 166, 254
  degradative 166
  fatty acid biosynthesis 254
Peptides 90, 95, 130, 200
  antibacterial 95
Phages 75
  lysogenic 75
  lytic 75
Phagocytosis 79
Phenotype, mutant 47
Phosphate solubilizing microbes 11
Photosynthesis 263
Plant 2, 14, 19, 43, 57, 60, 61, 106, 107, 108, 109, 110, 111, 116, 117, 118, 119, 122, 123, 124, 126, 127, 128, 186, 222
  -associated pathogens 119
  biotechnology 19
  disease resistance 57, 117, 126
  diseases 14, 43, 61, 111
  fitness 118
  genotype-microbiota-environment 109
  hormones 117
  immunity 107, 127
  infections 186
  microbiome composition 106
  microbiota 106, 107, 108, 110, 123
  -pathogen interaction 57, 60, 111, 116, 119, 124
  pathosystems 116
  signaling mechanisms 124
  -soil bacteria interaction 2
  transcripts 122
  -virus interactions 128

wastewater treatment 222
Plant growth 12, 106, 107, 112, 115, 125
   -promoting microbes (PGPM) 106, 107, 112, 115, 125
   promotion 12
Plasmid curing method 291
Pollutants 150, 151, 152, 154, 156, 159, 165, 169, 171, 172, 189, 197, 225, 226, 227, 228
  harmful 152
  hydrophobic 165
  inorganic 226
  toxic 225
Polluted soils 240
Potato virus 129
Process 157, 189
  bioreactor 157
  chemoorganotrophic 189
Proteins, fusion 290

R

Radiation therapy 97
Reactive oxygen species (ROS) 44, 114, 198
Restriction 45, 46, 263, 293
   enzyme (REs) 45, 46, 293
   inhibitors 263
Rice stripe mosaic virus (RSMV) 129
RNA 111, 268
   binding protein 268
   -dependent DNA methylation 111
RNA polymerase 73, 93, 94, 267, 268
   and DNA binding proteins 93

S

Single nucleotide polymorphisms (SNPs) 128, 134, 204
Soil fertility 184
Stress 7, 20, 110, 112, 113, 114, 115, 116, 186
  chemical 186
  draught 7
  heavy-metal 110
  osmotic 20, 115
  oxidative 112, 113, 114
  tolerance, moisture 116

T

Tobacco mosaic virus (TMV) 129

Toxicity, organic acid 266
Transcription factors 134, 163, 249
Transcriptional activation 267
Transport proteins 204

V

Viral transposases 76
Virulence 77, 78, 79, 80, 81, 82, 84, 85, 95, 126, 230
  factors 77, 80, 81, 82, 84, 95
  gene products 78
Virus(s) 3, 43, 60, 87, 94, 95, 96, 97, 125, 128, 129, 130, 186, 223
  cucumber mosaic 129
  tobacco mosaic 129
  tumorigenic 96
Vitamin A deficiency (VAD) 21
Volatile fatty acids (VFAs) 233, 236, 237, 239

W

Wastes 3, 4, 22, 153, 164, 165, 166, 187, 193, 194, 206, 236
  agricultural 22, 187
  industrial 194
  organic 3, 4, 153, 206
  toxic 193
Wastewater 3, 157, 166, 192, 193, 194, 196, 197, 223, 224, 227, 228
  environment 227
  treatment 3, 196, 224, 228
Wheat dwarf virus (WDV) 129

Z

Zinc finger 44, 45, 46, 47, 48, 49, 51, 54, 97, 120, 161, 163, 249, 250
  nucleases (ZFNs) 44, 45, 46, 47, 48, 49, 51, 54, 97, 120, 161, 163, 249, 250
  proteins (ZFPs) 46, 163

www.ingramcontent.com/pod-product-compliance
Lightning Source LLC
Chambersburg PA
CBHW051143220526
45473CB00003B/643